Machine Learning for Complex and Unmanned Systems

This book highlights applications that include machine learning methods to enhance new developments in complex and unmanned systems. The contents are organized from the applications requiring few methods to the ones combining different methods and discussing their development and hardware/software implementation. The book includes two parts: the first one collects machine learning applications in complex systems, mainly discussing developments highlighting their modeling and simulation, and hardware implementation. The second part collects applications of machine learning in unmanned systems including optimization and case studies in submarines, drones, and robots. The chapters discuss miscellaneous applications required by both complex and unmanned systems, in the areas of artificial intelligence, cryptography, embedded hardware, electronics, the Internet of Things, and healthcare. Each chapter provides guidelines and details of different methods that can be reproduced in hardware/software and discusses future research.

Features

- Provides details of applications using machine learning methods to solve real problems in engineering
- Discusses new developments in the areas of complex and unmanned systems
- Includes details of hardware/software implementation of machine learning methods
- Includes examples of applications of different machine learning methods for future lines for research in the hot topic areas of submarines, drones, robots, cryptography, electronics, healthcare, and the Internet of Things

This book can be used by graduate students, industrial and academic professionals to examine real case studies in applying machine learning in the areas of modeling, simulation, and optimization of complex systems, cryptography, electronics, healthcare, control systems, Internet of Things, security, and unmanned systems such as submarines, drones, and robots.

Machine Learning for Complex and Unmanned Systems

Edited by
Jose Martinez-Carranza
Everardo Inzunza-González
Enrique Efrén García-Guerrero
Esteban Tlelo-Cuautle

CRC Press
Taylor & Francis Group
Boca Raton London New York

CRC Press is an imprint of the
Taylor & Francis Group, an **Informa** business

Designed cover image: ©Shutterstock

First edition published 2024
by CRC Press
2385 NW Executive Center Drive, Suite 320, Boca Raton FL 33431

and by CRC Press
4 Park Square, Milton Park, Abingdon, Oxon, OX14 4RN

CRC Press is an imprint of Taylor & Francis Group, LLC

ISBN: 978-1-032-47224-9 (hbk)
ISBN: 978-1-032-47330-7 (pbk)
ISBN: 978-1-003-38561-5 (ebk)

DOI: 10.1201/9781003385615

Typeset in Nimbus font
by KnowledgeWorks Global Ltd.

Publisher's note: This book has been prepared from camera-ready copy provided by the authors.

Contents

PART I Machine Learning for Complex Systems

PART II   Machine Learning for Unmanned Systems

Preface

Nowadays, in a very huge number of engineering applications one can infer the use of machine learning methods. From the sensing of data, one can infer the necessity of signal conditioning techniques, data classification algorithms, signal processing, and control of devices or loads. Obviously, not all applications can be developed using the same machine learning methods, so this book includes a collection of recent works, which focus on special cases in the areas of modeling, simulation, optimization and hardware implementation of complex and unmanned systems. These topic areas require special machine learning methods, as for example for the classification of data acquired by different types of sensors and video-cameras, signal processing and the control of motors, wheels, and other loads or actuators. In this manner, the sixteen chapters in this book emphasize the usefulness of machine learning methods, not only for theoretical problems but also for real case studies with real experiments.

Part 1 is devoted to complex systems and includes seven chapters. Chapter 1 introduces a neural network model for machine learning tasks, but the authors found that simple echo state networks (ESNs) can be used instead. In ESNs, only the output layer weights are calculated by linear regression, which has a low computational cost. The study shows how ESNs with improved reservoir creation can solve classification tasks using simulated data and public databases. Chapter 2 discusses the challenge of monocular camera localization due to constant changes and proposes a review of state-of-the-art methodologies for this task. Traditional methods use end-to-end training with a large dataset, but this study presents Continual Learning (CL) strategies such as distillation, weights optimization, dynamic architectures, incremental labels, and latent replay to avoid the need for a large dataset and catastrophic forgetting. The aim is to describe the benefits and limitations of these methods for camera localization tasks from a single image. Chapter 3 proposes a dataset of images of three distinct fish species, each with a unique number of images, which are trained in different convolutional neural networks (CNN) for species identification applications. The performance metrics obtained from the proposed dataset show that the YOLOV5 and RESNET models achieved a precision of 95% and 92%, respectively, and were evaluated on edge computing devices. The study plans to increase the images and species in the database to train algorithms for identifying more fish species. Chapter 4 presents a regression model that utilizes regression tree, random forest, and convolutional neural networks to model the nonlinear behavior of power amplifiers. The use of recursive subset partitioning with the regression tree model provides a fast runtime and highly accurate output estimates without overfitting. The ensemble model RT offers a single-step iteration solution for modeling sparse AM/AM behavior in a cascade setup that drives a Class AB power amplifier, which can address residual distortion present in the direct-conversion transmitter by tuning the parameters. Experimental results demonstrate the effectiveness of the proposed model for a wideband of LTE signals with varying input power levels. Chapter 5 highlights the increased use of video surveillance for security and operational efficiency and the need for effective data analysis techniques. The chapter focuses on anomaly detection in video surveillance systems, which requires unsupervised or semi-supervised learning techniques. The study describes the use of convolutional autoencoders, variational autoencoders, Long Short-Term Memory networks, and Generative Adversarial Networks for anomaly detection in video surveillance, including theoretical aspects and practical examples. Chapter 6 presents tuberculosis (TB) as a leading cause of mortality globally, with a 30% increase in the mortality rate. Reliable, fast, and sensitive diagnostic tests are needed for TB and/or COVID-19 to control disease transmission. Artificial intelligence and deep learning have been used to develop computer-assisted diagnosis systems for pulmonary diseases, offering high sensitivity and specificity values that surpass those of the human eye. Chapter 7 explores the variability present in frequencies of analog hardware and applies a charge-controlled memristor model for hardware security. The study focuses on using memristors in physical unclonable functions based on ring oscillators and evaluates the performance of the PUF response using various metrics for hardware security. The study demonstrates that memristive PUFs exhibit excellent metrics under diverse conditions of system complexity.

Part 2 is devoted to unmanned systems and includes nine chapters. Chapter 8 discusses the challenge of developing an artificial pilot capable of autonomously flying a drone in autonomous drone racing. Learning-based methods propose training an artificial pilot to associate sensorial data with fly commands, but current approaches cannot consider future events, such as an incoming turn. The study proposes augmenting the temporal analysis to consider data from future events observed through visual data and associated with flight commands optimized to follow a flight trajectory. The study discusses this approach in the context of the state of the art and an experimental framework to assess its effectiveness. Chapter 9 proposes optimizing UAV flight control's performance using metaheuristics and comparing two relevant AI techniques, Differential Evolution and Accelerated Particle Swarm Optimization algorithms. The study also applies two control techniques, Sliding Mode Control and Proportional-Integral-Derivative control. The study presents UAV model development, trajectory methodology, and the application of metaheuristics to adjust controller parameters, followed by the results and discussion of tests for adjusting controller parameters and trajectory tracking. Chapter 10 focuses on object classification in aerial navigation using texture as an important feature. The study presents a classification model that uses transfer learning and wavelet-based features as an additional feature extraction method, achieving a 53% accuracy using the Describable Texture Database. The study validates the results using a virtual world in the Gazebo simulator and creates a new Synthetic Aerial Dataset of Textured Objects, showing generalization of knowledge for some classes of the database. Chapter 11 explores using UAVs for air-ground communications to improve coverage in areas with poor or nonexistent coverage, with a focus on fifth-generation mobile communication (5G). The behavior of Air Base Stations (ABS) is analyzed to determine the coverage percentage over a simulation time, taking into account random disturbances. The analysis is conducted by simulating a quadrotor-type UAV and measuring power levels received by ground users to determine the effective coverage area. Chapter 12 presents a review of the literature on the topic of noise produced by UAVs. The aim of the review is to describe the effects of noise on society, summarize past audio analysis of noise produced by popular UAV models, and revise recent techniques on noise mitigation. The chapter concludes with potential research avenues and insights extracted from the review. Chapter 13 reviews the Neural Radiance Fields (NeRF) methodology and its latest works in computer vision and robotics, with a focus on aerial robotics applications. NeRF has become a state-of-the-art technique for generating new synthetic views due to its deep learning-based approach. The chapter also briefly discusses other potential applications and works proposing improvements to the original formulation, as well as potential opportunities for using NeRF in Unmanned Aerial Vehicles (UAVs) or drones. In Chapter 14, a small drone equipped with an OAK-D smart camera is used to perform autonomous flight and spatial AI for warehouse inspection. The OAK-D computes depth estimation and neural network inference on its chip, allowing for package scanning by detecting QR and bar codes, and a person detector for safe flight. The drone localization is performed using RGB-D ORB-SLAM, enabling autonomous flight in a GPS-denied environment, and the experiments show that the inspection task can be easily distributed using multiple drones. Chapter 15 discusses the use of Cognitive Dynamic Systems (CDS) as a framework for designing complex and intelligent cyber-physical systems. CDS is inspired by human cognition, mimicking the human brain and consisting of perception-action cycles, memory, attention, and intelligence. The approach can provide adaptation and self-modification, making it suitable for proactive unmanned vehicle systems and other applications that require machines to interact with the physical world and operate autonomously. Finally, Chapter 16 shows a study that analyzes the classification of EEG signals using machine learning algorithms, such as: discriminant line analysis, decision trees, k-nearest neighbors, naive Bayes, and support vector machines. The study presents the different extracted characteristics related to motor and imaginary movements to train the different signal classification algorithms selected. The study concludes with exemplary performance in the classification of EEG signals that allows them to be used in scenarios such as robotic movement control, prosthetics, robotics, and electronic wheelchairs.

About the Editors

Jose Martinez-Carranza is a Full-Time Principal Researcher B (equivalent to Associate Professor) in the Computer Science Department at the Instituto Nacional de Astrofisica Optica y Electronica (INAOE). In 2015, he was awarded the Newton Advanced Fellowship granted by the Newton Fund and the Royal Society in the UK. Currently, he holds an Honorary Senior Research Fellowship in the Computer Science Department at the University of Bristol in the UK. He leads a research team that has won international competitions such as 1st Place in the IEEE IROS 2017 Autonomous Drone Racing competition and 1st Place in the Regional Prize of the OpenCV AI Competition 2021. He also served as General Chair of the International Micro Air Vehicle conference, the IMAV 2021. In 2022, he joined the editorial board of the journal "Unmanned Systems". His research focuses on vision-based methods for robotics with applications in autonomous and intelligent drones.

Everardo Inzunza-González received his Ph.D. degree in Electrical Sciences from UABC Mexico in 2013, and the M.Sc. degree in Electronics and Telecommunications from the Scientific Research and Advanced Studies Center of Ensenada (CICESE) in 2001, the B.Sc. degree in Electronics Engineering from Culiacan Institute of Technology, in 1999. He is currently a full-time Professor and Researcher of Electronics Engineering at Universidad Autónoma de Baja California (UABC-FIAD) Mexico. He is currently a reviewer for several prestigious journals. His research interest includes the Internet of things, Network Security, Data Science, Artificial Intelligence, Machine-Learning and Deep-Learning, Wireless Communication, Image Processing, WSN, Pattern Recognition, Wearable Devices, Embedded Systems, FPGA, SoC, Microcontrollers, Chaotic encryption, Image encryption, Image enhancement, Image processing, Chaotic oscillators, and Applied Cryptography.

Enrique Efrén García-Guerrero studied physics engineering at the University Autonomous Metropolitana, Mexico, and received the Ph.D. and M.Sc. degree in optical physics from the Scientific Research and Advanced Studies Center of Ensenada (CICESE) Mexico. He has been with the Facultad de Ingeniería, Arquitectura y Diseño of the Universidad Autónoma de Baja California (UABC-FIAD) Mexico since 2004. His current research interest includes Image enhancement, embedded systems, chaotic cryptography, artificial intelligence, machine-learning, deep-learning, neural networks, digital image processing, and optical systems.

Esteban Tlelo-Cuautle received a B.Sc. degree from Instituto Tecnológico de Puebla (ITP) México in 1993. He then received both M.Sc. and Ph.D. degrees from Instituto Nacional de Astrofísica, Óptica y Electrónica (INAOE), México in 1995 and 2000, respectively. During 1995-2000 he was with the electronics-engineering department at ITP. In 2001 he was appointed as Professor-Researcher at INAOE.

He has been Visiting Researcher in the department of Electrical Engineering at University of California Riverside, USA (2009-2010), in the department of Computer Science at CINVESTAV, México City, México (2016-2017), and Visiting Lecturer at University of Electronic Science and Technology of China (UESTC, Chengdu 2014-2019). He has authored five books, edited 12 books and more than 300 works published in book chapters, international journals and conferences. He is member in the National System for Researchers (SNI-CONACyT-México). His research interests include integrated circuit design, optimization by metaheuristics, fractional-order chaotic systems, artificial intelligence, security in Internet of Things, and analog/RF and mixed-signal design automation tools.

Contributor

Aguila-Torres, Daniel Santiago

Tecnológico Nacional de México, IT de Tijuana, Tijuana, 22435, Mexico
0000-0001-8164-0963

Aguirre-Castro, Oscar Adrian

Universidad Autónoma de Baja California
0000-0002-8000-2043

Cabrera-Ponce, Aldrich A.

Benemerita Universidad Autonoma de Puebla (BUAP)
0000-0002-9998-7444

Cárdenas-Valdez, José Ricardo

Tecnológico Nacional de México, IT de Tijuana, Tijuana, 22435, Mexico
0000-0002-5437-8215

Guerrero-Chevannier, Miguel Angel

Universidad Autónoma de Baja California

Colores-Vargas, Juan Miguel

Universidad Autónoma de Baja California
0000-0001-9336-2470

Cruz-Esquivel, Ernesto

Universidad de Las Americas Puebla (UDLAP)
0000-0002-5314-6738

Cureño Ramírez, Andres

Cinvestav, Computer Science Department

de la Fraga, Luis Gerardo

Cinvestav, Computer Science Department
0000-0002-9373-9837

Espinosa Flores-Verdad, Guillermo

INAOE, Department of Electronics, Mexico
0000-0002-3182-6353

Flores, Dora-Luz

Universidad Autónoma de Baja California

Fortuna-Cervantes, J. Manuel

Tecnológico Nacional de México, Instituto Tecnológico de San Luis Potosí
0000-0002-9229-3159

Galaviz-Aguilar, Jose Alejandro

Tecnologico de Monterrey, School of Engineering and Sciences
0000-0002-6550-124X

Galván-Tejada, Giselle Monserrat

Cinvestav, Laboratorio Franco Mexicano de Informática y Automático

García-Guerrero, Enrique Efrén

Universidad Autónoma de Baja California
0000-0001-5052-6850

Gonzalez-Landaeta, Rafael

Universidad Autónoma de Baja California

Gonzalez-Zapata, Astrid Maritza

INAOE, Department of Electronics
0000-0001-6398-5802

Gutierrez-Giles, Alejandro
Instituto Nacional de Astrofisica Optica
    y Electronica (INAOE)
0000-0002-2166-2463

Guzman-Zavaleta, Zobeida J.
Universidad de Las Americas Puebla
    (UDLAP)
0000-0002-3163-8862

Inzunza-González, Everardo
Universidad Autónoma de Baja
    California
0000-0002-7994-9774

Lara-Barrón, Manuel Mauricio
Cinvestav, Laboratorio Franco Mexicano
    de Informática y Automático

López-Bonilla, Oscar Roberto
Universidad Autónoma de Baja
    California
0000-0003-4635-2813

Martin-Ortiz, Manuel
Benemerita Universidad Autonoma de
    Puebla (BUAP)
0000-0002-4725-2059

Martinez-Carranza, Jose
Instituto Nacional de Astrofisica Optica
    y Electronica (INAOE)
0000-0002-8914-1904

Martinez-Torres, Cesar
Universidad de Las Americas Puebla
    (UDLAP)
0000-0003-3791-663X

Mejía-Carlos, Marcela
Universidad Autónoma de San Luis
    Potosí
0000-0003-2872-9461

Mérida-Rubio, Jován Oseas
Universidad Autónoma de Baja
    California
0000-0002-9355-4787

Mitchell-Moreno, Joseph Herbert
INAOE, Department of Electronics,
    Mexico
0000-0003-2304-6944

Muniz-Salazar, Raquel
Universidad Autónoma de Baja
    California

Perea-Jacobo, Ricardo
Universidad Autónoma de Baja
    California

Ramírez-Arias, Francisco Javier
Universidad Autónoma de Baja
    California
0000-0001-8222-5673

Ramírez-Torres, Marco T.
Universidad Autónoma de San Luis
    Potosí
0000-0002-7457-7318

Rascon, Caleb
Universidad Nacional Autonoma de
    Mexico
0000-0002-1176-6365

Rojas-Perez, Leticia Oyuki
Instituto Nacional de Astrofisica Optica
    y Electronica (INAOE)
0000-0002-9440-9944

Rosas-Ordaz, Luis Fernando
Universidad de Las Americas Puebla
    (UDLAP)
0009-0000-0423-4806

Sarmiento-Reyes, Arturo
INAOE, Department of Electronics,
    Mexico
0000-0002-7619-4994

Tamayo-Pérez, Ulises Jesús
Universidad Autónoma de Baja
California
0000-0002-2800-9694

Tlelo-Coyotecatl, Esteban
Cinvestav, Laboratorio Franco Mexicano
de Informática y Automático

Torres Huitzil, Cesar
Tecnologico de Monterrey
0000-0002-8980-0615

Trujillo-Toledo, Diego Armando
Universidad Autónoma de Baja
California
0000-0003-1482-8581

Vargas-Rosales, Cesar
Tecnologico de Monterrey, School of
Engineering and Sciences
0000-0003-1770-471X

Part I

Machine Learning for Complex Systems

1 Echo State Networks to Solve Classification Tasks

Luis Gerardo de la Fraga
Cinvestav, Computer Science Department, Mexico City, México.

Astrid Maritza Gonzalez-Zapata
INAOE, Department of Electronics, Luis Enrique Erro No. 1,
Tonantzintla, Puebla, Mexico

Andres Cureño Ramírez
Cinvestav, Computer Science Department, Mexico City, México

CONTENTS

1.1 INTRODUCTION

An echo state network (ESN) is a kind of recurrent neural network. The ESN has a reservoir for mapping the inputs into a high-dimensional space and a readout for pattern analysis from the high-dimensional states in the reservoir [591, 602]. According to [167, 602], ESNs can be used in machine learning applications of supervised learning such as classification and regression. The reservoir is fixed in an ESN, and only the readout is trained with a simple method such as linear regression, which has the cost of a single matrix inversion. Thus, the major advantage of reservoir computing compared to other models of neural networks is its fast learning, resulting in a low training cost [602].

In the recent work [167], an ESN and an extreme learning machine (ELM) are presented. ELM is a training algorithm for a feed-forward neural network with a single hidden layer [257, 656]. Although the training algorithm is similar in an ESN as in single-hidden layer feed-forward neural network (SLFN), in this work only a single ESN is used to solve supervised classification problems. This article shows in

DOI: 10.1201/9781003385615-1

a practical test on how a single ESN, with its common training method, can be used to solve a supervised classification problem successfully.

This work is organized as follows. Section 1.2 is explained an ESN. In Section 1.3, the experiments with four examples with artificial data to learn a bi-classification problem, which are very descriptive for themselves, are described, and also a second experiment with ten databases with real data is shown. In Section 1.4, a brief discussion about the obtained results is presented. Finally, in Section 1.5 the conclusion of this work is drawn.

1.2 ECHO STATE NEURAL NETWORK

ESN belongs to the set of recurrent neural networks with the special characteristic of a sparsely connected hidden layer. The weights of hidden neurons are assigned randomly. Also, the connectivity of hidden neurons is kept fixed and is assigned randomly in the network initialization. Therefore, the weights of the output neurons can be learned so that the network can reproduce specific temporal patterns. An ESN network is a dynamic system and not a function, this feature allows us to use it as a predictor for chaotic signals [246, 468, 554]. The behavior of an ESN for prediction is described in Eq. (1.1).

$$
\begin{aligned}
u_{t+1} &= \tanh\left(W_{\text{in}} \cdot [1, x^{\text{T}}]^{\text{T}} + W u_t\right), \\
o &= [1, x^{\text{T}}, u_{t+1}^{\text{T}}] \cdot w_{\text{out}},
\end{aligned}
\tag{1.1}
$$

where vector $x \in \mathbb{R}^n$ is the input vector, $o \in \mathbb{R}$ is the single output, W_{in}, W, are the weight matrices of the input and hidden layers, respectively, and w_{out} is the weight vector of the output layer. Here, there is an output weight vector instead of an output weight matrix because there is a single output. $W_{\text{in}} \in \mathbb{R}^{m \times (n+1)}$, $W \in \mathbb{R}^{m \times m}$, and $w_{\text{out}} \in \mathbb{R}^{m+n+1}$, m is called the reservoir size. The implementation of (1.1) is shown in Algorithm 1. Figure 1.1 shows the diagram of ESN behavior.

In Algorithm 1, as the used network has a single output, instead of an output matrix an output vector, w_{out}, is used. The first row of input matrix W_{in} and the first element of output vector w_{out} form the neuron's bias values. When the network is trained, at the initialization step, input and reservoir weights (and bias) are assigned randomly, and only the output vector values are modified.

For the network training phase, we must have a set of data pairs with the input and the corresponding output. For s pairs of data input x_i and output y_i, for $i = \{1, 2, \ldots, k\}$, it is calculated the vector u_{t+1} using Eq. (1.1), and this result is stacked in matrix (here we suppose that we start at $t = 0$)

$$
X = \begin{bmatrix}
1 & x_1^{\text{T}} & u_1^{\text{T}} \\
1 & x_2^{\text{T}} & u_2^{\text{T}} \\
\vdots & \vdots & \vdots \\
1 & x_k^{\text{T}} & u_k^{\text{T}}
\end{bmatrix}.
\tag{1.2}
$$

X has size $k \times (n + m + 1)$. And the output weights vector w_{out} is calculated as

$$
w_{\text{out}} = X^+ y,
$$

where X^+ denotes the Moore-Penrose's pseudo-inverse of matrix X.

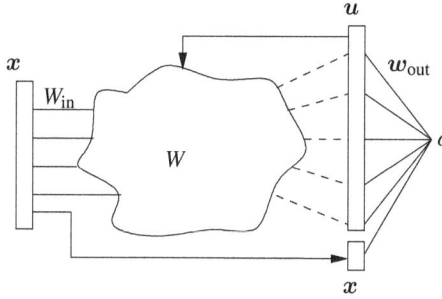

Figure 1.1   Diagram of the behavior of an ESN with a single output.

Algorithm 1 Implementation of the ESN network applied to solve a prediction problem.

Require: $\{x\}$, $x \in \mathbb{R}^n$, a set of input vectors,
   n input samples, m reservoir size,
   $W_{\text{in}} \in \mathbb{R}^{m \times (n+1)}$ input weights matrix,
   $w_{\text{out}} \in \mathbb{R}^{n+m+1}$ output weights vector,
   $W \in \mathbb{R}^{m \times m}$ reservoir weights matrix.

Ensure: One output value o for the predicted output.

1:   $u, v \in \mathbb{R}^m$ $\triangleright$ u and v are auxiliary vectors of size m
2:   $u \leftarrow [0]$ $\triangleright$ Vector u is initialized to zero
3:   **for each** input vector x **do**
4:       $v \leftarrow W_{\text{in}} \cdot [1, x^{\text{T}}]^{\text{T}}$ $\triangleright$ Input layer calculation
5:       $v \leftarrow v + Wv$ $\triangleright$ Hidden layer calculation
6:       $u \leftarrow \tanh(v)$ $\triangleright$ Calculation of the neurons' activation function tanh
7:       $o \leftarrow [1, x^{\text{T}}, u^{\text{T}}] \cdot w_{\text{out}}$ $\triangleright$ Single output calculation
8:   **end for**

1.2.1   ESN FOR CLASSIFICATION TASKS

The ESN is used bit differently to solve classification tasks. Algorithm 2 shows the pseudocode of function predict(), which is an ESN to solve a classification problem. The input vector x is of size n, where n is also the number of features of the input vector. The input weight matrix is of size $W_{\text{in}} \in \mathbb{R}^{n \times m}$, $W \in \mathbb{R}^{m \times m}$ is the reservoir, and $W_{\text{out}} \in \mathbb{R}^{n \times n+m}$ is the output matrix, where m is the reservoir size. The implementation of the class assigned to an input vector x, is described in Algorithm 2.

   The procedure to train the output matrix W_{out} for a classification task is a little different to predict, as explained at the beginning of Section 1.2.

   The function inverse_activation in line 12 of pseudocode 3 performs the inverse of the sigmoid function, and it is coded as

$$f(x) = \ln \frac{0.01 + 0.98x}{0.99 - 0.98x}. \tag{1.3}$$

Algorithm 2 An ESN to solve a classification task. The reservoir matrix is of size $m \times m$.

Require: A trained ESN: input, reservoir, and output weights, in matrices W_{in}, W, and W_{out}, respectively.
Require: Input vector x with n features.
Ensure: The class in $[0, 1, \ldots, n-1]$ to which x belongs

 1: **procedure** DECISION(x)
 2:      $V \leftarrow$ zeros($1 + $ n_input $+ m$, n) $\triangleright$ n_input $= 1$
 3:      $v \leftarrow [0]$ $\triangleright v \in R^m$
 4:      **for** $i = 0 : n-1$ **do**
 5:          $u \leftarrow x[i, 0]$
 6:          $v \leftarrow \tanh(W_{in}[1, u]^T + Wv)$
 7:          $V[:, i] \leftarrow [1, u, v^T]^T$
 8:      **end for**
 9:      $Y = W_{out}V$ $\triangleright Y \in R^{n \times n}$
10:      $z = \text{mean}(Y, 1)$ $\triangleright z$ is a vector with the mean by each row of Y
11:      **return** argmax(z)
12: **end procedure**

In this manner, Eq. (1.3) gives the value of -4.6 when $x = 0$.

1.2.2   RESERVOIR CALCULATION

The GitHub of the project EasyESN [735] has a naive form to generate the weights of the reservoir, as described in pseudocode 4. The rand() function in line 1 of pseudocode 4 returns a pseudo-random variable within the interval $[0, 1]$ with a uniform distribution.

As it is possible to see in line 2 of pseudocode 4, d is not a value of the density of the values in matrix d. The values of d represent a threshold for the random values in matrix A, which is below that value and set equal to zero. In the simulations with this naive reservoir calculation, it often produces a singular matrix, where at least a single column or row is filled with zeros.

To avoid the generation of singular reservoirs, we propose a new form of generate the reservoir. The minimum density is with a single value in each column or row in the matrix that represents the reservoir; this is equivalent to hidden neurons with a single input. The proposed way to generate the reservoir matrix is shown in pseudocode 5.

An ESN must have the echo state property, whereby it asymptotically eliminates any information from the initial conditions. It is empirically observed that the echo state property is obtained for any input if the spectral radius (i.e. the maximum absolute eigenvalue of W) is adjusted to be smaller than unity [602].

1.3 EXPERIMENTS

Two experiments were performed. The first one with four synthetic bidimensional data and two classes. And the second one with 10 real data of the University of California in Irvine Machine Learning Repository [162] as used in [167].

For the first experiment, four synthetic sets of data were selected to train an ESN. Three sets are the same as the ones presented in the software Scikit Learn [2] to show the performance of the different algorithms for classification available in the software package (see Figure 1.2). The fourth data set was created to show two classes with a not linear and very clear separation between the two classes.

Algorithm 3 Procedure to calculate the weights of the output layer of an ESN to solve a classification task.

Require: Input matrix X of size $s \times n$, for s samples, and the vector y of target classes for each sample.
Require: Input and reservoir matrices W_{in}, and W, already initialized.
Ensure: The weights of output matrix W_{out}.

1: **procedure** FIT(X, y)
2:    $v \leftarrow [0]$ $\triangleright\ v \in \mathbb{R}^{m \times 1}$
3:    $V = [0]$ $\triangleright\ V \in \mathbb{R}^{1 + \text{n\_input} + m \times s \cdot n}$
4:    $Y_{\text{target}} = [0]$ $\triangleright\ Y_{\text{target}} \in \mathbb{R}^{o \times s \cdot n}$
5:    **for** $i = 0 : s - 1$ **do** $\triangleright\ s$ samples
6:      $v = [0]$ $\triangleright$ Clean vector v
7:      **for** $j = 0 : n - 1$ **do**
8:        $u \leftarrow X[i, j]$
9:        $v \leftarrow \tanh(W_{\text{in}}[1, u]^{\text{T}} + W v)$
10:       $V[:, i \cdot n + j] = [1, u, v^{\text{T}}]^{T}$
11:                                 $\triangleright$ Set the target values:
12:        $Y_{\text{target}}[:, i \cdot n + j] = \text{inverse\_activation}(y[i, :])^{\text{T}}$
13:      **end for**
14:    **end for**
15:    $W_{\text{out}} = Y_{\text{target}} V^{-1}$
16:    **return** W_{out}
17: **end procedure**

Algorithm 4 Naive creation of the reservoir matrix of size $m \times m$.

Require: Size of the reservoir m and its density value d.
Ensure: Reservoir matrix W of size $m \times m$

1: $A \leftarrow \text{rand}() - 0.5$ $\triangleright$ Values in matrix A are in interval $[-0.5, 0.5]$
2: $\text{Mask} \leftarrow \text{threshold}(A, d)$ $\triangleright$ Values in $A_{i,j} < b$ are set to 0.0
3: $W \leftarrow \text{apply}(A, \text{Mask})$
4: $e \leftarrow \text{maximum\_eigen\_value}(W)$
5: $W \leftarrow W / e$

Algorithm 5 Proposed generation of the reservoir matrix of size $m \times m$ with the minimum density.

Require: Size of the reservoir m
Ensure: Reservoir matrix W of size $m \times m$

1: $a \leftarrow \text{rand}() - 0.5$ ▷ Vector a is of size m
2: $p \leftarrow [0, 1, \ldots, m-1]$ ▷ Vector p is also of size m
3: $W \leftarrow \text{zeros}(m, m)$
4: $p \leftarrow \text{shuffle}(p)$
5: **for** $i = 0 : m - 1$ **do**
6:     $W[i, p[i]] \leftarrow a[p[i]]$
7: **end for**
8: $e \leftarrow \text{maximum\_eigen\_value}(W)$
9: $W \leftarrow W / e$

Hundred samples are generated for each data set. For the first three data sets, 60% of data is used for training and 40% for testing. For the last set, 75% and 25% of data are used for training and testing, respectively. Input data was normalized in the range $[-1, 1]$. The hyperbolic tangent function was used at the output in the neurons at the hidden layer, as it is possible to observe in Algorithm 2.

The results of the four data sets with the algorithms nearest neighbors, linear support vector machine (SVM), radial basic function SVM, Gaussian process, and feed-forward neural network are shown in Figure 1.2. The column at the left in

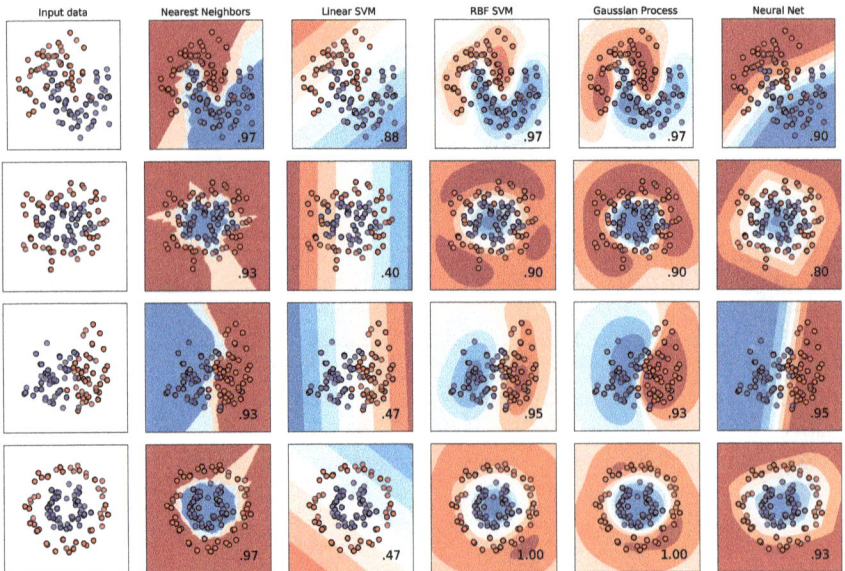

Figure 1.2 The four data sets and the results of some classifiers available in SciKit Learn [2]. The plots show training points in solid colors and testing points semi-transparent. The lower right number shows the classification accuracy on the test set.

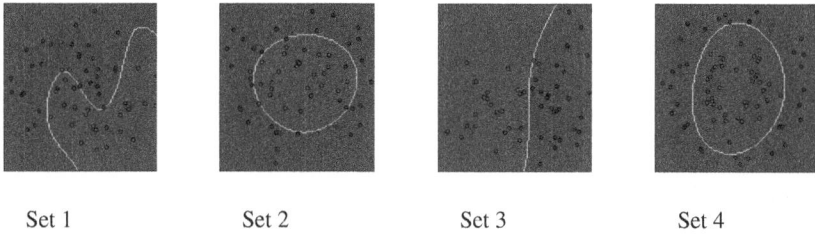

Set 1   Set 2   Set 3   Set 4

Figure 1.3 The surface of the decision obtained with the ESN on the four synthetic data sets. Training samples for each class are represented with solid circles, and testing samples are marked with "x".

Figure 1.2 shows the four data sets, and on each of the other columns are the results with each classification algorithm. The numbers on the graphs in this figure show the accuracy obtained with the test data subset.

For our test with the ESN, a five-fold validation was used. Twenty times were iterated with a new reservoir using Algorithm 5, and the best result was reported.

The results of the ESN in the four data sets are shown in Figure 1.3. The variables that can be used in the training phase are the reservoir size, and the seed of the random number generator to initialize all the input and reservoir weights. The used values in the reservoir size and in the resulted training and test accuracy with a five-fold cross-validation are shown in Table 1.1.

The optimum reservoir size was searched for sizes 5, 10, 15, 20, and 25. The results are shown in the four graphs in Figure 1.4. For example, for the first data set, observing the graph in the left upper part of Figure 1.4, the optimum size for the container is 20 that is when the test accuracy value is the biggest.

Comparing the accuracy in Figure 1.2 with our obtained results with ESN in Table 1.1, we can see that the results for sets 3 and 4 are similar to the ones obtained with the best algorithms in Figure 1.2: the nearest neighbors and SVM using radial basic functions (RBF SVM). The accuracy obtained with ESN for sets 1 and 2 (0.93 and

---

Table 1.1

Values in reservoir size and training were used, and the accuracy was tested for each synthetic data set with a five-fold cross-validation.

Data set	Reservoir size	Training accuracy	Test accuracy
1	20	94.72%	92.53%
2	5	87.50%	87.80%
3	10	95.04%	94.99%
4	5	98.83%	99.00%

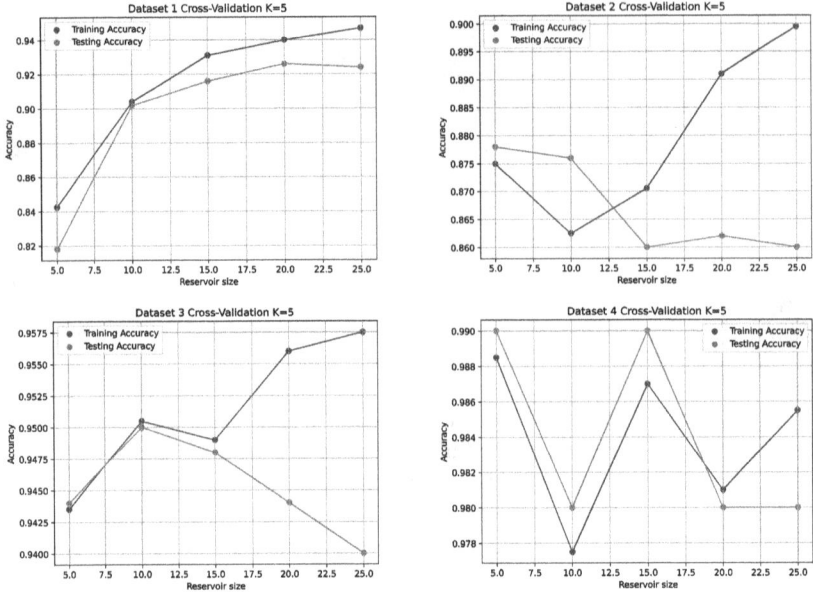

Figure 1.4   The search for the best reservoir size for the four synthetic data sets.

0.88) is greater than the ones obtained using feed-forward neural networks (0.90 and 0.80 in Figure 1.2).

All Python scripts used in the experiments are publicly available at `https://cs.cinvestav.mx/~fraga/ESN1.tar.gz`

The results for the second experiment with 10 real databases are shown in the Table 1.2. In four databases, we obtained better results in Breast Tissue, Seeds, Balance scale, and Diabetic retinopathy. In datasets Image Segmentation and Banknote authentication, we obtained similar results (see Table 1.2). And in the rest of the four databases, the results obtained in [167] are clearly better than our results. These last differences should be studied in the near future.

The search for the best reservoir size for each real data was also performed, as in the case of the first experiment with synthetic data. Results are summarized in graphs in Figure 1.5. Authors in [167] do not show with which reservoir size obtained their best results also shown in Table 1.2, thus we cannot compare directly our results. In [167] also use an ESN with a reservoir with more neurons linked from the input layer with the reservoir. Remember that we use the minimum possible link with a single neuron (see Algorithm 5). Therefore, we increase the density in the reservoir for the cases of the Banana and Wine databases, these results are shown in Figure 1.6. As we can see in Figure 1.6, changing the reservoir density does not increase significantly the already obtained accuracy.

Table 1.2

Results to apply ESN to ten real classification problems. F represents the number of features, C is the number of classes, S is the number of samples in the database, Tr is the training accuracy, Te is the test accuracy. Res is the obtained reservoir size. In bold is marked the best testing accuracy in comparison with [167].

Dataset	F	C	S	Tr [167]	Te [167]	Tr	Te	Res
Image Segmentation	19	7	2315	95.87	**89.50**	89.60	88.90	1000
Pima Indian Diabetes	8	2	762	84.23	**77.79**	73.13	70.20	300
Wine	13	3	178	100.00	**97.44**	62.50	40.20	300
Breast Tissue	9	6	106	93.65	22.43	76.80	**67.10**	100
Cardiotocography	22	3	2126	95.90	**92.89**	89.20	84.54	1000
Seeds	7	3	210	99.40	93.33	94.65	**93.47**	50
Banknote authentication	4	2	1372	100.00	**100.00**	99.47	99.34	250
Balance scale	4	3	625	94.72	89.76	90.91	**90.93**	150
Banana	2	2	5300	90.86	**84.32**	74.50	74.52	10
Diabetic retinopathy	19	2	1151	79.39	73.35	80.25	**77.70**	1000

1.4 DISCUSSION

It is very easy to use an ESN, indeed the Python package that was used to preform the experiments is called *easyesn* [735].

It is well known that a SLFN can be used to separate two classes if the separation region is not linear between them [268]. A neural network without a hidden layer can separate two classes if the region between them is linear [268]. The decision surface in the problem with two classes and synthetic data in Figure 1.3 is non-linear; thus the ESN produces a non-linear decision surface.

An ESN is normally used to classify or in regression problems depending time data. As it is possible to see in this work, ESNs can be used to solve classification tasks.

There is not yet an algorithm to obtain the best reservoir size and its density. The results obtained here were obtained by trying several combinations. The result depends also directly on the value of the seed for the random number generator. Here, simple algorithm is used to avoid the creation of singular reservoirs when the hidden weights result in a singular X matrix in (1.2), and then it cannot be inverted.

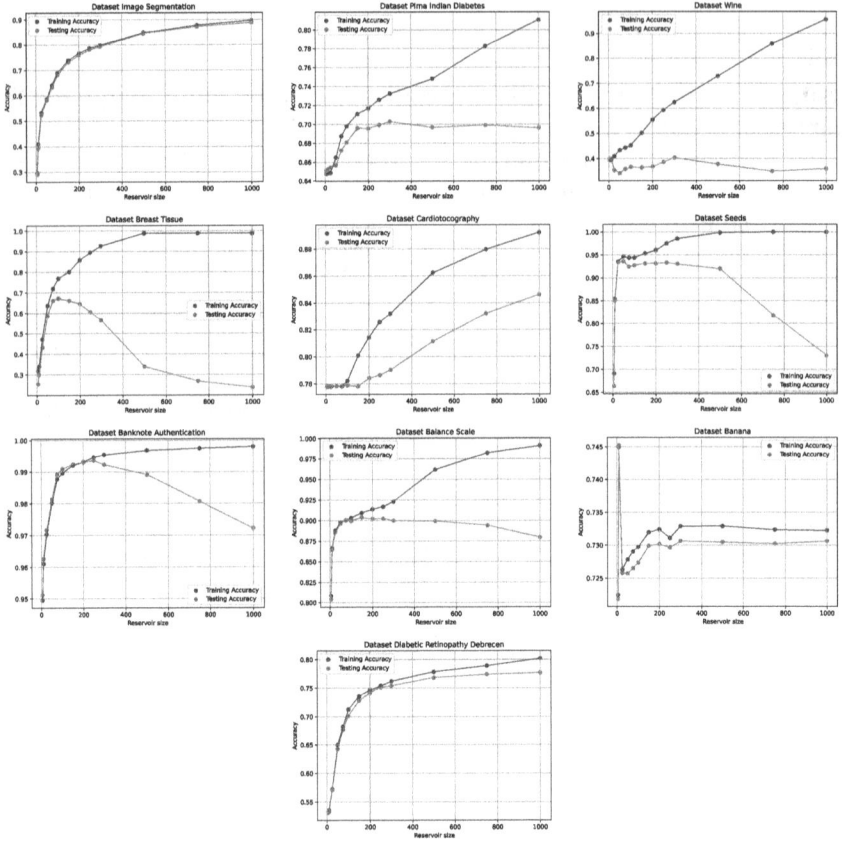

Figure 1.5 The search for the reservoir size for the ten databases with real data.

As the proposed reservoir has the minimum size of hidden neurons with a single input, the model is very suitable for a hardware implementation of the classifier model. The implementation of hardware models of ESN [305] is a research topic in the near future.

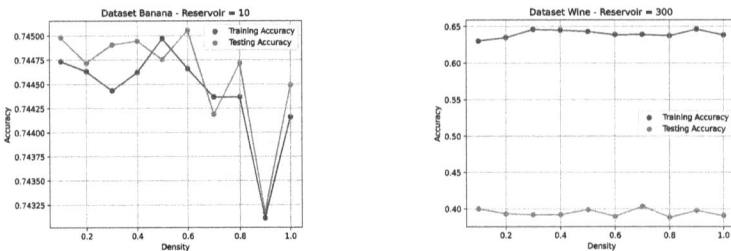

Figure 1.6 Searching for the best reservoir density for the Banana and Wine databases.

Also, as a future work, the automatic search for the size of the reservoir, searching for the best permutation of the reservoir neurons, and the best seed for the pseudo random generator, could be performed using a genetic algorithm.

1.5 CONCLUSION

This work showed that an ESN can be used to solve classification tasks with non-temporal data. The advantage of using an ESN is the training time: it is equivalent to the computational effort of inverting a matrix, and it is not necessary to do long calculations as in a forward neural network. It is necessary for more research to offer an algorithm to select automatically the values for the hyperparameters used in an ESN to solve a specific task.

2 Continual Learning for Camera Localization

Aldrich A. Cabrera-Ponce, Manuel Martin-Ortiz
Faculty of Computer Science, Benemérita Universidad Autónoma de Puebla (BUAP), Puebla 72592, Mexico

Jose Martinez-Carranza
Department of Computational Science, Instituto Nacional de Astrofísica, Óptica y Electrónica (INAOE), Puebla 72840, Mexico

CONTENTS

2.1 INTRODUCTION

Camera localization has been a computational challenge within the vision and robotics tasks, such as tracking, detection, and navigation within a scenario. Hence, it is necessary to determine the camera's pose based on an image captured by a robot while developing an action. In contrast to alternative sensors, cameras offer more information that enables more robust processing and facilitates the acquisition of localization through visual descriptors, landmarks, patterns, and other visual data present in the surroundings [414]. Moreover, visual information allows us to develop novel manoeuvres of multiple tasks involving ground-based and aerial robots, thus serving as a helpful tool for localization [632].

DOI: 10.1201/9781003385615-2

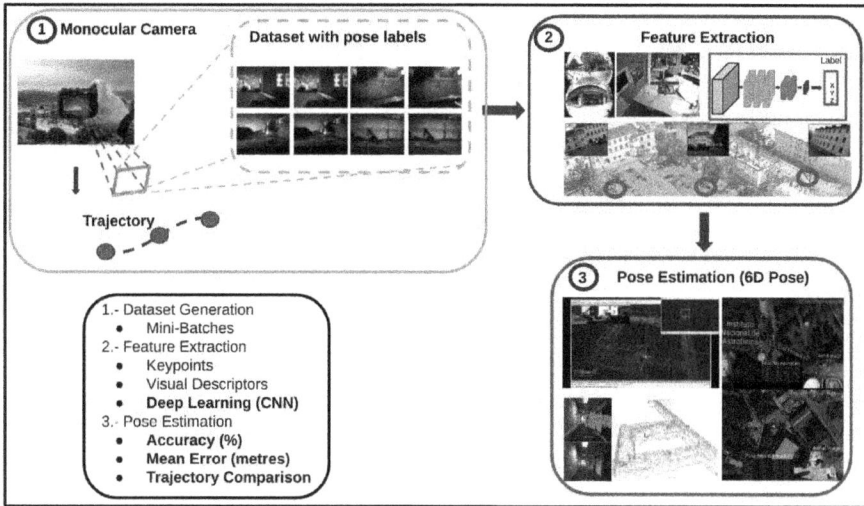

Figure 2.1 Methodological process used to obtain the poses and localization of the camera from a single image.

Motivated by the latter, the scientific community in the field of robotics has explored several methodologies to estimate the camera's poses, such as nearest neighbours [414], visual descriptors, simultaneous localization and mapping (SLAM) [437], and visual odometry (VO) [425]. These methods aim to address the challenge of camera localization using feature extraction and visual information to obtain the image's 3D coordinates. In this way, it is possible to achieve localization without relying on additional information such as depth, triangulation calculations, or metric measurements. Nevertheless, these methods are limited in scenarios with sparse features, resulting in low accuracy localization .

In order to tackle this challenge, deep learning emerges as a tool for extracting information using convolutional neural networks (CNNs) from a dataset. Specifically, architectures such as PoseNet have been developed to leverage images of the environment and corresponding coordinate labels for monocular camera-based pose estimation [292]. Figure 2.1 shows the overall process for dataset generation, the application of deep learning and other methods to extract visual information, and obtaining the camera's localization . Due to the diverse techniques for camera localization available, we propose reviewing the state-of-the-art methodologies focused on pose estimation in indoor and outdoor scenarios.

Furthermore, we incorporate the advances in continual learning techniques applied to vision problems and localization , exploring the strategies developed to enable incremental learning in the lack of a predefined dataset. Additionally, we discuss the relevant works employed to address the same challenge, showing their contributions to the learning and vision domain.

In order to present the findings of this study, we have structured our work as follows: Section 2.2 explains the importance of conducting this review and summarises the literature search. Section 2.3 provides an overview of the datasets utilised for the different techniques. Section 2.4 presents the existing methods and strategies for computer vision and localization tasks using deep learning. The principal methodologies utilised in this review are present in Section 2.5. Section 2.6 presents an analysis and discussion of the relevant works reviewed. Finally, Section 2.7 shows the conclusion derived from this review and our viewpoint.

2.2 PURPOSE OF THE REVIEW

This literature review aims to present those methodologies that focus on image processing for tasks such as segmentation, detection, and localization , with a specific emphasis on the latter in order to obtain camera poses using computer vision, mapping systems, and deep learning. The latter has greatly interested the scientific community due to networks' practicality and ability to be trained using image datasets. Convolutional Neural Networks (CNNs) extract and learn from visual information to create a model to predict the expected position. Consequently, a learning model can acquire knowledge of poses from a dataset and localise a camera based on a single evaluation image of the given scenario.

Nevertheless, these learning methodologies require a large dataset and multiple training iterations to acquire the camera's pose, which requires substantial computational resources for constructing the learning model. In this context, this study also reviews state-of-the-art methodologies focused on continual learning using images for classification, detection, and localization tasks. Our objective is to identify and provide the reader with an overview of methods concentrating on camera localization , enabling a network to learn continuously without needing a pre-existing dataset or extensive resource consumption. Furthermore, we prioritize works that exhibit sufficient precision and contribution to prevent catastrophic forgetting, thus safeguarding the final localization outcome. Consequently, for this review, we identified 125 articles classified into four methodologies: computer vision, visual odometry (VO) and simultaneous localization and mapping (SLAM), end-to-end learning, and continual learning (Figure 2.2).

2.2.1 SEARCHING

The search for relevant works followed an organised scheme of a systematic review, considering the problem statement, methodology, and results. Initially, we conducted a search focusing on image processing within the field of robotics, using various computer vision techniques to address specific tasks. These techniques encompassed detection, extraction, tracking, segmentation, and localization . Examples of such methodologies include ORB-SLAM2 [438] and PoseNet [292], which concentrate on estimating camera poses through feature extraction, mapping, and deep learning approaches [7, 414]. However, continual learning strategies still require further improvement, given the ongoing advancements in these techniques. Consequently, we identified an average of 15 papers that attempt to tackle the localization problem within this domain.

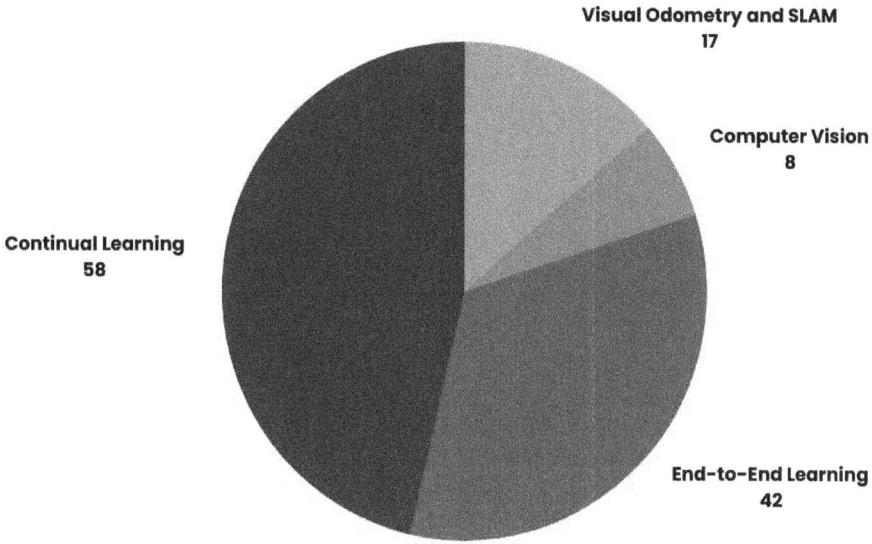

Figure 2.2   Number of papers found divided into methodologies to be reviewed.

Based on this search, we conducted a comprehensive analysis of these works to identify the methodologies, metrics, evaluations, datasets, and results that successfully address the problem. Additionally, our search encompassed over 100 papers sourced from conferences and impact journals within the fields of computer vision and robotics. We utilised relevant keywords related to the topic and feature extraction, such as descriptors, edges, lines, blobs, and abstract information obtained through CNNs. Through this approach, we identified the best papers that aligned with the objective of this review. Figure 2.3 illustrates a concise flow diagram outlining the phases of our search process to select state-of-the-art papers.

We prioritised papers that have had a significant impact on the scientific community, including those that are still in the development phase. We used various tools and platforms to conduct our searches, such as Google Scholar, Scopus, Research-Gate, IEEE Xplore, and ELSEVIER. We also used the INAOE Institute's digital

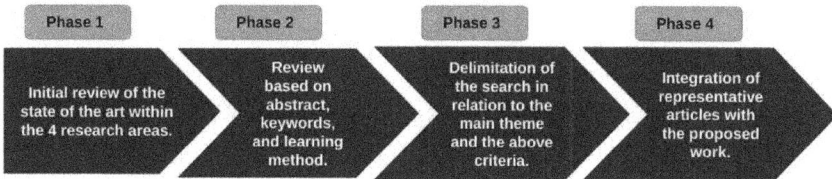

Figure 2.3   Phases carried out to search and review the state-of-the-art papers.

Table 2.1

Number of papers reviewed in conferences, indexed journals, and ArXiv.

Publication	Number of papers
Conferences	52
Journals	40
ArXiv	33
Total	125

library and the BUAP. We also explored recognised conferences like IROS, ICRA, IMAV, CPVR, and MCPR, which often feature works focused on robotics and computer vision. Furthermore, we explored forums and websites such as arXiv, where the research community shares unpublished articles and scientific reports related to localization to foster further advancements and research. Table 2.1 provides an overview of the number of papers we found in conferences and journals, where conference papers often propose novel methods that are subsequently published in journals.

In addition, we thoroughly examined the supplementary content provided in each paper to enhance their methodology. This includes datasets, source code, demonstration videos, applications for tablets or mobile phones, and web pages that augment the applied methodology. Table 2.2 illustrates the amount of additional information we found in these papers. This supplementary material is crucial for understanding the data utilised in experiments and the corresponding results. Finally, we proceed to present a detailed review, description, and comparison of selected works focused on camera localization tasks in the following sections.

Table 2.2

Number of papers that provide: Demonstration videos, dataset links, source code, web pages, and applications.

Additional content	Number of papers
Demonstration videos	19
Dataset	28
Source code	45
Web pages	26
Applications	7
Total	125

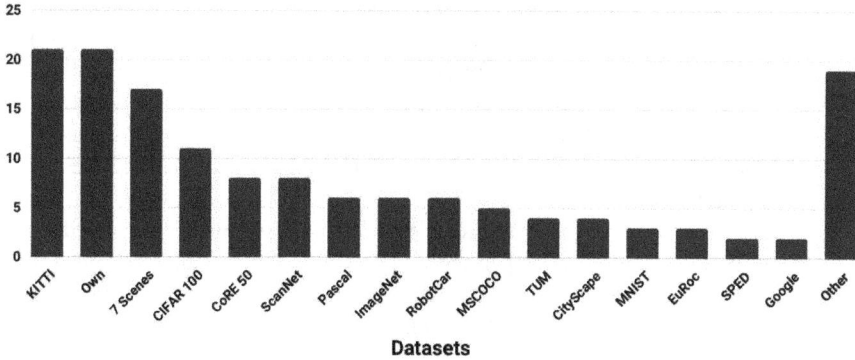

Figure 2.4 Datasets used in the different methodologies for detection, classification, segmentation, and localization tasks.

2.3 DATASETS

This section focuses on the datasets utilised in the reviewed works to conduct their experiments and achieve the desired outcomes. These experiments encompass various tasks, including object segmentation, tracking, classification, and recognition, using CNNs and learning models. Moreover, some of these datasets were employed for camera localization tasks, allowing for the comparison of experimental frameworks and trajectories with other approaches. The works that utilise these datasets for camera localization typically involve visual odometry, SLAM, and neural networks for pose estimation. In this session, we aim to provide an overview of the most commonly used datasets in state-of-the-art research and describe the information and labels they contain, contributing to the final outcomes.

In Figure 2.4, we can observe the number of works that have utilised various datasets for localization tasks, including KITTI, 7 Scenes, and custom-created datasets. KITTI has emerged as one of the most popular datasets for urban scenarios, offering a comprehensive collection of stereo images, point clouds, 3D maps, and position annotations. This dataset has been extensively used in visual odometry and SLAM systems such as [88,437,661]. Its extensive image set provides sufficient data for trajectory evaluation and autonomous navigation.

Another dataset, 7 Scenes, and its variations (14 Scenes and 21 Scenes), feature locations captured in Oxford, encompassing RGB images, depth data, 3D models, and camera orientation and position annotations that precisely determine the localization where the image was captured. This dataset has found application in deep learning works for creating learning models and estimating the camera's poses with respect to a Ground Truth, as demonstrated by [290–292].

However, some work such as [38,79,345,380] created their dataset using monocular cameras in indoor and outdoor scenarios, within laboratories, institutes, and even on the streets of a city to carry out their experimental frameworks. Also, they capture multi-view for better information about the scenario, store the position annotations for each image, and in some cases, apply synthetic data for data augmentation. For instance, they use simulators such as AirLoop [191], Gazebo, or 3D environment models created to generate more information about the place. They argue that the synthetic data expand the dataset and samples to train the CNN, improving the accuracy's results in the evaluation step. Besides, for comparison, some datasets are available for evaluating the output with a novel methodology and testing the results with the Ground Truth.

Likewise, other authors prefer to create their datasets for more specialised tasks, such as aerial geo-localization [80] and robot localization in agricultural fields [105]. These non-public datasets also include RGB, RGB-D images, pose and orientation annotations, and segmentation [7, 407]. Others are AirLoop [191], DeMoN, ROA, TAO, ACM, DAG, and ISPRS Potsdam, including cartography, perception, infrared and information from Photogrammetry and Remote Sensing (ISPRS). They created these datasets to solve a problem that cannot be achieved with public datasets, even for evaluating a methodology using known data of their scenarios.

Other datasets for localization tasks are EuRoc and TUM, developing for autonomous navigation and localization in indoor and outdoor scenarios using visual odometry and SLAM systems. These datasets have RGB images, point clouds, and depth images for mapping and trajectory comparison. In contrast, the datasets Scan-Net, RobotCar, and CityScape are used for deep learning methodologies to pose estimation from a learning model [731]. On the one hand, we also find datasets such as CIFAR 100, Pacal, ImageNet, SPED, and MS COCO allow the authors to recognise and classify objects based on the methodologies of detection and segmentation to track an object in frames [705, 724]. On the other hand, these datasets are essential to finding regions in certain parts of an object and thus to acquire the shape, colour, position, and orientation. We discuss these datasets due to processing to find an object, which we can use to localise a monocular camera from object detection.

Finally, we found two datasets for continual learning tasks used for object detection and classification in an incremental learning way. For example, CoRE50 [376] is a new dataset for continuous object recognition split into 11 sessions for training and seven for evaluation. This dataset considers 50 object classes and a learning model that trains every five without losing prior knowledge. MNIST is another dataset for digit and letter classification using typographic images of multiple examples. This dataset applies permutations in pixels, changing the pixel's organisation and allowing to train of a new task with different pixels configuration [303]. Figure 2.5 presents an example of all these datasets described in this session.

In summary, we found public and created datasets by the authors for specific tasks such as classification, detection, navigation, and localization . In this way, most datasets used for classifications task are CIFAR100, ImageNet, and MSCOCO for the number of objects and samples in them. In contrast, for mapping, localization , and pose estimation, the dataset used are KITTI, 7 Scenes, TUM, EuRoc, and ScanNet. This is due to pose annotations, RGB-D data, and images captured in rural scenarios and indoor and outdoor places. One of the main objectives of obtaining the localization of an object or camera is to know the scenario through datasets to support final

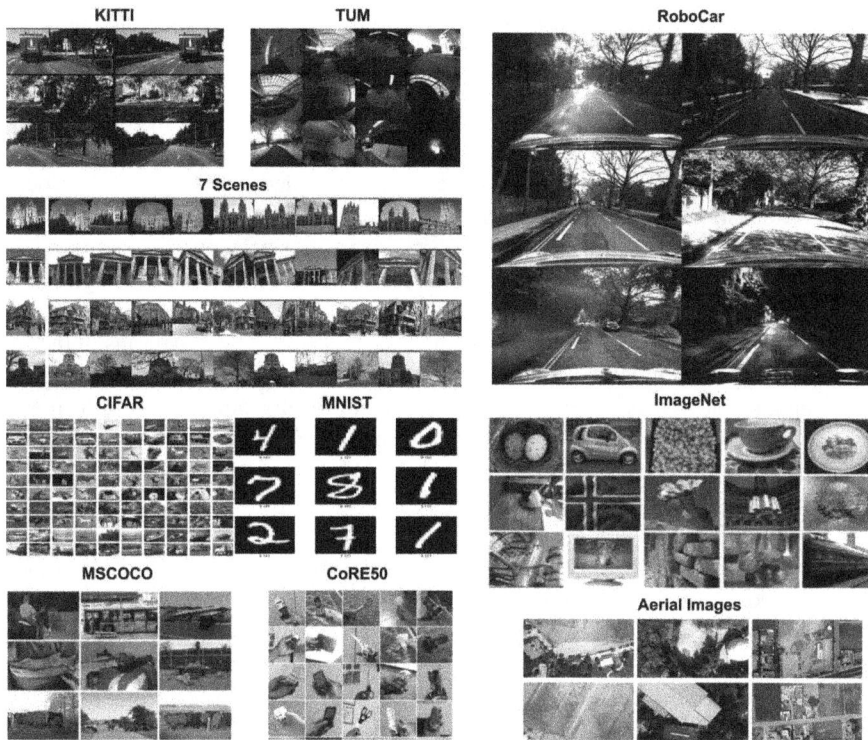

Figure 2.5 Datasets found and used in some of the papers reviewed.

results. Finally, for continual learning methods, there are existing datasets for object detection, like CoRE50. However, no dataset exists for the camera localization task, thus creating datasets for this purpose. This can contribute to the next generation of works that try to solve localization using continual learning methods and training a model on the fly.

2.4 TRAINING STRATEGIES

This section presents architectures, networks, and continual learning strategies for camera localization from a single image and other tasks. PoseNet [292] was one of the main architectures presented for camera pose estimation and localization . This architecture consists of 23 convolutional layers, nine inception modules and three linear regressions for pose estimation. Inspired by this work, several authors have tried to improve the accuracy and reduce the pose error with changes within the original architecture. For example, [547] use a theoretical model for camera pose regression and their model to show the closest pose approximation via 3D structure. In [86] presents a combination of a CNN with a GPR to propose a probabilistic framework to model the uncertainty in the regression of 6DoF camera pose.

Other works employ a lightweight CNN called SqueezeNet to adapt the network for pose regression [436] and add LSTM modules in the final layer to reduce the parameterisation of weights [555, 651]. Furthermore, VLocNet is a network for pose regression and odometry estimation using consecutive monocular images, incorporating hard parameter sharing for learning inter-task correlation [630]. Finally, an improved PoseNet architecture is presented in [712] called PoseNet++. This modification consists of a VGG-16 network to complicate out-of-image plane regression problems, replacing the last softmax classification layer with one fully connected layer with 2048 neurons and adding an output layer with seven output regression units. On the other hand, [82] presents a CompactPoseNet, in which the architecture consists of 3 convolutional layers, 3 Inception Modules, and three neurons in the final layer for pose regression. These methods aim to outperform PoseNet accuracy, ensuring a stable convergence for real-time camera localization . Figure 2.6 shows some architecture mentioned in this section using PoseNet-like based to build new architectures.

For deeper strategies, [679] presents a study of neural fields in visual computing by identifying standard components of these methods. Here, they expose five classes of techniques, such as hybrid approaches, mapping, conditioning by concatenation, hyper-networks, and meta-learning, which are important to solve computer vision tasks. Besides, he mentions using share weights, formulation, conditioning, and representations developed for neural field progress and training modalities. Some works like [74] and [119] demonstrate the importance of using depth maps, colour maps and panoramas captured with Google Street View, generating shadow and brightness to enhance the learning as well as symmetrically splitting the neurons.

On the other hand, combining two or more methods to solve the same problem is possible. For example, in [350, 425, 661] analyse the problem of monocular visual odometry using a CNN based on recurrent convolutional neural network (RCNN) architecture. This provides knowledge of the environment using visual odometry to estimate the 6-DoF pose and the depth with RCNN. Other works use depth estimation combined with pose using an end-to-end cycle optimisation and Kalman filter to reduce the pose noise from PoseNet [216]. MaskNet is an unsupervised learning framework for depth and ego-motion estimation to predict dynamic objects in the environment and position [219].

We argue that the influence of a depth image for visual odometry and the Kalman Filter leverages spatial losses and temporal losses between image sequences, performing a better estimation of the pose and depth map. Besides, it can improve the trajectory and localization into a known environment from a single image captured with a monocular camera. Figure 2.7 shows the methodology applied for depth estimation and camera localization .

Finally, another strategy is reinforcement learning [360], which considers an egocentric map and global exploration map to speed up learning and improve different size environments, allowing one to explore an environment without prior mapping. However, they must adopt deep and continual learning approaches for camera localization on the fly. In this case, [467] presents a study of the localization strategies

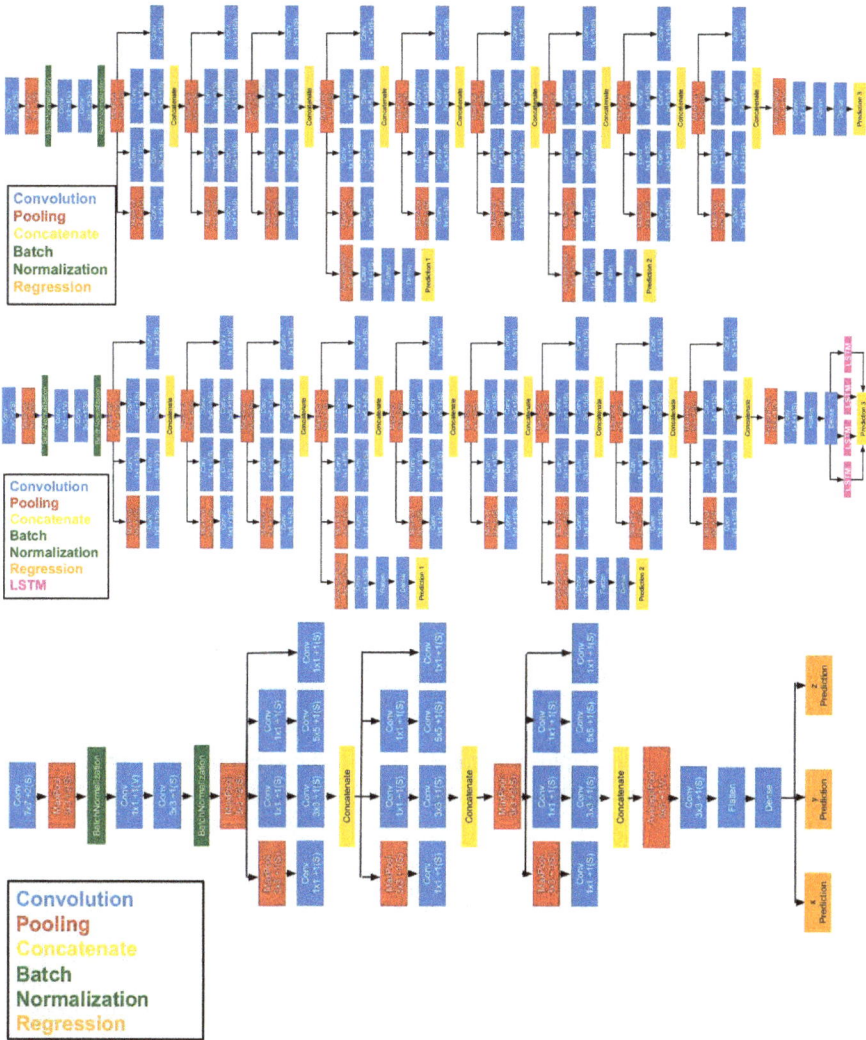

Figure 2.6 In top, the original PoseNet architecture is presented in [292], in the middle PoseNet + LSTM modules, and at the bottom the CompactPN showed in [82].

for autonomous robots, showing the main problem, principles, and approaches with Markov localization , Kalman filter, SLAM system, probabilistic schemes, and comparison of principal works for estimation and localization . Nonetheless, a continual learning approach is needed to solve this task and analyse ideas that encourage and apply these methodologies.

2.4.1 CONTINUAL TRAINING

After briefly discussing and describing some deep learning techniques, we present the continual learning techniques in this section. Although most of these strategies

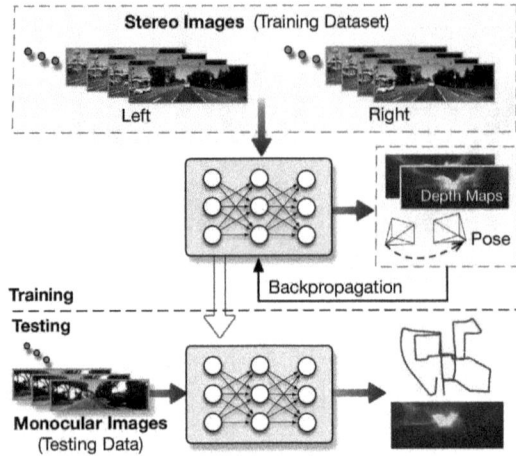

Figure 2.7 Visual odometry and depth estimation with monocular images for pose estimation and camera localization .

focus on object segmentation, detection, and classification tasks, we will mention them for the reader's knowledge. Likewise, we will present some works important for continual learning, considering that learning can be applied in a localization task or pose estimation from a single image. Firstly, we define continual learning as a branch of deep learning whose process is to continuously or incrementally acquire information over time and avoid forgetting previous knowledge. Some of the continual learning strategies are divided into three parts:

> **Architectural Strategies:** Techniques that modify the structure of the CNN, adapting it to acquire new data through weight freezing, dual memories, hybrid models, weights sharing, and hippocampus.
>
> **Regularisation Strategies:** Techniques that modify the loss functions using penalties and rules, thus avoiding the loss of knowledge and prior learning. In other words, it accesses the loss function to extend the selection time allowing it to consolidate the learning and regularize the specification of the weights.
>
> **Rehearsal Strategies:** Techniques that use subsets stored in external memories to reproduce them during the following training steps. This means that the past information is periodically repeated to update the model with new information but maintains the previous one through connections that enrich learning. Some known techniques here are using separate and external memories where the previous weights are stored, preventing catastrophic forgetting.

In this way, we show the three strategies divided in a Venn diagram in Figure 2.8. Also, we can combine 2 or 3 techniques for a better result and reject forgetting, maintaining learning for longer. This idea comes from the plasticity-stability dilemma

Architectural Strategies

CWR   PNN

ARI*

EWC

LwF   iCaRL

GEM

SynInt

Regularization Strategies   Rehearsal Strategies

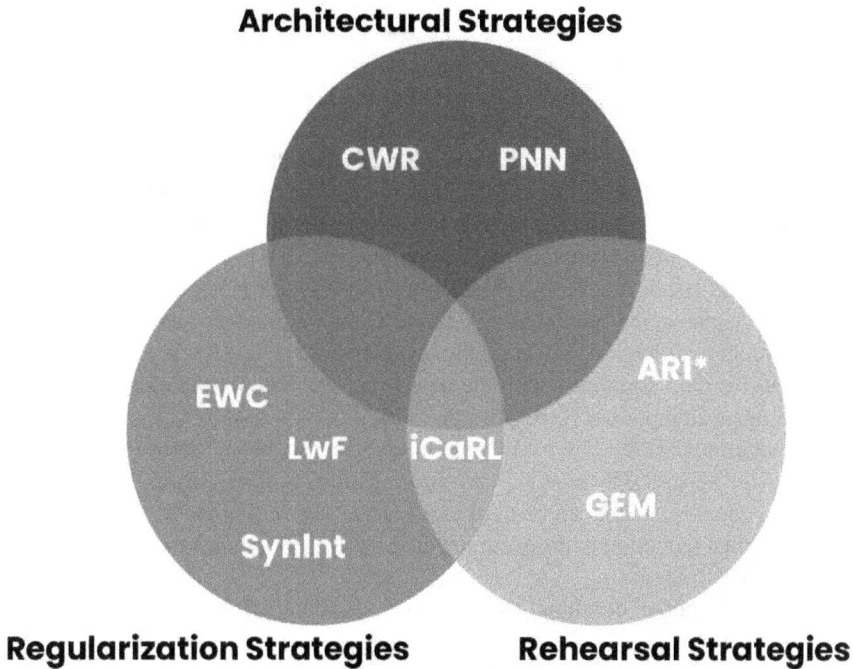

Figure 2.8   Continual learning strategies divided into three categories.

within deep learning, where a network has neural plasticity if the learning is adapted to the new data. However, if the network has sufficient stability, it can preserve previously learned information, avoiding forgetting it, but if further information arrives, it loses that experience when learning the new one. For this reason, continual learning focuses on this dilemma, maintaining neural plasticity and its stability in a network, allowing us to train continually and avoiding catastrophic forgetting.

One of the strategies presented in [377] replaces Batch Normalisation for Renormalisation, freezing deep layers, restricting weights, and reducing computational consumption when learning long data sequences. In [226] proposes a regularisation strategy on a subset of branches, thus assembling the learning for multi-level tracking representations. Another method is knowledge distillation, a technique used to transfer knowledge from a large model to a smaller one, reducing the model's size to estimate a dataset from the large model. In this way, the model's parameters are initialised instead randomised, maintaining knowledge based on regularisations and modeling rules [100]. For the latter, in [626, 716] propose a modality-specific selection of CNN pre-trained to fine-tune in combination with classifier Mondrian Forest (MF) the incremental class learning, maintaining the discrimination within old classes.

In [112, 416] use Incremental Label Propagation with a distillation method to retain the information about previously learned classes. Besides, an Auxiliary

Distillation method preserves knowledge to reconstruct previous models, allowing keep the knowledge of the last data [724]. We notice that the distillation-based techniques are used with the ideology of having two models, a teacher and student, sharing the knowledge with a student to maintain the learning. In this way, some work leverages the distillation strategy to learn a large stream of unlabeled data and increase the classes' learning while avoiding overfitting [259, 302, 337]. In addition, a distillation methodology divides the capacity into several parts and utilises them to input training, retaining the previous learning and reducing the conflict in pseudo labels by combining the previous and current models [705].

Finally, the last strategies are the rehearsal techniques using external memories to apply a latent replay methodology to the CNN architecture. This methodology consists of slow-down learning at all the layers below the latent layer where weights are stored for then combined with the new ones, maintaining the essential patterns for each instance input [478]. Besides, in [376] proposes to store activation volumes at some intermediate layer to a continual learning approach and shows the technique improves the accuracy by keeping activation features in the following training data. Also, they mention that it can be used as a baseline for further studies than the naïve solution with a reduction in the accuracy drops w.r.t the cumulative approach is large [376, 378, 479].

We present another continual learning method called lifelong learning which aims to train and adapt multi-task knowledge to iterative policies, improving the performance of knowledge to new tasks [208]. Besides, use visual perception to develop new models with several agents such as objects, dynamic changes, and annotations to navigate and refine the training [44, 450]. This method consists of a regularisation strategy but can be considered an architectural strategy, updating the model and using more scenario information to improve the knowledge in a multi-task approach. Although considering other methods used for different tasks in Table 2.3, these works are designed to solve computer vision tasks such as classification and detection using objects. Nonetheless, we consider them important due to the reproducibility of the experiments and their application to a camera localization problem. Thus, we can adapt the method to a structure or architecture and evaluate it with camera localization experiments.

2.5 METHODOLOGIES FOR CAMERA LOCALIZATION

In this session, we analyse the methodologies used in state-of-the-art works, focusing on papers that use deep learning and continual learning techniques. Also, we briefly overview the most relevant reports and the architectures or networks used to perform computer vision tasks. Finally, we present some papers that show the necessary foundations for developing and acquiring the camera pose and localization into a scenario applying these methodologies.

2.5.1 COMPUTATIONAL VISION TECHNIQUES

We present a brief description of the computer vision techniques for camera localization ; some of these methodologies are applied in tasks such as detection, autonomous navigation, and pose estimation. For example, in [672, 695] presents a

Table 2.3

Continual learning strategies for computer vision tasks such as classification, segmentation, detection, and localization .

Work	Strategy	Task
[340]	Pseudo-labelling, Regularisations	Segmentation
[296]	Regularisation and Replay methods	Depth
[608]	Regularisation	Classification
[622]	Regularisation	Segmentation
[155]	Self-labeling	Detection
[156]	Knowledge restore and transfer	Class incremental
[673]	Spatial jittering method	Object Detection
[660]	Architectural, Regularisation	Classification
[256]	A domain-shared encoder	Depth estimation
[574]	Multi-label annotation	Autonomous Driving
[646]	domain-mixing strategy	Segmentation
[35]	A contrastive regularization	Segmentation
[601]	Compositional Replay	Classification
[723]	Regularisation, Architectural, Replay	Class incremental
[614]	Generative memory	Classification
[38]	hyper networks	Manipulation
[643]	Regularisation	Classification

camera localization methodology using scale features and point cloud with noisy annotations. This method allows for a better pose estimate by considering the background and image noise. On the other hand, in [89, 386] presents a vision method for UAV navigation in unordered scenarios using features obtained from a point cloud and a laser sensor. However, there are multiple methods to acquire information from the environment and thus get the localization of a UAV for navigation tasks. The techniques mostly use extra sensors or visual information extracted from some mapping produced by a camera or laser, and it is optional to acquire the location of an image by relying on other data sources.

In addition, we describe five computer vision techniques in Table 2.4, where they seek the position from overlapping views, sensor fusion, and extension of scenes with feature trees. With this, we can realise the applicability of the techniques to develop and obtain the camera pose with them. However, large consumption of computational resources is required to get an adequate result for the camera localization so that a deep learning methodology can give us a better result.

2.5.2 VISUAL ODOMETRY AND SLAM

Another scenario in which camera localization is used is with Visual Odometry (VO) and SLAM system. These methods allow obtaining a robot's position with an onboard camera within a scenario, thus determining the pose and orientation of the

Table 2.4

Camera localization using computer vision techniques.

Work	Technique
[717]	Overlapping views for pose estimation
[407]	Fusing 6D camera's pose and visual odometry
[588]	A vision-based system with an infra-red and RGB camera
[361]	Estimation using RGB camera in 3D environments
[414]	Extend scene with regression forests using features

robot. Works have performed localization by combining estimation with deep learning and features obtained from a map [350, 366, 425, 661]. Other approaches have employed building a SLAM map of the scenario and thus locating the camera on board a robot. Under this scenario, features of the site are extracted based on Landmarks to provide essential site information, to then build an area map.

With the above, a SLAM system exploits visual information and finds matches between a previous image and the current image to project points onto a 3D plane that will shape the map of the location. Visual descriptors are information composed of landmarks and a description at a point in that image and are thus used to recognise paths based on the match of that point to that of the scene [418]. The best known SLAM-based works are [88, 437, 438] presenting elaborate systems of a SLAM to localise a monocular camera, stereo, RGB-D, as well as IMU sensors in combination with an image. On the other hand, in [270, 721] uses a Lidar sensor to map implicit representations and leverage binary cross-entropy loss to optimise the global feature within the scenario map. Other works use visual odometry in combination with CNN for pose estimation and robot localization [210, 462], whose purpose is autonomous navigation in different scenarios through a mapping approach.

Finally, existing works employ the slam system and visual odometry in combination with deep learning methodologies for multi-scale localization and camera localization using monocular, stereo, and RGB-D cameras [499, 610, 719]. The novel combination allows the interpretability of the scenario through a map and thus estimates the position using deep learning approaches. We consider these works novel for tasks requiring autonomous navigation as well as constant localization of the robot within competitive scenarios. However, it is still necessary to have a large dataset or large features to build the learning model. In addition, we summarised some works in Table 2.5 for visual odometry and SLAM system for camera localization tasks.

2.5.3 END-TO-END LEARNING

For deep learning methods, we found work combining cross-information between images captured with a drone and satellite imagery to retrieve the position [563]. However, such approaches still rely on georeferencing, coordinate systems, and

Table 2.5

Camera localization works using visual odometry and SLAM techniques.

Work	Technique
[437]	ORB-SLAM using ORB descriptor
[438]	ORB-SLAM2 using 3 cameras types for localization
[88]	ORB-SLAM3 with 4 kinds of camera and IMU initialisation
[334]	3D mapping and localization system for agricultural robots
[418]	A visual navigation under changing conditions scenarios
[425]	A prior knowledge of the environment using a Visual Odometry
[350]	A novel deep end-to-end networks for long-term 6-DoF VO task
[661]	A DL-based monocular VO algorithm
[691]	Improve ORB-SLAM2 for camera localization and Mapping

ground planes where the presence of the information may not be available; also, the re-localization of a camera from a single image may need to be corrected. For the latter, the use of neural networks to solve the problem of camera localization has grown in the last decade. Mainly by using the captured database that can predict position using regression architectures such as [547].

In this way, the estimation of a camera's pose is based on the number of samples in the training dataset used to train the network, one of the most commonly used architectures being the PoseNet [292]. Sometimes these systems are combined with other architectures to create a hybrid model capable of re-localising a camera in multiple scenarios [86]. This has provided an important factor that allows the architectural basis of neural networks to be explored and exploited to recognise where we can capture the images and pose estimation. On the other hand, some authors use this architecture for geo-localization purposes using GPS-like annotation and aerial images to train a learning model. Thus, the learning model gives a close geo-localization to create planned routes with geographic information that determine the locations of the images that have been found.

In addition to the previously mentioned works, SPED [109] is a known work for scenario recognition using visual descriptors and images taken from Google Street View. This work mentions that having shots from different angles and perspectives allows localising the photograph's place, city, and country using the geographical coordinates estimate. Camera localization using deep learning is beneficial for robotics, using only a dataset with images and either 3D or GPS annotations [358]. Furthermore, deep learning can estimate a possible position in outdoor scenarios where the difficulty with multiple factors such as lighting, change of location, perspective, and scale has presented a challenge to localising a robot with an onboard camera using traditional visual methods [62, 399].

Finally, [451] presents two outdoor localization methods based on deep learning and landmark detection, successfully determining the robot's coordinates and orientation. Similarly, a geospatial deep neural network (CNN+LSTM) estimates

Table 2.6

Camera localization works using end-to-end learning.

Work	Technique
[82]	Learning over small and highly batches
[292]	Regression CNN for pose estimation
[290]	Loss modeling for CNN and pose estimation.
[291]	A geometry method for improve the estimation
[80]	A CNN for geo-localization with aerial images
[605]	Scene agnostic camera localization using DSM
[563]	An image-based cross-view geo-localization method
[366]	A end-to-end networks for long-term 6-DoF VO task
[547]	Theoretical model for camera pose regression and 3D structure
[86]	Combine a CNN and GPR to regression of 6DoF camera pose
[543]	A solution to end-to-end learning of pose estimation

precise geographic locations using only ground-based [597] imagery. Others use un-supervised learning for depth estimation and re-localization using video frames in indoor scenarios [600] and in combination with visual odometry [210, 462]. As a summary of the review of deep learning methods for localization , we present in Table 2.6 the main works for estimating and localising a monocular camera from a single image.

2.5.4   CONTINUAL LEARNING

There are some works for camera localization using incremental learning methods, such as iMAP [589] and continual SLAM [644]. These have developed novel techniques based on continual learning camera localization using some methods based on mapping in unseen areas. Other works [236, 460, 461] use a modified cross-distillation loss with a two-step learning technique for incremental learning as well as distribution probabilistic to improve the learning in different scenarios. Concurrently, a siamese network is used to obtain the camera's poses in both indoor and outdoor scenarios [301], and a structure-aware method for direct pose estimation using PoseNet and explicit 3D constraints into the network [63].

On the other hand, a metric exists to evaluate the pose estimation and trajectory using neural nets. Relocnet [43] is a CNN for camera pose based on nearest neighbour matching between pairs of images to optimise the estimations. In contrast, two works present a continual learning method based on BioSLAM [701], and pigeons' behaviour to detect trajectories and plane an autonomous navigation [552]. These works aim to place recognition and then find the localization using rehearsal strategies with external memory for storing the knowledge about the place. Besides, incremental learning is suitable to extend the information about the areas for

Table 2.7

Camera localization works using continual learning.

Work	Technique
[377]	Learn over long sequences of small and highly batches
[85]	Multi-Model Approach with continual learning
[416]	Approach working on the output logits and features
[705]	Incremental learning for semantic segmentation
[264]	Automatic annotation for object detection
[259]	Incremental learning for semantic segmentation
[398]	Architectural and regularisation methods
[632]	Use triplet loss for place recognition and localization
[690]	Replay approach and distillation method
[662]	Sample images that provides an improved scene coverage
[478]	Latent replay tecnique to store activations volumes
[376]	Latent Replay method for object detection
[180]	Convolutional representations based on nearest neighbour
[589]	A MLP to serve as the only scene representation
[79]	Multi-Camera continual learning based on Replay strategy
[107]	Orthogonal projection and loop-closure detection
[693]	Learning-based loop closure based on 3-D features
[725]	Knowledge-Distillation for pose regression
[54]	Combination of multiple stereo with continual learning
[645]	Sampling strategy to maximize image diversity
[30]	CNN with a regressor to output the drone steering commands
[553]	LiDAR-based odometry
[63]	PoseNet structure-based and 3D incorporation

navigation purposes using drones; thus, a drone learns continually under different scenarios updating the learning model for localization [236, 419, 533]. Finally, the continual learning techniques presented in these works and analyses demonstrate the successful estimation of monocular camera position. However, methods are still being developed to find a methodology capable of training on the fly, avoiding forgetting the previous information, and obtaining the camera localization on a robot or drone in real time.

In addition, we summarised some works and presented others in Table 2.7 for continual learning methods and camera localization . We argue that several of these works present novel techniques applying architectural strategies. However, the presence of rehearsal strategies is notable because it is better to manage previous information within memory and thus present it when new information enters. Thus parameters and weights are combined to preserve learning while consolidating further information and avoiding the catastrophic forgetting of the previous data.

2.6 DISCUSSIONS

Through this comprehensive review of the state-of-the-art, we have observed the various techniques employed to address computer vision tasks. These techniques encompass using sensors to gather abundant information from the surrounding environment, enabling the achievement of classification, segmentation, detection, and pose estimation results. In addition, feature matching plays a crucial role within computer vision, identifying corresponding features and determining the nearest neighbours between a query image and a scene image.

In parallel, mapping techniques rely on feature extraction from images to generate a point cloud by projecting the feature positions onto a plane. This process enables the creation of an environment map with associated metrics, which subsequently facilitates camera localization . These mapping techniques form the foundation for determining the camera's position within the environment.

In the realm of deep learning, new techniques and architectures continually emerge, reflecting the dynamic nature of the field. One famous architecture in the current state-of-the-art is PoseNet [290, 292]. This architecture trains a dataset consisting of images and corresponding 3D coordinate annotations. By leveraging this training process, a learning model is constructed to estimate poses that closely align with the Ground Truth. Consequently, this facilitates the localization of objects or cameras within a given scenario without necessitating extensive computational resources [80, 82].

However, it is worth noting that the training of PoseNet or its variants, despite their compact nature, typically requires a substantial dataset to develop a robust learning model. Therefore, adequate dataset sizes are essential to achieve satisfactory performance and accuracy in the localization task.

Motivated by these considerations, our review focuses on the latest advancements in continual learning, addressing various tasks such as classification, object detection, and camera localization across diverse scenarios. Among the prominent techniques in this field are rehearsal techniques, which involve the management of past information by replaying its weights in an external memory located within a latent layer [376, 378, 478]. This approach ensures that previous knowledge is continually updated, preventing the learning model from forgetting previously acquired data while simultaneously learning new information.

It is worth highlighting that these methodologies have found particular applications in aerial localization tasks using images captured by drones and GPS positions [85,589]. Consequently, this review is valuable for comprehending cutting-edge localization and position estimation strategies using monocular cameras. Furthermore, by examining the works presented, readers will gain insights into this field's state-of-the-art methodologies and techniques.

2.7   CONCLUSION

Upon completing this review, it is evident that numerous works have proposed diverse solutions for estimating the position and localization of a camera within its environment. It is important to highlight that while these techniques showcase reliable and replicable methodologies, their practical application may be limited to computers equipped with GPU processors. Nevertheless, adopting a continuous learning approach aims to address the problem using minimal data and low-energy consumption resources. This consideration becomes particularly crucial when implementing such solutions on platforms or mobile robots, where computational constraints and power efficiency are paramount. By striving for efficient and resource-conscious methodologies, these works aim to optimise the performance and applicability of camera localization techniques in real-world scenarios.

After conducting this analysis, it is evident that ongoing research is still needed to identify new strategies that can accurately estimate the position of a camera in dynamic, indoor, outdoor, and unexplored scenarios. This highlights the complexity of the localization problem and the need for further advancements in the field.

Furthermore, it is important to acknowledge the role of conferences and journals in facilitating scientific contributions and providing a platform for researchers to share innovative results with the broader scientific community. These platforms play a crucial role in fostering collaboration, knowledge exchange, and the dissemination of research findings in localization , simultaneous localization , and mapping (SLAM), and robotics. By actively participating in these scientific events, researchers can contribute to the collective effort to advance the field and address the challenges of camera localization .

In future work, we aim to deepen our understanding of recent and upcoming advancements in the field, focusing on improving learning techniques for camera localization . Also, we aim to develop an approach that combines deep learning methodologies with continual learning strategies to enable the localization of drones, mobile robots, or cameras in challenging environments, ultimately facilitating autonomous navigation.

By leveraging the power of deep learning and continual learning, we anticipate overcoming existing limitations in camera localization and achieving more robust and accurate results. Our goal is to contribute to develop intelligent systems capable of autonomously navigating and localising themselves in complex and dynamic scenarios. We recognise the importance of this research, enabling various applications such as robotics, autonomous vehicles, and surveillance systems.

3 Classifying Ornamental Fish Using Deep Learning Algorithms and Edge Computing Devices

O. A. Aguirre-Castro, Everardo Inzunza-González,
O. R. López-Bonilla, U. J. Tamayo-Pérez,
Enrique Efrén García-Guerrero

Facultad de Ingeniería Arquitectura y Diseño, Universidad Autónoma de Baja California, Carretera Transpeninsular Ensenada-Tijuana No. 3917, Ensenada C.P. 22860, Baja California, México

D. A. Trujillo-Toledo

Facultad de Ciencias Químicas e Ingeniería, Universidad Atónoma de Baja California, Calzada Universidad No. 14418, Tijuana C.P. 22390, Baja California, México

CONTENTS

DOI: 10.1201/9781003385615-3

3.1   INTRODUCTION

Today ornamental fishes have a diverse variety of fish species in underwater environments, each of which is identified by different visual characteristics. The classification of ornamental fish provides valuable information to researchers to carry out various tasks or applications, one of which is the identification of different species of ornamental fish, as well as the optimal ecological study for the growth and development of different species [494, 560, 718]. Ornamental fishes are a variety of aquarium-bred fishes, which are primarily intended for decoration and display rather than consumption or use in scientific research. These ornamental fish can vary in size, shape, color, and pattern. These types of fish have become a popular hobby for animal lovers and aquarists. Freshwater marine fish in terms of volume and value of specimens traded. and value of specimens traded in the ornamental market. In terms of volume, the global trade in marine ornamental species represents less than 10% of total ornamental fish traded; however, due to their high unit value, the proportion is higher when calculated as a percentage of global trade by species.The percentage of global trade-in value, with trade growing strongly in recent years [371]. Authors such as [347] mention that aquaculture has a promising future due to the fact that in different countries the ornamental fish aquaculture industry has spread to a global level, due to its high export demand, in [343] shows an application of Deep learning for the detection of ornamental fish diseases, being this one of the main applications in the aquaculture area as shown by the authors [255, 538] since identifying diseases at an early age has lower losses and higher quantity of ornamental fish. Moreover, the information provided by the proposed dataset is to understand the grouping of fish learn the behavior and cooperation between various fish species in a natural culture condition [181, 648]. In the field of machine learning [687, 688] and computer vision [442, 538], classification [172], and counting fish species from images [164,241,568], is a research field topic for aquaculturists in multi-class recognition [121, 692]. Automated fish classification is very important in ornamental fish farming research as it helps in the automated monitoring of fish species activities in ponds, development, feeding, and disease behavior. Some of the most popular ornamental fish include Guppies, Bettas, Goldfish, Goldfish, Angelfish, Tetras, Discus, African Cichlids, South American Cichlids, Angelfish, and Pufferfish. For the classification of ornamental fish, Zebra fish were used, shown in Figure 3.1, also known as Zebra danios or striped Zebra (scientific name: Danio rerio) [331], and are a species of small, active fish that are popular as pets in home aquariums due to their attractive appearance and ease of care. These fish species are native to India, Nepal, Bangladesh, Burma, and Pakistan. Small in size, they can reach approximately 3 to 5 cm in length when fully grown. These species are schooling fish and feel most comfortable and secure when kept in groups of at least five individuals. They are omnivorous and eat both live foods such as mosquito larvae, daphnia, and artemia as well as commercial foods, in [252] mentions that this particular species prefers live foods. Zebra fish have a life expectancy of approximately 2–3 years in captivity. Being known to be very prolific and can reproduce quickly in the home aquarium, because of these characteristics they are one of the most in-demand species in ornamental

(a) Green Zebra   (b) Pink Zebra   (c) Yellow Zebra

Figure 3.1 Example of Zebra fish with different colors. (a) Green Zebra, (b) Pink Zebra, and (c) Yellow Zebra.

fish farms. In recent years [265], zebrafish have also become a common model for scientific research due to their ability to regenerate tissues, their well-developed immune system and their fully sequenced genome, which has led to important advances in biology and medicine. Another of the species mentioned in this work is shown in the Figure 3.2, this is the Mollys fish species.(scientific name: Poecilia sphenops), a species of freshwater fish native to Central and South America. This species is commonly popular as aquarium fish, due to its particular coloring and shape. This species is one of the most demanded freshwater ornamental fish due to its ease of care and reproduction. This species of ornamental fish is known primarily for its veil-like fins and bright coloring. Males have larger fins and more intense colors than females. Molly fish are ovoviviparous, which means they give birth to live fry instead of laying eggs. Molly fish are omnivores that eat a variety of foods, such as vegetables, live or frozen foods. This species of Molly fish can grow up to about 8 cm in length and have a life expectancy of about 3-5 years in ideal aquarium conditions. There are several varieties of Molly fish, including black Mollies, gold Mollies, and velifera Mollies. Molly fish are a good choice of ornamental fish species for beginning aquarists because of their ease of care and hardy nature. As long as aquarists have the proper water conditions and cleaning care, because this species is very prone to disease. One of the most popular fish species in the world for its attractive coloring and easy care is the Goldfish (Carassius auratus). As shown in Figure 3.3, this ornamental fish species is a freshwater fish species native to Asia [671]. This species consists of a wide variety of striking colors that can range from orange and red to black and white. Goldfish are omnivorous, they can eat a variety of foods, including food. Goldfish reach a size of up to about 30 cm in length and have a life expectancy of up to 20 years in ideal aquarium conditions. Goldfish, like mollies, are an ornamental fish species that due to their easy care and hardy nature can breed readily

(a) Yellow Molly   (b) Black Molly   (c) Bicolor Molly

Figure 3.2 Example of Molly fish with different colors. (a) Yellow Molly, (b) Black Molly, and (c) Bicolor Molly.

(a) Common Goldfish   (b) Oranda Goldfish   (c) Oranda RedHat Goldfish

Figure 3.3 Example of Goldfish found in the proposed dataset. (a) Goldfish Common, (b) Goldfish Oranda RedHat, and (c) Goldfish Oranda.

in aquarists' farms, which should keep the aquarium clean and well-maintained to prevent disease. There are a great variety of goldfish, among which the goldfish, red hat fish, and other varieties stand out. Some authors such as [697] use artificial neural networks such as Yolo V3 for the detection of 11 species of fish, with more than 2200 images. In this work, the aforementioned species were taken into account with the detection of fish with deep learning algorithms such as Yolo V5. The test dataset has 1200 images to identify the aforementioned species. Image data acquired in the normal way is called standard format data. The VGG-GoogLeNet model adapted to the dataset of this work was obtained by improving the network structure based on VGG and GoogLeNet. In the fish detection test, the data set is fed directly into the Yolo V5 for identification and detection. The remainder of this work is organized as follows: Section 2 describes materials and methods; Section 3 presents the results; and Section 4 concludes this work.

3.2   MATERIALS AND METHODS

In this section, the materials and methods used for the training of the deep learning models will be described, as well as the acquisition of the set of images created to create the dataset, as well as the training models for the classification and identification of ornamental fish species. This section also describes the hardware used for training and testing the deep learning models.

3.2.1   DATASET FOR TRAINING

The dataset used consists of capturing images of the ornamental fish species to be identified in an environment in which the aquarists have the optimal conditions for their preservation. To create the dataset, a digital action camera was used, which provides high-quality images. Initially, 400 images per species were taken in a natural environment where the necessary information for their identification was not available. The dataset was collected using a digital camera (VEMONT) equipped with a 2x wide-angle lens, 12 Mega pixels of resolution, and 1080P video recording of up to 30 frames per second. In video camera mode, a resolution of 640×480 pixels was employed, and because it is an action camera, it provides useful information on how to work with underwater shots. This is the size that is used throughout the entire image collection. By capturing a variety of images, it is possible to monitor

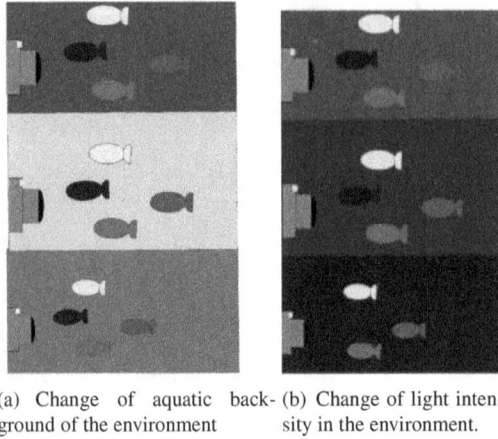

(a) Change of aquatic back- (b) Change of light inten-
ground of the environment   sity in the environment.

Figure 3.4   Image acquisition methodology to perform the dataset.a) Change background b) Change the light intensity.

the behavior of various species, in their natural environment. Researchers can collect images of the fish from various angles, allowing them to extract traits for use in several applications, thus giving the aquaculturist a better solution to the different problems that can affect their environment. Figure 3.4 shows the methodology used for the acquisition of images, this in an environment controlled by the aquarists, in Figure 3.4 a) shows the methodology used for the acquisition of images is shown by applying variations from the bottom of the environment of the species of Fish will provide more information for identification, another change that was established in the methodology is the variation of light intensity shown in Figure 3.4 b) these changes were found to simulate different environments in which the other species of ornamental fish provided by the acuarist. This methodology provides us with information so that the models that have been trained can be tested in different environments such as funds with little light, funds with cloudy water, funds with low visibility. Although there are different algorithms to attack these natural phenomena as presented by [18], these authors apply digital image processing to recover information from images in low light based on retinex. Although this image processing provides us with more information for an improvement in the detection of species, having a dataset with these characteristics will provide us with a trained model to identify under these conditions, this will provide a lower demand for computational cost making just the detection process.

3.2.2   TRAINING METHODS

In this section, the training methods are discussed based on convolutional neural networks (CNN) such as Yolo V5, Resnet 502. These models are used for different applications as mentioned by the authors in [224]. These artificial intelligence algorithms have an application in this area of interest where classifiers based on SVM

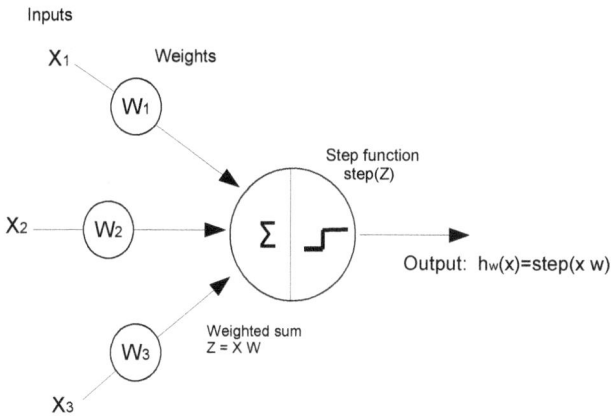

Figure 3.5 A binary classification model called the perceptron.

or KNN have a lower detection capacity. Some of those used in this work are mentioned below. For the detection and identification of species, we will start from the base of the system, what we call the neural network, this is the core of the object detection. An artificial neuron is made up of a series of inputs and outputs. The inputs are associated with some weights so that the TLU (threshold logic unit) calculates the weighted sum of its inputs to obtain an output, as we can see in Figure 3.5 a perceptron is a layer connected to all the inputs of the hidden layers. The result of each perceptron is given by its activation function, this will be activated depending on whether the weighted sum of the inputs has a value greater or less than an estimated threshold. The result provides values between 0 and 1, -1 and 1. The result can be null or -1. The response to the activation function is part of the output layer. In MLP models (multilayer perceptron), the outputs are interconnected to the input of another hidden layer, and this model is shown in Figure 3.6 These models are the basis for some classifiers and deep learning models based on convolutional artificial neural networks, used in different applications for detection and identification of species and objects as mentioned by the authors [52, 241]. These models were used in this work for the identification of ornamental fish.

3.2.3   CLASSIFICATION LEARNER

Learning classifiers are machine learning algorithms used to predict the class or category to which an object belongs based on its characteristics or attributes. The classifiers in which the best results were obtained for this work are those of support vector machines (SVM) and decision trees. SVM classifiers are used to classify objects into two or more classes using a linear or non-linear decision function, while decision tree

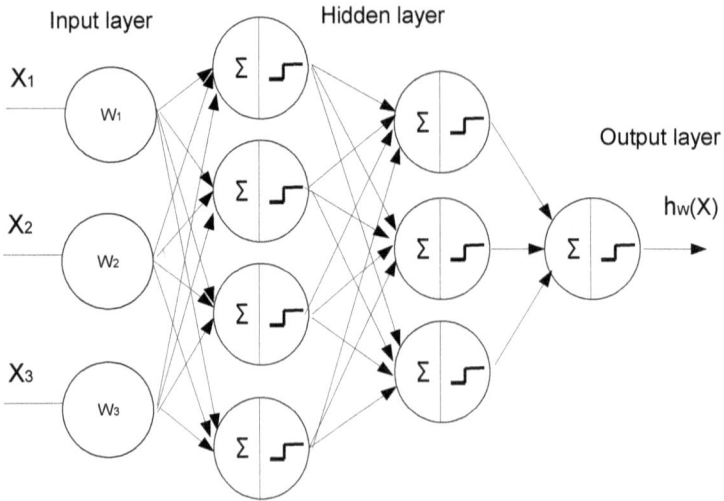

Figure 3.6   Neural network with hidden layer structure.

classifiers are based on the construction of a decision tree that divides the training data into larger groups. small until reaching a final decision. These classifiers have to have enough information for their results to be of a high probability, which is why the dataset must provide a sufficient number of images to obtain an identification with a higher probability of being correct. These type of classifiers shown in Figure 3.7 provide us with an initial point in the use of neural networks, having a point of comparison for the more developed models such as convolutional neural networks.

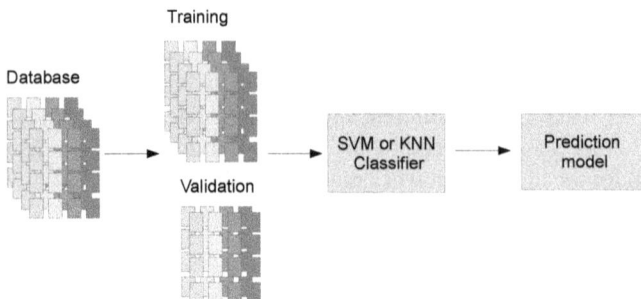

Figure 3.7   Methodology used for the use of classifiers SVM and KNN.

These classifiers were used by means of MATLAB tools and functions and provide us with some metrics of the already trained models. As mentioned above, it is a reference point for more complex models based on CNN, and this in the application identification of ornamental fish.

3.2.4 MODELS OF DEEP LEARNING WITH ARCHITECTURE CNN

Convolutional neural networks (CNNs) are a type of artificial neural network architecture that has become popular in the field of machine learning and computer vision. CNNs are used in classification tasks, object detection, image segmentation, and other types of visual analysis. These neural networks automatically learn image characteristics, such as edges, shapes, textures, and colors, through a process called convolution. Convolution refers to the application of filters or kernels to images to extract important features from them. These filters are small numerical matrices that move around the image and perform mathematical operations on the pixels at each position, producing a new image with more abstract characteristics. In addition to convolution, CNNs also include other layers such as pooling, which reduces the size of the image and the amount of information that is processed, and the fully connected layers, which ultimately perform the classification or detection of objects. CNNs have proven to be very effective in the task of recognizing objects in images and videos, outperforming other traditional techniques in many cases. In this work, two architectures will be used which use the CNN as the core in their internal architecture, such as Yolo V5 and Resnet, using a technique called transfer learning, this is an automatic learning technique in which the knowledge acquired by a model that is previously trained for a given task and is used to solve a different or related task.

3.2.5 CNN MODELS WITH YOLO-BASED ALGORITHMS

Yolo (You Only Look Once) [275] is a popular real-time object detection system that uses deep learning algorithms to identify and locate objects in images or videos. Figure 3.8 SHOWS The architecture of model Yolo v5, which is the latest version of Yolo, developed by Ultralytics. Compared to previous versions of YOLO, YOLO v5 is faster, more accurate, and more flexible. It uses a novel architecture called CSPNet (Cross Stage Partial Network) that combines the advantages of both residual and dense connections to improve performance. YOLO v5 also incorporates a number of optimization techniques, such as dynamic scaling and filter pruning, to reduce model size and improve inference speed. One of the key advantages of YOLO v5 is its flexibility. It can be trained on a variety of datasets, including custom datasets, using a simple configuration file. This makes it a popular choice for a wide range of computer vision applications, such as self-driving cars, robotics, and surveillance. Overall, YOLO v5 represents a significant improvement over previous versions of YOLO and is a powerful tool for object detection and recognition.

3.2.6 RESNET

ResNet (Residual Networks) is a deep convolutional neural network (CNN) architecture proposed by Microsoft Research in 2015. This architecture shown in Figure 3.9,

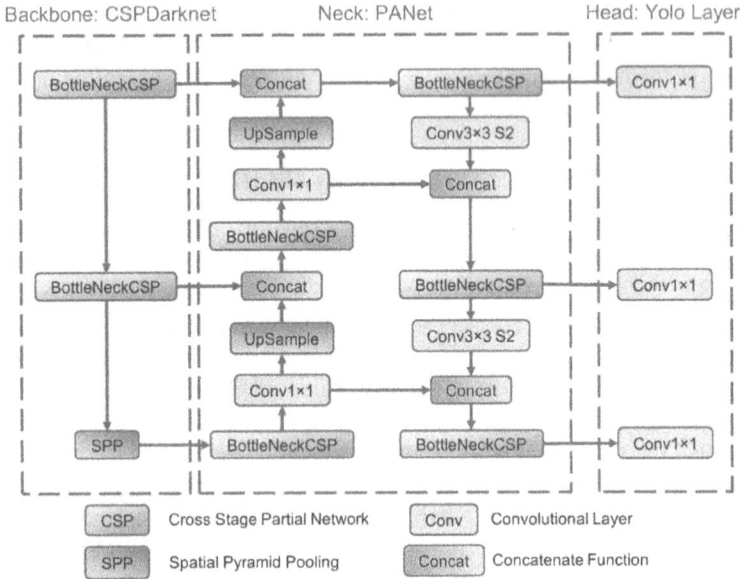

Figure 3.8 The architecture of model Yolov5 is described in [686].

addressed the vanishing gradient problem in deep neural networks. The performance degradation problem refers to the decrease in model accuracy as the number of neural network layers increases. ResNet solves this problem by using "residual" blocks, which allow information to flow through the network more effectively. Instead of stacking layers on top of each other, as in conventional CNNs, ResNet uses residual blocks that skip layers and allow information to flow directly through them. This allows input and output signals to be combined in the model more effectively and helps prevent performance degradation. ResNet has been widely used in image classification tasks and has proven to be very effective in object identification in complex images. In addition, variants of ResNet have been proposed for other tasks, such as image segmentation and real-time object recognition.

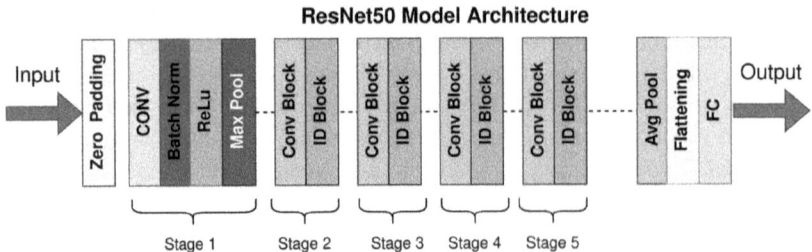

Figure 3.9 Architecture of model ResNet 50 [237].

Table 3.1

Hardware for training deep learning models [18, 19].

Hardware	Jetson Nano	Jetson Tx2	Personal Computer (PC)
GPU	MPCore processor Quad-Core	Denver 2, 2.5 GHz, Quad-Core ARM	Intel Core i7-8750
GPU	128-Core NVIDIA Maxwell GPU	256-Core NVIDIA Pascal GPU	768-Core NVIDIA Pascal GPU
RAM	4 GB DDR4	8 GB DDR4	16 GB DDR4
Connectivity	Wifi, Ethernet	Wifi, Ethernet	Wifi, Ethernet
Storage	32 GB SD-Card	32GB eMMC 5.1	1 TB SSD
Energy consumption	6.5 W ~10 W	7.5 W~15 W	180 W ~ 200 W
Video output	HDMI	HDMI	HDMI
Operating voltage	5 V	19 V	19 V
Operating System	Ubuntu	Ubuntu	Ubuntu/Windows
Cost (USD)	$108.00	$479.00	$979.00

3.2.7 TRAINING HARDWARE

This section describes the hardware used to train the YoloV5 and ResNET algorithms and another classifier model based on convolutional neural networks or CNN. This hardware is described in Table 3.1, where a Jetson Nano is shown in Figure 3.10, a Jetson Tx2 is shown in Figure 3.11, and a personal computer is described in Table 3.1. Although the beginning of this work was done with this hardware, it was necessary to migrate to the use of the Google Colab platform to carry out the training and the embedded systems and personal computer to execute the training result.

3.2.8 JETSON NANO

Jetson Nano is an NVIDIA-designed embedded system for artificial intelligence (AI) and machine learning applications. It's a low-cost system that packs a quad-core ARM CPU, 128-core NVIDIA Maxwell GPU, 4GB of RAM, and multiple I/O ports as shown in Table 3.1. This embedded system can run AI applications in real-time and is ideal for computer vision, voice processing, and robot control applications for robotics and Internet of Things (IoT) projects. The Jetson Nano embedded system comes pre-installed with a software package that includes the Linux operating system, necessary hardware drivers, the NVIDIA SDK for AI and machine learning application development, and a wide range of libraries and tools. This embedded system is shown in Figure 3.10.

3.2.9 JETSON TX2

Jetson TX2 is a computing platform designed by NVIDIA for artificial intelligence (AI) and machine learning applications in industrial and commercial environments. It

Figure 3.10 Jetson NANO hardware [455].

is more powerful than Jetson Nano and can handle complex AI and machine learning applications. The Jetson TX2 features a quad-core ARM Cortex-A57 CPU, a 256-core NVIDIA Pascal GPU, 8 GB of RAM, and multiple input and output ports, shown in Table 3.1. It also includes an inference engine for Integrated AI to speed up the inference of deep learning models. This platform is ideal for applications in the field of robotics, industrial automation, and high-resolution image processing. It is also famous for security and surveillance and autonomous vehicle applications. This embedded system is shown in Figure 3.11. Although the personal computer is listed in Table 3.1, the following results section will compare the performance of the proposed embedded systems, jetson nano and jetson tx2, which will be used as a reference while evaluating the outcomes of the trained models. This allows for a cost-benefit analysis of the hardware used in this project.

3.3   RESULTS

This section shows the results of the models trained for the detection of ornamental fish in controlled environments, in which the obtaining of the dataset will be detailed and how the classification of the species was generated with the different identification models mentioned as Yolo v5, v8 and ResNet50.

3.3.1   DATASET RESULTS

Two different types of datasets were applied to deep learning models, in which the choice of the dataset to be used depends on the application. In the previous section

Figure 3.11   JetsonTX2 hardware [455].

on materials and methods, the methodology to be used for the collection of information was described. In this section, the characteristics of these datasets used will
be detailed. The URL https://data.mendeley.com/datasets/tdn9cw7mrm/1
dataset contains the first edition used in different models made in MATLAB applications and other models developed with Python libraries such as TensorFlow. This
first dataset contains the fish species separated into folders in which the species to
be classified were in different folders regardless of the color. This type of dataset
provides us with various information on the species. However, the images provided
by the action camera contain different data for detection and stop noise caused by
foul shots. These images may present malformations of the species and the environment in which the ornamental fish species to be detected are found. Figure 3.12
shows an example of the images found in the folders of the different ornamental fish
species. These images are used for classifiers based on KNN or SVM implemented
in training applications such as MATLAB classification learner. This type of dataset
proposes a simple answer used on machine learning algorithms executed on MAT-
LAB. However, it was observed that it was not the best for identifying the various
species, given that the images have too many alterations, making too much noise at
the time of detection. As shown in Figure 3.4 it was proposed to increase the dataset
to have more information on the fishes at the time of collecting their characteristics
to obtain more and in different factual situations such as the lack of lighting and
background of the environment in which to be found. This dataset is increased with
specific images, making cuts of the images, leaving images of different sizes, and

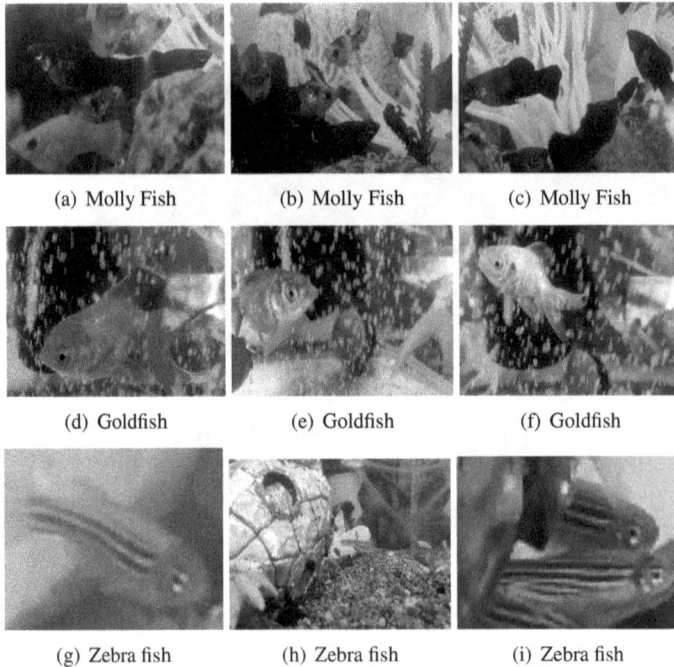

(a) Molly Fish   (b) Molly Fish   (c) Molly Fish

(d) Goldfish   (e) Goldfish   (f) Goldfish

(g) Zebra fish   (h) Zebra fish   (i) Zebra fish

Figure 3.12   Example of species ornamental fish found in the proposed dataset.

delimiting the fish figures. Figure 3.13 shows the initial distribution, in which it is observed that the dataset is not balanced. Although some models were made with this dataset, the results are shown in the following paragraphs, showing the importance of a balanced dataset. Data generation is important because it allows you to balance the dataset, resulting in accurate and dependable AI models. The quality of the data used to train deep learning models has a significant impact on the model's performance and accuracy. There are various methods for generating data, including collecting existing data through additional imaging or through the literature, which provides existing data sets and images; it is important to note that the data obtained be relevant and correctly classified. There are synthetic data generation and random data augmentation; these data can be generated by synthetic data that simulates the characteristics of the actual data. However, it implies the generation of new instances because the data is generated through rotations, scaling, and deformation of images. This generation can help improve the performance of the models as long as they properly belong to a correct labeling, indicating the category in which they belong. These high-quality data generators take time to make them relevant and representative of the actual data. This dataset was implemented in classifying models; although the species were separated by means of folders that contained the different species to organize, the images of each one were of low quality, and they did not have the proper balance. The characteristics of each species were extracted from these images and attached to a model that divides the information into two parts, one to train the

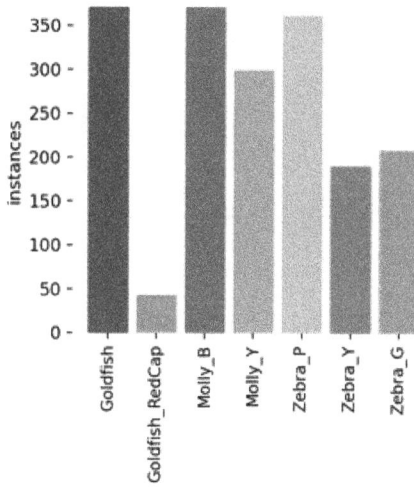

Figure 3.13 Dataset not balanced.

classifier model and the other for validation. This methodology is shown in Figure 3.14 and was previously described. The images of the species to be identified were increased to improve the dataset. The dataset was modified to obtain a more significant number of species and extract their characteristics from different perspectives, obtaining important characteristics for the development of the models to be used. The software labelimg was used to identify the species in an image to carry out this image increase. There are currently tools to increase our dataset from the photos in

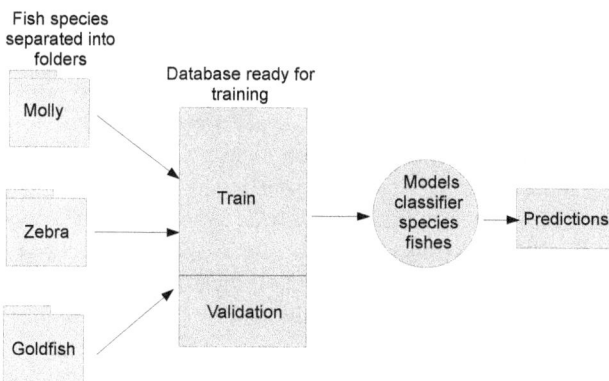

Figure 3.14 Dataset implemented into the classifier model, SVM and KNN.

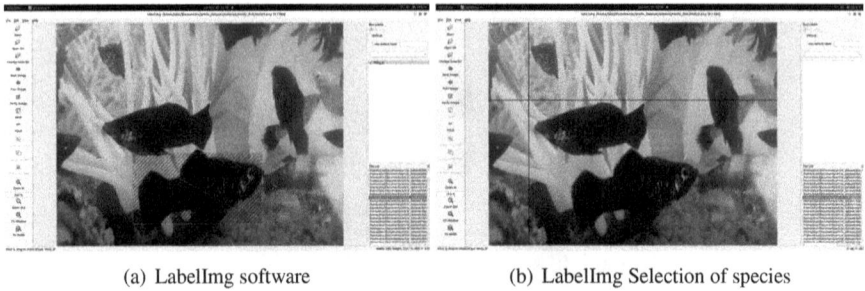

(a) LabelImg software   (b) LabelImg Selection of species

Figure 3.15 Example of classification of species ornamental fish with labelimg.

the dataset. A significant difference compared to the first dataset is that the classification is based exactly on the selected species, decreasing the error due to the fact that the characteristics extracted are only from the fish and not from the entire environment in which the species is found. A species classification example utilizing the labelimg software is shown in Figure 3.15. This image classification program offers a selection option to choose the area to be classified, and this option also contains the extraction of the species or object's detection criteria. Only the sections to be detected in the photos were chosen, unlike the prior dataset. With this software for the classification, an improvement was obtained in the labeling of species, selecting images where two or three different species were found in one image; this labeling provided us with a more balanced dataset, as shown in Figure 3.16. In this dataset, it was intended that the species had a balance of 200 samples per species, unlike the dataset in Figure 3.13; this imbalance is notable in the goldfish_redcap species compared to the other species. For the increase of this species, images were added, and software implemented an increase in samples. In order to train detection models like YOLOV5 and Yolo V8, the labelimg software creates a text file with the name and position of the species to be detected.

3.3.2 CLASSIFICATION LEARNER MODELS

Using MATLAB and its Classification Learner application, three models were required for species identification. Based on architectures such as SVM, KNN, and bilayered neural networks, these models generated the best results, as shown in Table 3.2. the models trained with the highest accuracy are SVM and bilayer neural networks, as seen by the results provided in Table 3.2 at 92.7% SVM model and the KNN model at 90.7%. Although these models have superb accuracy when working with them in real-time, they provide another option for the detection of species due to the fact that the dataset used large images with different environments, taking into account a large amount of noise. This is one of the main differences concerning models, in addition to the fact that these models could only be validated on the personal computer; another limitation of this model, it provides a reference for models with larger architectures, such as those based on YOLO and ResNET50.

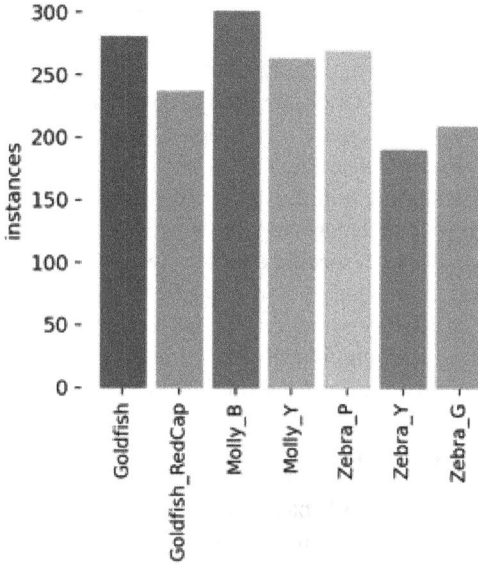

Figure 3.16   Dataset with balanced species.

3.3.3   TRANSFER LEARNING MODELS

As was already noted, the Ultralytics-developed YOLOv5 object detection system is built on a deep convolutional neural network and is intended to recognize objects in photos in real time. To use YOLOv5 requires Hardware capable of running deep learning models efficiently. This includes a graphics card (GPU) with support for NVIDIA CUDA and an adequate amount of RAM. A programming environment with Python 3.6 or higher is required, including the following packages: PyTorch, OpenCV and NumPy. To use YOLOv5, you must download the pre-trained model corresponding to the object detection task you want to perform. Adjustments and customizations can be made to the model configuration to adapt it to the specific needs of each lesson. A method known as transfer learning is the foundation for

Table 3.2

Classification learned models with MATLAB validation results.

Model	Accuracy (%)
SVM	92.7
KNN	90.7
Bilayered Neural Network ResNet50	92.7

task-specific adaptation. A pre-trained model is used as the foundation for a new learning challenge in machine learning. A pre-trained model is utilized on a comparable task and tuned to a current study instead of training a model from scratch, which can considerably cut down on the time and resources needed to train an effective model. The pre-trained model was able to understand general patterns that are helpful in various machine-learning tasks since it had been pre-trained on a vast and varied data set. By starting with the pre-trained model, you can use previously discovered patterns to your advantage and fine-tune the model to a particular task by training on a fresh, smaller task-specific data set. Several machine learning applications, such as object identification, picture classification, natural language processing, and others, have effectively exploited transfer learning. The method has made it possible to develop deep learning models that are more effective and accurate while also drastically reducing the time and resources needed to train them.

3.3.4 MODELS WITH YOLO

Two convolutional neural-network based deep learning models, YOLO V5, and YOLO V8, each used this learning transfer technique. The learning transfer function operates in the final layer of the architecture in Figure 3.8, where the YOLO architecture is remembered, making this technique well-known for its optimization in object recognition. The following sections discuss the results and comparisons of the trained models. The model trained by means of transfer learning in the yolo v5 models was taken into account at different times to see if it had variations at the time of training and detection. these results are shown in the confusion matrices shown in Figure 3.17. The confusion matrix is used to evaluate machine learning models to visualize and summarize the accuracy of a model's predictions. Figure 3.17 (a) shows the confusion matrix of the Yolo v5 model trained with 20 epochs, it shows that each species to be classified has a percentage greater than 60 %, which has a precision index greater than 87%. In this model, in the species Zebra_G and Zebra_P, the best results of the classifier are obtained, being the Goldfish_RedCap the one with the lowest detection percentage. Figure 3.17(b) shows the Yolo v5 model with 50 epochs. It is observed that some species increased their probability of detection. In this model, the species that is most difficult to detect continues to be Goldfish_RedHat with a probability of 52%, this being the only one species that lowered its probability of detection; however, the model had an improvement in the precision index to 89%. Figure 3.18 (a) shows a model with 100 epochs, although the 20 epoch model had better results in some species, lower than one percent compared to the previous one due to overtraining of the model, this model was trained to have a comparison with the version of the Yolo V8 because this version was only trained in 100 epochs. Figure 3.18 (b) shows the results of yolov5x; this is the last improved version of YOLOv5. It has a more significant number of layers and parameters than YOLOv5, which allows a more substantial learning capacity of the model and, in theory, a greater precision in detecting objects. This model has improved the detection of Zebras, goldfish, and black Mollie's species with percentages more significant than 80 specific metrics such as precision and recall are used to evaluate the quality of the

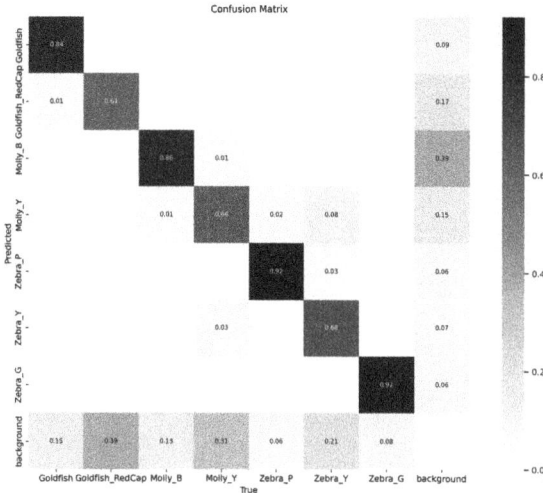

(a) Matrix confusion 20 epoch with YOLO v5s.

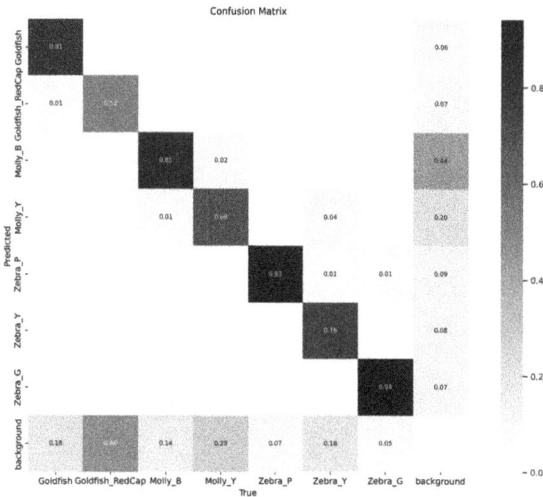

(b) Matrix confusion 50 epoch with YOLO v5.

Figure 3.17   Yolo model confusion matrix with different epochs.

deep learning model and especially in classification tasks. Precision is a metric that provides how accurate the model's optimistic predictions are. This metric measures the proportion of positive cases the model correctly classifies in relation to the total number of positive predictions. This metric of the trained model is shown in Figure 3.19. This figure shows the precision generated by the trained models. In blue, a light version of the YoloV5 is demonstrated. This version was trained with 20 epochs,

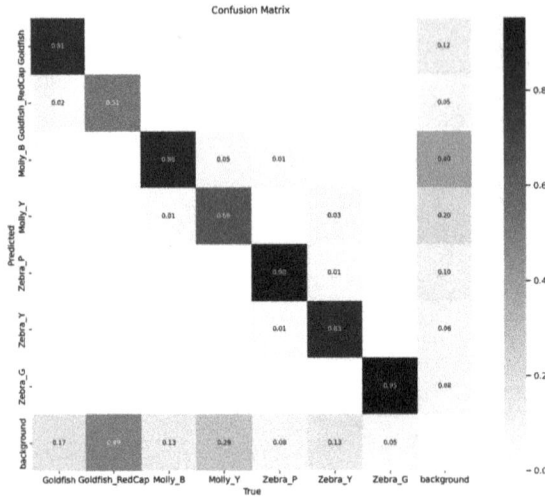

(a) Matrix confusion 100 epoch with YOLO v8.

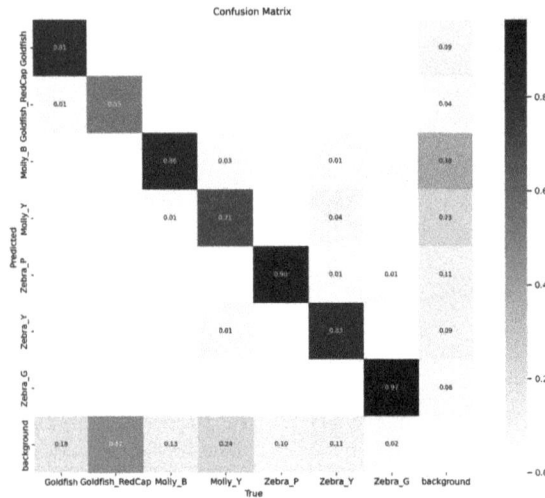

(b) Matrix confusion 100 epoch with YOLO5x.

Figure 3.18   Yolo model confusion matrix with different epochs.

giving a precision of 80%. In orange, the Yolo V5 model has 50 epochs, obtaining a precision of 83%; in gray Yolo V8 model trained at 100 times with a model that had a precision of 85% and finally, in yellow, the most extensive version of the YoloV5, obtaining a precision of 83% its best performance. The recall gauges the model's ability to recognize affirmative examples. In other words, it counts how many positive examples the model correctly categorizes with respect to all positive cases; the

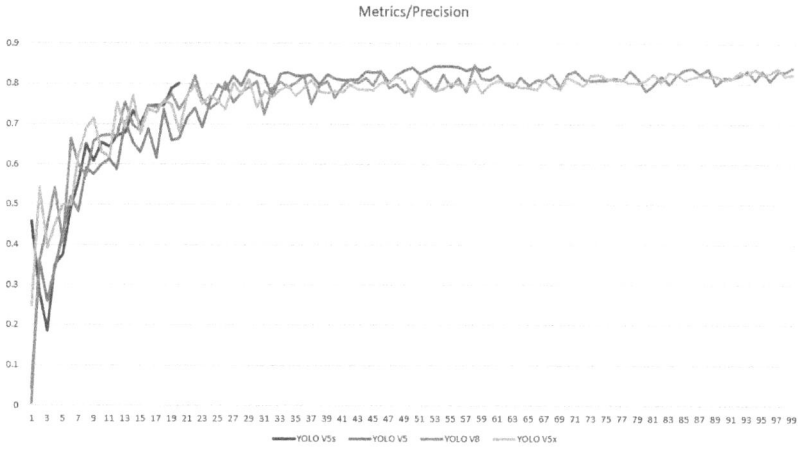

Figure 3.19 The precision metric of models trained

confusion matrix provides the data for this calculation. The results of the models' comparison are shown in Figure 3.20. According to the graph, YoloV8 achieves a recall of 80% and YoloV5x a recall of 79%, while the other two models fall short of these percentages. The F1 score is a measure of evaluating the performance of a machine learning model in binary classification tasks. It is calculated as the harmonic mean of the precision and the recall and provides an overall measure of the quality of the model. The results shown in Figure 3.21 show the performance of each model; this shows that the models with more epochs, such as YoloV8 and YoloV5X, obtain a stable average of 80%, while the other models YoloV5s and YoloV5 have a considerable variation having models with a high probability of being wrong. Three distinct

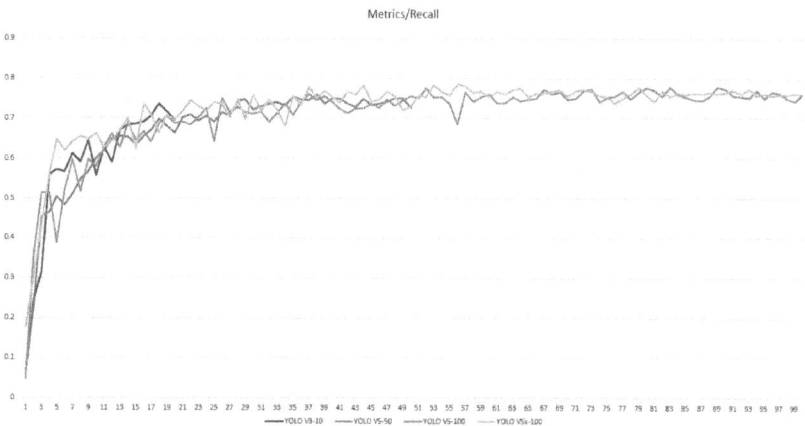

Figure 3.20 The recall metric of models trained

films that included the ornamental fish species that the aquaculturist was going to classify in an appropriate setting were used to validate the trained models. When discussing the findings, we can see that the species are categorized with a precision above 80%; the trained models produce these results. Figure 3.22 shows the results of the trained models. It can be seen that ornamental fish species are classified in an environment controlled by the bodybuilder. Figure 3.22 (a) and Figure 3.22 (b) show the classification obtained by Yolo V5s. This model detects two types of goldfish and some zebrafish. Although this model showed lower metrics than the other models, it performed very well in the rankings. The Yolo 5 model was used in the results shown in Figure 3.22 (c) and Figure 3.22 (d). These images show the results of the model for the detection of ornamental fish; this model was validated with molly fish and zebrafish, and these results show a higher probability of detection of the species, obtaining an improvement concerning the previous model. In Figures 3.22 (e) and 3.22 (f), the results of the yolov8 model are shown; in these images, they are shown with species of fish mollys and zebras; these results show an improvement in detection, in Figure 3.22 (f), it is demonstrated that the detection of fish Molly is invertive in size and even with objects that are interfering with the species to be detected. These models generally present the detection of ornamental fish species in environments controlled by aquaculturists; However, these models have a good performance in the classification, and the performance will be measured in the different embedded systems, Jetson nano, Jetson tx2 and PC. This will make it possible to compare the necessary resources to obtain the best model that has the best cost-benefit result for this application with the models indicated. Table 3.3 shows the performance of the Yolo v5s model with different resolutions, this table shows that the Jetson NANO has a performance of 30 FPS with a resolution of 240×120, the Jetson TX2 embedded system has a medium resolution of 480×240 has a performance of 35 FPS, this obtaining a performance similar to that obtained by the PC.

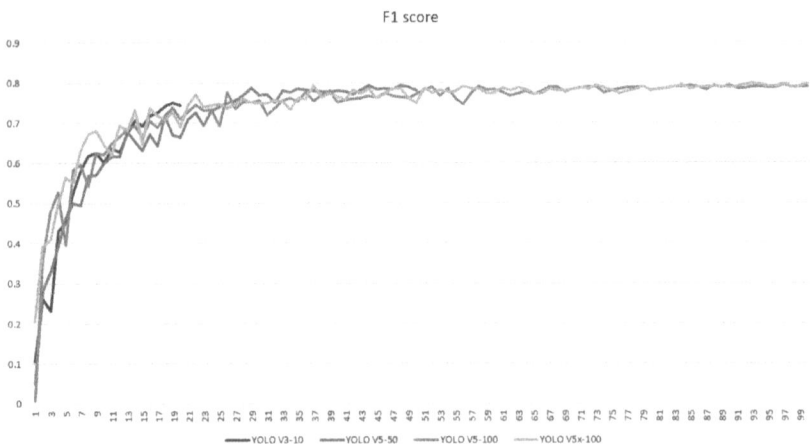

Figure 3.21   The F1 score metric of models trained

(a) YoloV5s model

(b) YoloV5s model

(c) YoloV5 Model

(d) YoloV5 model

(e) YoloV8 model

(f) YoloV8 model

Figure 3.22 Example of species ornamental fish found in the proposed dataset.

The results shown for YOLO V5 in Table 3.4 are the light model; this is because robust models need greater computational capacity, and this Yolov5s model obtained competitive results in comparison with the Yolo V5s model. In this model, the Jetson

Table 3.3

Comparison of the Yolo V5 model with different resolutions.

Embedded Systems	FSP		
	240×120	480×240	640×480
Jetson Nano	30	24	18
Jetson Tx2	48	27	20
PC	49	35	30

Table 3.4

Comparison of the Yolo V5 model with different resolutions.

Embedded	FSP		
Systems	240×120	480×240	640×480
Jetson Nano	25	12	8
Jetson Tx2	46	25	12
PC	52	29	21

NANO obtained a good performance in the same way in 240×120 images with 25 FPS and the Jetson TX2 in medium size images.

Although the YoloV8 model obtained a model with higher detection accuracy in performance, the model has a lower performance in the embedded systems where the tests were performed. The results obtained in Table 3.5 compared to the previous models contain a lower amount of FPS in the jetson nano and jetson Tx2 embedded systems, obtaining the best results in smaller images.

3.4 CONCLUSIONS

In this work, the detection of ornamental fish was presented by applying different machine learning models through convolutional artificial neural networks. Although the use of residual neural networks obtained good precision in the evaluation metrics of the trained models, the model did not obtain the expected results when classifying ornamental fish, adding computational cost to apply it to high-performance embedded systems. It was also observed that the YOLO V5 model trained with 50 epochs obtains a better detection result and provides in embedded systems like the Jetson NANO 18 FPS with a resolution of 640x480 pixels on the Jetson TX2. This model reached 20 FPS with a similar resolution of 640x480 pixels. The YOLO V5 model has an accuracy of 90%; the YOLO V8 has an accuracy of 87%, these models being the best in the classification of ornamental species.

Table 3.5

Comparison of the Yolo V8 model with different resolutions.

Embedded	FSP		
Systems	240×120	480×240	640×480
Jetson Nano	12	9	5
Jetson Tx2	25	12	8
PC	42	28	19

4 Power Amplifier Modeling Comparison for Highly and Sparse Nonlinear Behavior Based on Regression Tree, Random Forest, and CNN for Wideband Systems

J. A. Galaviz-Aguilar
School of Engineering and Sciences, Tecnologico de Monterrey,
Monterrey 64849, Mexico

C. Vargas-Rosales
School of Engineering and Sciences, Tecnologico de Monterrey,
Monterrey 64849, Mexico

D. S. Aguila-Torres
IT de Tijuana, Tecnológico Nacional de México, Tijuana 22435,
Mexico

J. R. Cárdenas-Valdez
IT de Tijuana, Tecnológico Nacional de México, Tijuana 22435,
Mexico

CONTENTS

DOI: 10.1201/9781003385615-4

4.1   INTRODUCTION

Although behavioral models of power amplifiers (PAs), such as higher-order polynomial models that consider short-term memory effects, can provide an excellent
implementation option, the computational requirements, and coefficient calculations
that come with regression-based methods may pose challenges for wireless systems
that use high-order modulation techniques toward 5G. This is because such techniques can increase the peak-to-average power ratio (PAPR) for high-bandwidth signals and lead to greater non-linearities in the transmitter path, potentially impeding
the construction of a modeling function [136], [189].

Modern communication systems require high efficiency and linearity to meet the
requirements to avoid the effect of spectral expansion and decrease the adjacent channel power ratio (ACPR) levels. The main issue of the PA performance comprises to
operate the device in its linear region, another critical challenge is due to the high
levels of PAPR, which excite the PA in its saturation zone when used with high
transmission schemes data. The bandwidths used in 5G involve stronger memory effects in the transmitter stage, implying a more efficient digital predistortion (DPD)
process [684]. Various efforts based on machine learning (ML) use DPD with direct
learning architecture (DLA), reducing the polynomial expansion [50].

ML techniques have been increasingly employed to optimize the modeling of
highly nonlinear systems. A crucial aspect in achieving successful behavior modeling is the accurate prediction of desired outputs while minimizing the number of
coefficients required [49]. Random forest technique is used for behavioral modeling component selection applied to a commercial PA operating under 5G-NR signal [737]. In [599], a residual-fitting modeling method applied for DPD is presented
by reducing the number of coefficients and ACPR compared to existing works; alternative techniques have used a tangent of the midpoint of the piecewise for modeling
stage [168].

The selection of the primary method in the modeling and DPD stages is crucial
for the signal under analysis. Models based on ML have been explored as a technique that can effectively handle multivariate data for both modeling and linearization purposes. Compared to considering memory effects, ML-based models may also
provide better performance, particularly with regard to polynomial models [298]. In
the exploration of nonlinear problems through the advancement of artificial intelligence (AI), specifically with the utilization of convolutional neural networks (CNN),
remarkable efficacy in nonlinear analysis has been demonstrated [681], [374].

The radio-frequency PA requires a proper characterization and measurement procedure to provide a linearization stage into a device under test (DUT) platform, this
is also crucial to model in-band transmissions. Additionally, is provided a saving
method to characterize radio-frequency PAs based on code division multiple access
(CDMA) [36]. In the literature related works are based on ML, such as weighted linearization method for highly nonlinear PAs [71], [153]. Similarly, applied to Class-
E device for amplitude to amplitude (AM/AM) conversion curve and third-order

intermodulation distortion (IMD3) modeling at simulation level [709]. In real time, closed loop ML based dual-input digital Doherty PA is introduced.

ML provides an ensemble model with enough fitting in extracting and monitoring the behavior of nonlinear devices as radio-frequency PAs. the derived model of ML as random forest is utilized to make classification decision for PA model driven wireless emitter identification method [384] and as behavioral modeling structure and DPD validated for PA operating for 5G signals with a proper approach [737], [75].

By employing regression tree, random forest, and CNNs, the system is devised to enhance the supervised learning phase, leveraging the advantages of offline training. Additionally, it facilitates versatile comparisons for validating diverse activation functions. The modeling stage achieves a high precision of -44.7908 dB for LTE 10-MHz and -43.8446 dB for LTE 20-MHz signals. This ensures the future feasibility of using digital DPD for spectral enhancement and reduction in radio frequency transmissions, thus enabling the attainment of requisite peak-to-average power ratio (PAPR) levels for 5G applications.

The radio-frequency PA, as one of the most essential components in any wireless system, suffers from inherent nonlinearities. DPD has been widely accepted as one of the fundamental units in modern and future wideband wireless systems.

L. Guan and A. Zhu
IEEE Microwave Magazine

4.2   LEARNING FRAMEWORK FOR PA BEHAVIOR MODELING

Over the years, modeling approaches have been developed to represent the PA behavior, such as polynomial-based models [422], the time-delay neural network [657], and the nonlinear auto-regressive moving-average (NARMA) models [147]. However, in search for a model order reduction, Random Forest shows a suitable technique that offers flexibility for extracting a memory PA model from a generalization. Similarly, linear methods such as multiple linear regression are often used to compute Volterra-series model parameters [685]. Thus, the input-output signal relationship is modeled through the AM-AM characteristics of the PA. Furthermore, by taking into account only the diagonal kernels, a conventional memory polynomial model (MPM) makes an initial approximation of the model. The MPM structure used is representative of the equation 4.1:

$$y(n) = \sum_{k=1}^{K} \sum_{m=0}^{M} = a_{m,k} x(n-m) |x(n-m)|^{k-1}. \tag{4.1}$$

where, $a_{m,k}$ represents the coefficients related to the static and dynamic part of the model, $x(n)$ the sampled and discretized input signal, $y(n)$ the output signal, M the short-term memory level of the system, K the order of the polynomial to represent the static part.

On the other hand, supervised learning techniques for behavior modeling can be beneficial in uncovering the intricacies of single-objective optimization problems that involve varying numbers of parameters and constraints. In such cases, polynomial or exponential time solutions can fall under primary categories that offer models that are feasible to solve.

Classification and regression trees perform repeated splits on a training dataset until a rule establishes a terminal node. At each terminal node, the predicted response value is constant [73]. Predicting requires traversing the tree from the root node to a leaf node. Since Decision Trees are approximately balanced, crossing through the tree to make a prediction requires roughly $O(log_2(m))$ nodes, where m is the number of training samples. It is noted that the order of growth for the prediction does not depend on the number of features and only increases with the number of observations. For simplicity, the model can be constructed with each sample of the radio-frequency PA input signal x is transformed into a feature vector X as in equation 4.2

$$X = \{x[n], x[n-1], ..., x[n-m]\} \in \mathbb{R}^{3m} \tag{4.2}$$

where $x[n]$ is a vector containing the real, imaginary, and absolute instantaneous values of the RF-PA. The output model is compared with the learning system and is defined by equation 4.3

$$y_{out} = y_1(x_{in}) + y_2(x_{in})i + \varepsilon_1(x_{in}) + \varepsilon_2(x_{in})i, \tag{4.3}$$

The complex signal I and Q components are calculated in an ensemble fashion, using trees ε_1 and ε_1.

$$\varepsilon_1(x_{in}) = y_1(x_{in}) - \text{Re}(y_{out}), \tag{4.4}$$

$$\varepsilon_2(x_{in}) = y_2(x_{in}) - \text{Im}(y_{out}) \tag{4.5}$$

The equations (4.4) and (4.5) represent the computation of the error between the predicted output and the actual output for each dimension. The variable x_{in} represents the input to the model, while y_1 and y_2 are the true outputs for the real and imaginary parts, respectively. The variable $\text{Re}(y_{out})$ represents the real part of the output, and $\text{Im}(y_{out})$ represents the imaginary part. The error for the real part and imaginary part are computed using Equations (4.4) and (4.5). These error terms are modeled to boost the accuracy of the predicted outputs.

ML models have demonstrated their efficacy in DPD, particularly in offline data applications where software implementations are easily available in popular languages. However, implementing these models on hardware devices like FPGAs can be challenging, due to timing and resource limitations. Recent efforts have been made to optimize one-dimensional CNNs for FPGA implementation, and they have been shown to outperform LSTM and multi-layer perceptron models in terms of prediction accuracy [145]. One advantage of these models is their modular design and

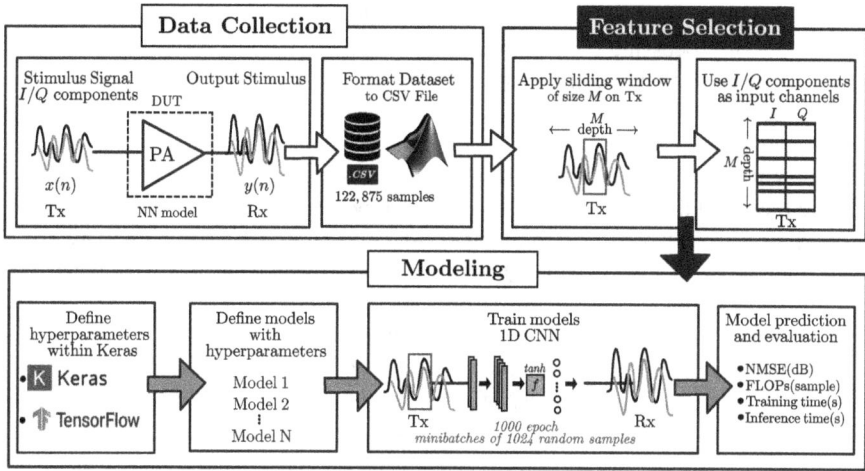

Figure 4.1   Acquisition process, feature selection, and model train modeling step.

parametric flexibility, which allows them to be adapted to different timing and re-source constraints. Furthermore, ongoing research is being conducted to improve the adaptability of DPD parameters in real-time applications. The weight sharing structure of the convolutional model serves to alleviate the influence of the strong memory effect, which is particularly beneficial for processing time series data, including DPD applications. Notably, when the PA's memory effects are augmented by utilizing wider signal bandwidths, the convolutional model's advantage is evident [681].

Figure 4.1 describes the general scheme of ML used in this research work; in the first stage, the generation of I/Q signals through the AD-FMCOMMS2-EBZ transceiver board is shown, as well as saving datasets. In the feature selection, the sliding window is made, and the signal is packaged as input channels. In the modeling stage, the performance of the system is validated, as well as the FLOP calculation using hyper-parameters within Keras and Tensor Flow, as well as the training of the random forest regression tree and CNN.

The PA 1D-CNN modeling process involves the use of a physical testbed equipped with specialized hardware to capture the inputs and outputs of the DUT, namely the PA. The DUT is subjected to a digitally generated LTE 64-QAM signal, which results in a complex signal encompassing the I and Q components. The signal is transmitted through the Tx channel, amplified by the PA, and received on the Rx channel, with the resulting data comprising $122,875$ samples of the LTE signal for both the input and output, stored in a CSV file. The collected data is significant as it offers valuable insight into the behavior of the DUT, which is essential for accurate modeling of the PA. Therefore, it is critical to emphasize the use of high-quality datasets to obtain optimal performance in PA modeling.

In the feature selection process of the neural network, the input (TX) and output (RX) signals of the PA are modeled, taking into account the complex nature of the signal with its real and imaginary components. Although some neural networks are capable of processing complex numbers, this work employs a traditional neural network approach that accommodates the inputs and outputs to a one-dimensional CNN structure. To adequately input data into the network, a feature selection process is employed, which takes into account the memory effects of the PA, resulting in nonlinearities and distortions in the output. A sliding window of size M is used to capture these nonlinearities, which are affected by memory effects, and to allow the network to learn the behavior. In this way, M samples are used to predict any given value, and the input signal's I and Q components are included as input channels. The resulting matrix has two columns for the I and Q components and M rows for the memory value, which assists in predicting memory effects.

Following the feature selection process, the modeling stage is initiated. The modeling pipeline may utilize a variety of libraries and tools from the field of programming languages. Python is selected as the modeling platform due to its open-source nature and the availability of numerous libraries for scientific computing and ML, thereby streamlining the modeling workflow. The Keras framework is utilized, along with TensorFlow serving as the backend, this combination enables efficient implementation of the modeling pipeline and allows for flexible and scalable development. At this stage, the definition of hyperparameters, as well as the selection of suitable metrics to assess the performance of the model, becomes paramount. Hyperparameters dictate the configuration of the model's architecture and the parameters governing the learning process. The choice of hyperparameters is crucial to achieve optimal model performance and ensure convergence. On the other hand, the selection of appropriate performance metrics ensures that the model's output accurately reflects the desired outcome of the task at hand, while providing insights into its strengths and limitations.

Within the modeling framework, hyperparameters are defined, including the loss function used to train the model. Selecting an appropriate loss function is crucial for successful training of the model, as it determines how the model's predictions are compared to the actual values during training. Keras offers several loss functions such as binary cross-entropy, categorical cross-entropy, mean absolute error (MAE), mean squared error (MSE), and others. The type of loss function selected depends on the problem being addressed. In order to accurately predict the behavior of the PA, a regression model is utilized, necessitating the use of a loss function suitable for regression tasks. Although MAE and MSE can be used for regression, NMSE is selected as a commonly used metric in regression analysis to evaluate the accuracy of PA predictive models. NMSE measures the in-band distortion between the ideal and predicted signals, and it is expressed as a mathematical formula in equation 4.6.

$$\text{NMSE}\,(n)_{\text{dB}} = 10\log_{10}\left(\frac{\sum_{n=1}^{N}|y(n)-\bar{y}(n)|^2}{\sum_{n=1}^{N}|y(n)|^2}\right) \tag{4.6}$$

where $y(n)$ is the actual data and $\bar{y}(n)$ is the predicted models in time convergence and accuracy of the estimated PA output. The equation utilizes the predicted signal $\bar{y}(n)$,

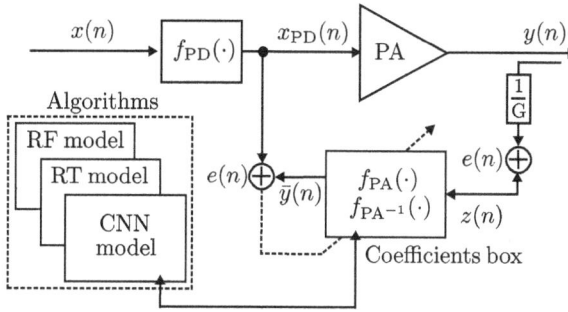

Figure 4.2 Block diagram for the indirect learning

and the actual signal $y(n)$, as inputs. The actual signal is obtained by amplifying the measured signal with the PA, whereas the predicted signal is generated using a model selected from Figure 4.2. The equation then calculates the difference between the predicted and ideal signals, which is squared and summed over all data points. This value is then divided by the sum of the squared ideal signal. The entire expression is then logarithmically scaled and multiplied by 10 to yield a value in decibels (dB). As the NMSE measures the distortion between the predicted and ideal signals, which are usually close in value, the resulting NMSE value in decibels will be negative.

The next step after defining the loss function in the modeling pipeline is the selection of the optimizer. In this study, the Adam optimizer was chosen over RMSprop, as it has demonstrated better convergence for the specific problem at hand, achieving lower error rates and requiring less training time. The default learning rate of 0.05 provided by Keras was used for the Adam optimizer. The training process was carried out using mini-batches of size 1024, with 1,000 epochs. Early stopping was implemented to speed up the training process and eliminate redundancy between the mini-batches, ensuring that sufficient information was learned from each mini-batch without overfitting on specific examples. A patience value of 5 was used to allow the network to learn the parameters from the example data, this means that training is stopped after 5 consecutive epochs with no improvement in validation loss, allowing the network to learn the optimal parameters without fitting too closely to the training data.

An array of configurations was defined to test the model's performance, which included the number of layers in the neural network and the type of activation functions used. The configurations were tested for one, two, and three layers of convolution with an additional dense layer to flatten the output for prediction. The goal was to identify the best architecture that yielded best results. The activation functions tested included elu, ReLU, LeakyReLU, sigmoid, and hyperbolic tangent functions. As the ReLU function does not have a negative part, it was omitted from the last layer of the network to prevent truncation of the output and ensure the accurate modeling of the PA behavior. Additionally, various memory values were tested for the network, which determined the size of the sliding window that served as the input. Values ranging from 1 to 11 were tested, and the one-dimensional neural convolutional network proved to be a suitable fit for this regression problem given its ability to process data based on previous samples.

During model evaluation, each configuration was trained and evaluated using the NMSE metric, which was defined as the loss function for training. Tensorflow utilities were used to provide a summary of the model's resource requirements, which included the number parameters, multiplications, additions, and the overall flop count at inference. This information was logged to compare the resources needed for the model with other state-of-the-art PA models. All the results were logged into a csv file for further analysis and data visualization. Additionally, the best-performing models for each activation function were saved for later visual comparison.

4.2.1 REGRESSION TREE

Classification and Regression Trees (CART) is an algorithm that performs repeated splits on a space of input vectors X into two descendant datasets. These sections on X represent nodes t of a tree T, and each split s is an element of the set S which contains all possible splits. The X space separated until it reaches a particular condition at terminal node t. At each terminal node t, the predicted value $\bar{y}(n)$ is a constant. It starts from a root node and ends in a leaf node that contains the response of the modeling system as can be depicted in Figure 4.3.

Tree construction for regression trees is a training process where the model is built using a set of training data. Once the process is completed, the resulting data structure can be used to make predictions based on conditions defined by the tree model. To make predictions, the input X is compared with one of the values s used to split the dataset, and based on the comparison, the procedure continues recursively on the left or right subtree until reaching a terminal node. At that point, a prediction is made based on the value associated with the terminal node. This prediction procedure can be mathematically described using equation (4.7), which defines a piecewise function

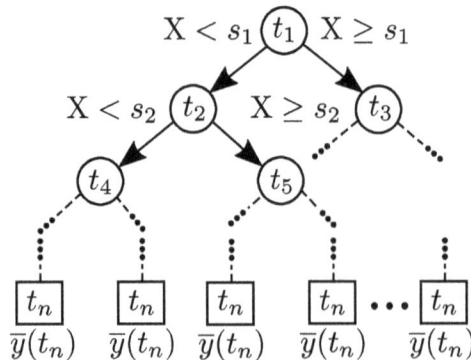

Figure 4.3 Block diagram of the RT architecture

that follows the aforementioned scheme.

$$f(x,t) = \begin{cases} \bar{y}(t_n) & t_L, t_R \in \emptyset \\ f(X, t_L) & X_n < s_i \\ f(X, t_R) & X_n \geq s_i \end{cases} \tag{4.7}$$

Regression Trees are widely used as they offer interpretability and ease of use, but over-fitting and poor generalization performance remain a concern. To address these issues, a proper approximation can be obtained by using a cost function to optimize the model. This flexible technique does not require extensive parametric adjustment and can be applied to highly non-linear models. The tree construction process may use different methods to determine the node splits and tree depth. Algorithm 6 describes the standard training procedure for Regression Trees, which maximizes the decrease in mean squared error (MSE) at each node split. In the context of predicting PA amplification, the accuracy of the models is evaluated using the NMSE, as defined by equation 4.6.

Notation: T, T_L, T_R, denote the training dataset, the left of, and the right subset of a split. The input and response values $X_{t,f}$ and y_t are contained within T. The S values correspond to the proposed splits and ε_t, ε_{t_L}, ε_{t_R}, $\Delta\varepsilon$ represent the MSE of the response values, the MSE of the left subset of a split, the MSE of the right subset of a split, and the MSE decrease of a split. The y_1, y_2, ε_1, and ε_2 are the Regression Trees obtained after the training procedure described in Algorithm 6.

4.2.2   RANDOM FOREST

The Random Forest algorithm is a popular ensemble learning method that involves constructing a collection of k decision trees. Each tree is dependent on a random vector Θ, which affects the numerical output of the tree predictor $h(X, \Theta)$ at a given input X. The overall prediction of the Random Forest is obtained by averaging the predictions of all k trees $h(X, \Theta_k)$, producing a more robust and accurate prediction. Unlike a single Decision Tree model, Random Forests use bagging methods that randomly select subsets of features and training samples to construct each tree. This approach reduces overfitting and improves the generalization performance of the model. As shown in Figure 4.4, a Random Forest model constructs k trees, each using random subsets of features, and then outputs the mean $\bar{y}$ of the predictions. The flexibility and interpretability of Random Forest models make them a popular choice in various applications, including classification and regression problems. For regression problems, the Random Forest predictor is the unweighted average over the Equation 4.8.

$$\bar{y} = \frac{1}{K} \sum_{k=1}^{K} h(X, \Theta_k). \tag{4.8}$$

Random Forest uses an ensemble of decision trees to improve the accuracy and robustness of the prediction. Unlike a single decision tree, Random Forest uses multiple trees to achieve more reliable predictions, albeit at a higher computational cost.

Algorithm 6 Training procedure for RT model.

1: **procedure** TRAIN_TREE(**T**)
2:     **if** CAN_SPLIT(T) **then**
3:        $\varepsilon_t \leftarrow \sum_{t \in T} (y_t - \text{MEAN}(y_T))$
4:        **for f** $\in$ **x$_t$ do**
5:           **sort x$_t$, y$_t$ in ascending order by x$_{t,f}$**
6:           $s^* \leftarrow$ **NULL**
7:           **for s** $\in$ **S do**
8:              **T$_L$, T$_R$** $\leftarrow$ **s**
9:              $\varepsilon_{t_L} \leftarrow \sum_{t \in T_L} (y_t - \text{MEAN}(y_{T_L}))$
10:             $\varepsilon_{t_R} \leftarrow \sum_{t \in T_R} (y_t - \text{MEAN}(y_{T_R}))$
11:             $\Delta \varepsilon \leftarrow \varepsilon_t - \varepsilon_{t_L} - \varepsilon_{t_R}$
12:             $s^* \leftarrow \text{MAX}(\Delta \varepsilon)$
13:           **end for**
14:        **end for**
15:        **t$_L$** $\leftarrow$ TRAIN(T_L)
16:        **t$_R$** $\leftarrow$ TRAIN(T_R)
17:        **return** y, **t$_L$, t$_R$**
18:     **else**
19:        **return** MEAN(y), **NULL, NULL**
20:     **end if**
21: **end procedure**
22: **procedure** TRAIN(**T**)
23:     $y_1 \leftarrow$ TRAIN_TREE(Re(x))
24:     $y_2 \leftarrow$ TRAIN_TREE(Im(x))
25:     $\bar{y} \leftarrow$ **y$_1$**(Re(**x**)) + **y$_2$**(Im(**x**))i
26:     **return** $\bar{y}$
27: Compute NMSE prediction $\bar{y}$ using Eq. (4.6).
28: **end procedure**

One of the main advantages of Random Forest is its ability to mitigate overfitting, a common problem in decision trees. In addition, it can handle missing values in the data set and performs well in high-dimensional spaces. Moreover, by using a bootstrap sample of the data set to build each decision tree, Random Forest is able to provide an unbiased estimate of the feature importance, which is useful for feature selection.

Figure 4.4 illustrates the architecture of a random forest, where the input vector X is partitioned and sent to different subtrees. Each subtree is trained on a random subset of features from X, which helps to prevent overfitting during the training process. However, this randomness also makes the training process non-deterministic, meaning that different sets of trees may be generated when training on the same data. Once the model is constructed, it becomes deterministic, giving consistent outputs for the same inputs. In the prediction phase, each subtree provides a prediction,

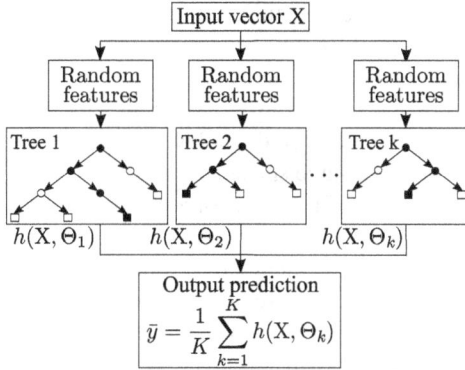

Figure 4.4   Block diagram of the RF architecture

which is then averaged to produce a robust estimate of the target variable. While Random Forest requires more computational resources than a single decision tree, the increased accuracy and robustness make it a valuable tool for many applications.

4.2.3 ONE-DIMENSIONAL CONVOLUTIONAL NEURAL NETWORK

One-dimensional convolutional neural networks (1D-CNNs) are a specialized subset of neural networks commonly used for processing sequential data. While traditional neural networks operate on images, videos, and audio in a straightforward manner, 1D-CNNs are designed to handle one-dimensional data such as time series, audio waveforms, and text data. One of the major benefits of 1D-CNNs is their ability to capture local features and patterns within the data, thanks to their use of convolutional filters. However, 1D-CNNs may not perform well in tasks that require global context, and their performance can be sensitive to hyperparameters. A 1D-CNN is a type of neural network that applies convolution on input data using a filter or kernel. The filter contains the weights of the convolutional layer, and it multiplies the input elements by the weights and sums the results. To increase the complexity of the network, multiple filters can be set per layer, and the filter size is determined by the size of the convolution window. However, the larger the filter size, the more computations are required by the network. To reduce computational resources, the stride value can be used to define the step size at which the filter is moved across the input data. The stride is simply the number of units by which the filter is shifted after each convolution operation. By using stride, the 1D-CNN can reduce the number of computations and thus the amount of processing power required to run the network. A larger stride value moves the filter more units, resulting in fewer convolutions and reducing the size of the output feature map, significantly reducing the amount of computation required by the network. Another useful technique to decrease the size of the feature map is the max pooling layer, which extracts the maximum value within a kernel size, effectively reducing the size of the feature map while retaining important features of the input data. In many cases, the reduction of dimensionality can result in the loss of important features at the edges of the input, so to preserve the spatial dimensionality of the input, padding is applied. It refers to the addition of extra elements to the input data, usually with a value of zero, before performing the convolution. The amount of padding is determined by the size of the filter and the desired output shape of the convolutional layer. Padding allows the output of the convolutional layer to have the same shape as the input, making it easier to connect the layers of the neural network. The output of a filter convolution is called a feature map, which is a matrix representing the filtered version of the input data.

A feature map is the output of a filter convolution, where each filter is responsible for producing one feature map. It represents the activation of the filter over the input, and its size is determined by the size of the input, the size of the filter, and the stride value. The output shape of a feature map is given by a formula that takes into account the size of the input, the size of the filter, and the stride value. The output shape of each layer can be calculated using the equation (4.9).

$$\text{Output shape} = \frac{(W - K + 2P)}{S} + 1 \qquad (4.9)$$

In this Equation 4.9, the input shape W refers to the shape of the input tensor to a convolutional layer, and kernel size K refers to the size of the filter kernel used in the convolution, S is the size of the stride of the convolution window and P is the padding size for the kernel. This equation is important in understanding the size of the feature map and how it changes as the input data is passed through the layers of a 1D-CNN. In a 1D-CNN, the input data is convolved with a filter, which is essentially a set of learnable weights that are used to extract features from the input data. The convolution operation involves multiplying the elements of the input by the corresponding weights in the filter and summing the results. An activation function is then applied to the result of this convolution, followed by the addition of a bias term. This process is repeated for each layer in the network, with the output of one layer becoming the input for the next. By chaining together these convolutional layers, the 1D-CNN can learn complex patterns and relationships within the input data. The definition of a convolutional layer is given in equation 4.10 where it uses the hyperbolic tangent as the activation function.

$$\omega_k = \sum \tanh(X(n) * W_i + b) \qquad (4.10)$$

Here, X is the input data, W is the filter weights, and b is the bias term. The algorithm loops through each position in the input data, and for each position, performs a convolution operation between the filter and the input data at that position, adds the bias term, and applies the activation function *tanh* to produce the corresponding feature map value ω_k.

In wireless communication systems, predicting the nonlinear distortion of a PA is a complex problem that can be formulated mathematically. To address this challenge, a 1D-CNN architecture is proposed, which includes two input channels representing the I and Q components of the complex input signal. To account for the memory effects of the PA, a sliding window of size 8 is applied, consisting of the present sample

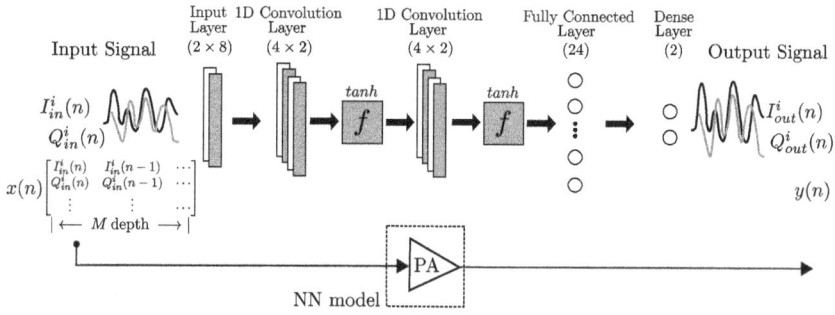

Figure 4.5 Block diagram of the CNN architecture

and 7 delayed samples. The resulting input layer has a sequence length of 8 and is processed by two convolutional layers with 4 filters each of kernel size 2, which are activated by the hyperbolic tangent function. These two layers are repeated to reduce the size of the output feature map, allowing the model to recognize relevant features and map them into a smaller space. The fully connected layer of the proposed 1D-CNN consists of 24 neurons that connect to the output layer of 2 neurons, one for each component of the output signal. Interestingly, during testing, adding an activation function to the output layer did not result in significant improvement, and therefore, no activation function was used in the output layer, and Figure 4.5 shows the proposed CNN architecture.

To further optimize the network's computational efficiency, 1D-CNNs share weights within their filters. This characteristic is a crucial aspect of the 1D-CNN architecture, because the filter acts as a sliding window that is applied to all the inputs. By sharing the same set of weights across all the inputs, the model can learn features that are useful for the task at hand and generalize better to new data. This property of weight sharing allows the model to be more efficient and have fewer parameters to train, which is particularly important in deep learning models that can have millions of parameters. Therefore, it is a fundamental characteristic of the 1D-CNN that the weights in the filters are shared across all the inputs, making it a powerful tool for various applications, such as image and speech recognition, natural language processing, digital signal processing, and many others. Overall, the proposed 1D-CNN architecture presents a promising solution to the challenge of predicting robust nonlinear distortion in PAs for wireless communication systems.

4.3 EXPERIMENTAL VALIDATION AND RESULTS

The validation models consist in a measurement experimental setup as shown in Figure 4.6. The FPGA provides a digital tuning of the signal under test to operate the transceiver system with efficient control of the output power levels. The DUT considers a cascade amplifier configuration, that consists in a driver amplifier ZX60-63GLN+ and a wide-band PA (ZX60−V63+) from Mini-Circuits, and

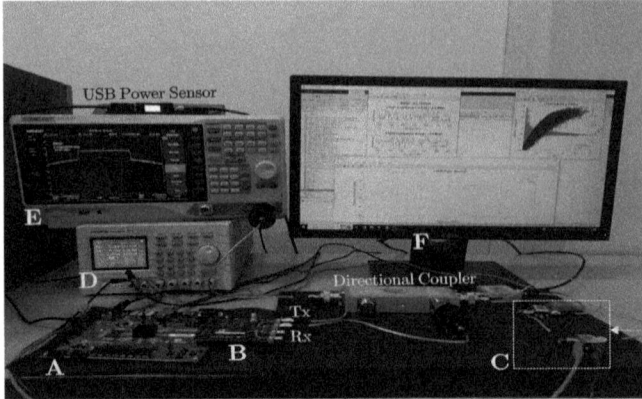

Figure 4.6 Experimental testbed platform. (A): Xilinx Zynq-7000 ARM/FPGA SoC Board. (B): AD9361 RF Transceiver at 2.45–GHz. (C): Cascade Amplifiers: pre-amplifier ZX60-63GLN+ & Wideband PA ZX60–V63+. (D): Power Supply GW Instek. (E): Spectrum Analyzer SIGLENT SSA3032X. (F): HOST-PC.

a DC power supply for PA biasing. The developed framework considers three key aspects: (*i*) The measurement system that includes the Xilinx Zynq-7000 ARM/F-PGA, RF transceiver dual-channel sub-6 GHz, a driver amplifier ZX60-63GLN+, a wideband Class-AB PA ZX60–V63+ with a typical gain for the cascade system of ≈ 45.3 dB at 2.45–GHz biasing with 5 V. (*ii*) A libiio library is run on this embedded Linux to provide a full-duplex communication between the data from MATLAB and the FPGA hardware. (*iii*) For the RF transceiver board AD9361; MATLAB sends the data to the FPGA. The data link between the ADI Hardware system and the FPGA-SoC platform is established through Ethernet, as described in [17]. The FPGA-SoC platform works by ADI Linux, which is similar to the operating system used on the ADI Hardware system. This enables efficient communication between the two systems, facilitating data transfer and resource sharing. The RF transceiver performance provides a narrow range from 2.4 to 2.5 GHz to support signals up to 56–MHz using dual 12-bit DACs with a synchronization clock set at 320–MHz. The up-conversion path uses an I/Q modulator driven by an internal local oscillator (LO) operating at the frequency of 2.45–GHz. At the feedback path, a receiver applies a down-conversion using an I/Q demodulator operated at a sampling rate of 640–MHz, while an interpolation filter is applied as data rate is set for a given bandwidth.

4.3.1 VALIDATION TESTS AND PROCEDURE

For the validation process, the system is provided with LTE signals of several bandwidths 10–MHz, 20–MHz a 6.75 dB PAPR is achieved. To enable accurate behavioral modeling, the associated estimation of the DUT delay is accomplished by calculating the maximum cross-correlation between the PA input and output samples signals. The validation presented is obtained from time-aligned data and a coefficient estimation extracted from a DUT testbed described in [136]. After that, the PA modeling framework is initially estimated using the real-valued data and training the algorithm to learn the parameters as shown in Figure 4.2. The observable data for the actual data and the predicted is computed with the NMSE degradation performance by offline training the predistorter based in an indirect learning architecture. The procedure is able to iterative validate the signal in terms of NMSE for the predicted output and the actual PA output behavior using and indirect learning approach.

In this work, the MATLAB `fitrtree` implementation of Regression Trees is used. The performed algorithm used considers a nonlinear behavioral modeling using an input vector $x(n)$ composed of delays of the input signal to take into account memory effects present in RF-PAs; that is $x_n = (x_n, x_{n-1}, x_{n-2}, ...)$. A well-known procedure based on an ILA architecture to identify the PA model is performed. As illustrated in Figure 4.3, a modeling stage identifies the predistorter through the Regression Tree model used to further implement an ILA linearization scheme, as well as coefficient extraction. The PA input complex signal is defined as $x(n)$, while $z(n)$ represents a normalized relationship between the maximum value of output power by attenuation $\frac{1}{G}$.

4.3.2 EXPERIMENTAL PERFORMANCE RESULTS

The RT model offers a good indicator of accuracy with a memory of 5, MinParentSize 1, MinLeafSize 1, Prunning off, MergeLeaves off for the LTE 10-MHz test, as shown in Figure 4.7 and Figure 4.9, for the AM/AM modeling and power spectrum, respectively. The sampling rate is 61.44 MSPS. The input data includes $122,880$ samples, while for the model training, the first $73,728$ samples are utilized. It should be noted that due to the limitation of the instantaneous bandwidth of ≤ 56 MHz in this platform, the DPD implementation is limited. Therefore, the experiments focus on a single-band LTE-20 MHz and a LTE-10 MHz signal with oversampling of 4, as depicted in Figure 4.10. It is worth noting that for the RT model prediction using an LTE 10 and 20-MHz an NMSE is obtained with an accuracy of -43.8446 and -44.7908 dB, respectively. The runtimes achieved for training and prediction of the LTE 10-MHz signal are 17.9932 s and 0.1797 s, respectively. Similarly, for the LTE 20-MHz signal, the runtimes are 22.3679 s for training and 0.8422 s for prediction. Note that for the LTE 20-MHz signal the quantization exhibits the limits for the AM/AM characteristics effects given the bandwidth filter limitation at the receiver. Figures 4.11 and 4.12, compare the complexity of the CNN model using a multicarrier LTE-3 & LTE-5 MHz signal, while the signal LTE-10 MHz is used in Figures 4.7 and 4.8, respectively.

Figure 4.11 depicts the competition of complexity of 1D-CNN in terms of floating point operations per second (FLOPS) at inference time versus the normalized mean squared error (NMSE) achieved, considering different activation functions. The figure illustrates that hyperbolic tangent performs significantly better than other activation functions, achieving the lowest NMSE score. The error of the hyperbolic tangent function levels off at around -37.5 dB, indicating that further improvements are unlikely with an increase in the number of FLOPS. The next best-performing

Table 4.1
Model accuracy comparison for the LTE-10 & LTE-20 MHz signals.

Model	Activation	NMSE (dB) LTE-10 MHz	NMSE (dB) LTE-20 MHz
RT	-	−44.7908	−43.8446
RF	-	−24.9254	−26.2404
CNN	ELU	−21.2706	−23.6074
CNN	LeakyReLU	−22.2914	−26.7572
CNN	ReLU	−22.2388	−26.0395
CNN	sigmoid	−21.9607	−24.4551
CNN	tanh	−22.2299	−25.6838

activation function is the exponential linear unit (ELU), which has comparable accuracy to other activation functions, but requires less computational resources. The sigmoid function is the worst-performing activation function, with a result that is comparable to the other activation functions. The figure demonstrates that as the number of FLOPS of computation allowed increases, the hyperbolic tangent and LeakyReLU functions perform better.

Figure 4.12 illustrates the performance of 1D-CNN in terms of the NMSE achieved versus the memory used (i.e., the length of the input sequence to the model), considering different activation functions. The figure demonstrates that the hyperbolic tangent activation function performs the best, achieving good results at memory values around 7. However, beyond that memory value, there is a trade-off between memory, network complexity, and performance. The next best-performing

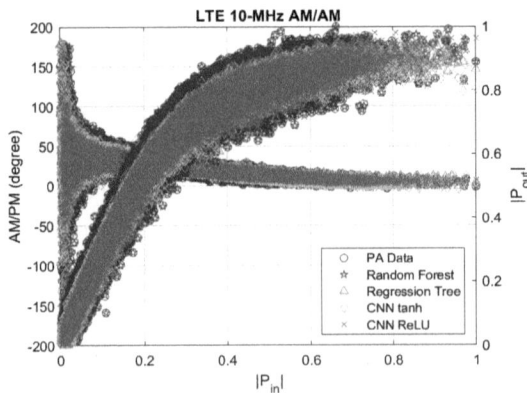

Figure 4.7 Normalized AM/AM and AM/PM characteristics with RF, RT, and CNN modeling results for an LTE 10-MHz signal.

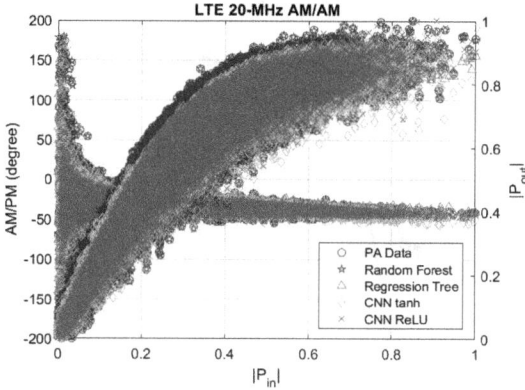

Figure 4.8 Normalized AM/AM and AM/PM characteristics with RF, RT, and CNN modeling results for an LTE 20-MHz signal.

activation function is the ELU, which follows a similar pattern to the hyperbolic tangent. In contrast, activation functions such as ReLU, LeakyReLU, and sigmoid seem to overfit the model beyond a certain memory threshold, leading to worse performance.

4.4   CONCLUDING REMARKS

In this work, a regression model with the application of regression tree, random forest and CNNs for the nonlinear behavior of PAs has been presented. The modeling results from experimental data show an accurate-enough feature in extraction with the target of the supervised technique. The framework benchmark shows a compact generalization through a learning structure for regressors when a hierarchical structure of a recursive subset is applied for long real-valued data.

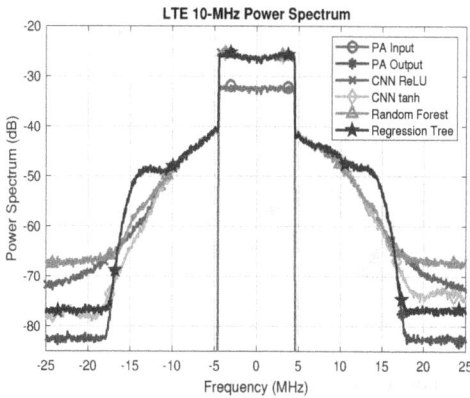

Figure 4.9 Output spectra with RT, RF, and CNN-based model prediction for an LTE 10-MHz signal.

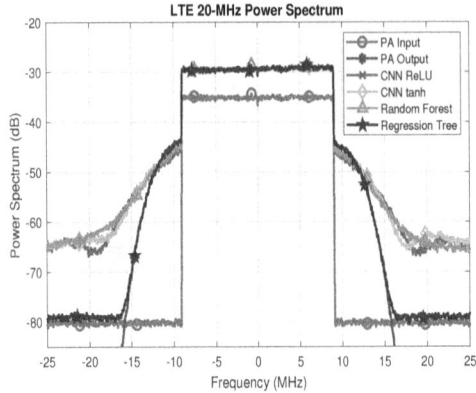

Figure 4.10 Output spectra with RT, RF, and CNN-based model prediction for an LTE 20-MHz signal.

Conventional methods for modeling often rely on sample segmentation. However, the use of the RT model offers an alternative approach. By recursively dividing input data into subsets, the RT model can produce a highly accurate estimated output model from the actual data in a fast runtime. Experimental results show the performance of RT algorithm yields an appropriate convergence with -44.7908 dB in NMSE for an LTE 20-MHz while has achieved an accurate model without overfitting. Furthermore, an ensemble PA behavioral modeling along power transmitter and impairments using the identification method for different signals and bandwidths OFDM-based 64-QAM modulated LTE signals.

The ensemble model RT offers a baseband single-step iteration solution for modeling sparse AM/AM behavior in a cascade setup that drives a Class AB PA. This

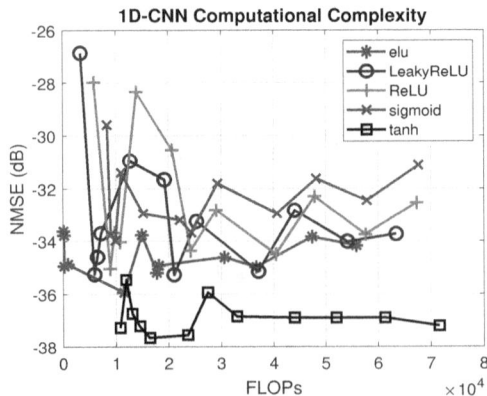

Figure 4.11 Comparison of activation functions in terms of complexity and accuracy.

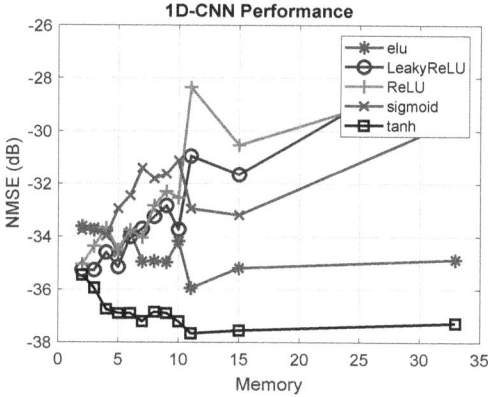

Figure 4.12   Relationship between memory and accuracy for different activation functions.

model takes into account the imperfections in the transmitter chain as well as PA non-linearity. To assess the effectiveness of the proposed model, the wideband for LTE 10 and 20-MHz signals with varying input power levels were tested. The measurement results, aided by ensemble methods, can provide flexibility in adaptation due to their robustness in achieving single-step compensation for high sparse data. This compensation addresses residual distortion present in the direct-conversion transmitter without changing the architecture, only by tuning the parameters.

5 Models and Methods for Anomaly Detection in Video Surveillance

Ernesto Cruz-Esquivel and Zobeida J. Guzman-Zavaleta
Department of Computing, Electronics, and Mechatronics,
Universidad de las Americas Puebla (UDLAP), Puebla 72810,
Mexico.

CONTENTS

5.1 VIDEO SURVEILLANCE SYSTEMS

A video surveillance system is a security system that uses video cameras and recording devices to monitor activities in a particular area allowing security personnel to view live or recorded footage. Video surveillance cameras are a powerful tool for security, especially in large public areas and big cities. For instance, monitoring could be in real time to detect and respond to potential security threats or incidents. Recording for later review allows the analysis of events for identifying, gathering evidence, or monitoring purposes. Alerting is based on sensors or event detection, helping to improve response times and reduce the risk of loss or damage. Video surveillance systems have grown worldwide; according to Cisco forecasts [122], connected city applications have a 26 percent compound movement rate. For instance, Mexico City has a network of over 63,000 video surveillance cameras in public spaces, and

DOI: 10.1201/9781003385615-5

public transportation [99]; moreover, almost four million seconds of video are generated daily. There is a similar amount of public video surveillance cameras in London [45] contrasting with a more significant number of cameras per person. The enormous amount of information generated is hard for human operators to process; therefore, computer-aided techniques are utterly required, among other tasks, to help operators in surveillance and detect situations of interest or anomalies automatically.

Anomalies in video surveillance systems are context-dependent. For instance, on streets, they can range from people running away and cars doing illegal U-turns to many other catastrophic scenes. Additionally, anomalies can be domain-dependent, classified as either spatial or temporal. In video data, spatial features define the appearance of objects at a pixel level, while temporal features represent the movement or flow of the objects. Therefore, spatial anomalies are related to unusual appearances, while temporal anomalies are associated with atypical movements of objects in the scene. An example of a spatial anomaly is a car on a pedestrian path, as cars are not supposed to be on pedestrian paths; therefore, any vehicle in that location is anomalous. In a similar scenario, a temporal anomaly would be a person running backward. Although people running on a pedestrian path might be expected, running backward is an unexpected movement for a person. Understanding the difference between the two domains of anomalies is crucial to comprehending why certain models excel at detecting spatial anomalies, while others perform better at detecting temporal anomalies. An ideal model would be capable of detecting both types of anomalies without difficulty.

5.2   STANDARD VIDEO SURVEILLANCE DATASETS

Several datasets have been presented to test the video surveillance anomaly detection techniques. Although these datasets still have some limitations, they have served as a standard benchmark. Initially, the datasets were bounded or synthetic. However, over time, datasets have become more challenging, encompassing different kinds of anomalies and offering more realistic scenarios. The following list includes the standard video datasets and benchmarks for assessing models detecting anomalies in video.

Subway. The Subway dataset [11] consists of two scenarios captured by surveillance cameras at an entrance and an exit of a public transportation system. The anomalies in this dataset include going in the opposite direction, unusual gestures, and running, among others. The entrance subset has 96 minutes of video data with 66 abnormal events. The exit subset has 43 minutes of video data, including 19 odd events. The main challenge for the Subway dataset is the detection of temporal anomalies like wrong-direction movement. Nevertheless, the area where abnormalities occur is limited to specific regions of the frames, and the number of anomalies in the dataset is small.

UCSD Ped. University of California San Diego Anomaly Detection Dataset (UCSD Ped) [357] includes two subsets, Ped1 and Ped2; both are pedestrian

Figure 5.1 Example of frames in UCSD Ped 1 and Ped 2. 5.1(a) is a frame with normal activity, only pedestrians are in the scene. 5.1(b) shows a frame with an anomaly: a car is crossing in the pedestrian path, both in Ped 1. In counterpart, in Ped 2 5.1(c) only pedestrians are in the scene. An atypical behavior is a bicycle crossing in the pedestrian path in the frame 5.1(d).

paths inside the UCSD campus as observed in Figures 5.1(a) and 5.1(c). In UCSD pedestrians are the only element considered normal, while anomalies include cars, bicycles, and skateboards. Examples of anomalies are observed in Figures 5.1(b) and 5.1(d). The UCSD Ped1 has 34/36 training/testing video clips, but Ped2 contains 16 training video clips and 12 testing video clips. The UCSD Ped1 contains more temporal anomalies while the UCSD Ped2 contains more spatial anomalies. However, this dataset has several drawbacks, including only low-resolution grayscale frames and the dataset only includes two different scenarios.

CUHK Avenue. The Chinese University of Hong Kong (CUHK) Avenue dataset [383] contains video clips from a pedestrian path inside the university. It includes 16 training video clips and 21 testing video clips. The anomalies in this dataset include people running or behaving strangely, like throwing bags into the air. This dataset only contains a scene, an avenue, and an entrance to the CUHK campus as observed in Figure 5.2(a), and the

(a)   (b)

Figure 5.2 Frames in Avenue dataset. The normal frame 5.2(a) shows a pedestrian path and an entrance. In Figure 5.2(b) an anomaly is a person throwing a backpack into the air.

anomalies are synthetic in some cases, as in the bag-throwing case that can be observed in Figure 5.2(b). Nevertheless, the scenario is more challenging as the angle makes the picture look crowded.

ShanghaiTech. The ShanghaiTech Campus dataset [388] contains different scenes of pedestrian paths inside the ShanghaiTech campus. The dataset contains 13 different scenes of pedestrians walking as normal behavior. Anomalous and normal frames from three different scenes can be observed in Figures 5.3(a) to 5.3(f). The difference between scenes is the camera's location, which has more variety of places on the ShanghaiTech campus and at different angles. ShanghaiTech's different scenarios make it a challenging dataset because one scenario might highly differ visually from the other; however, the expected behavior and the anomalies are the same. ShanghaiTech is still limited to pedestrian paths where bicycles and cars are considered anomalies.

StreetScene. The StreetScene dataset [501] is the most challenging of the listed datasets. This dataset includes videos from a busy street with a double-direction car lane and two pedestrian sidewalks; an example can be observed in Figure 5.4(a). This dataset includes bicycles, cars, and pedestrians as typical cases. The spatial anomalies are not as significant as in other datasets, as elements besides pedestrians are considered normal too. The most important in StreetScene dataset is how the objects behave, such as bicycles moving outside their lane, cars doing illegal u-turns, or pedestrians jaywalking. Therefore, StreetScene is a challenging dataset as most of its anomalies are temporal anomalies.

Table 5.1 shows the standard datasets with their main characteristics summarized.

Figure 5.3 Normal and anomalous frames from three different scenes from ShanghaiTech Campus dataset. Frames 5.3(a), 5.3(c), and 5.3(e) show only pedestrians as a normal behavior. Abnormal scenes are in Figure 5.3(b) with a bicycle; Figure 5.3(d) someone riding a bike with an umbrella; Figure 5.3(f) contains a skateboarding person.

5.3   DESCRIPTION OF MODELS FOR THE ANOMALY DETECTION

The supervised learning techniques to solve tasks in computer vision have increased since 2012 with the AlexNet architecture [26], and mainly for classification tasks. Nevertheless, the first technique applied for video surveillance anomaly detection

(a)   (b)

Figure 5.4 Example of frames in the StreetScene dataset. A frame labeled as normal is shown in Figure 5.4(a). In contrast, Figure 5.4(b) has a person crossing in the middle of the street as an abnormal behavior.

was in 2015 [682], and it was assessed with UCSD Ped 1 and 2 datasets listed in Section 5.2. The anomaly detection task is a binary classification problem where one of the classes is scarce. That is, anomalous samples in the available and standard datasets are not enough in most cases to train a standard classifier that is outstanding in other scenarios in the computer vision area. In contrast, the semi-supervised techniques combine labeled and unlabeled data to improve the model's accuracy; for instance, using labeled data to guide the learning process and using the unlabeled data to uncover patterns or structures in the data that can be useful in making predictions about unseen data. Moreover, unsupervised learning techniques analyze unlabeled data to find the patterns and structures representing the data in another space. Therefore, the following subsections describe the main pre-processing and learning techniques employed and examples of potential approaches to anomaly detection.

Table 5.1

Comparison of video surveillance anomaly detection datasets.

Dataset	Total frames	Train frames	Test frames	Frames' size
Subway Entrance [11]	144,249	22,538	121,711	512×384
Subway Exit [11]	64,900	7,546	57,354	512×384
UCSD Ped1 [357]	14,000	6,800	7,200	238×158.
UCSD Ped2 [357]	4,560	2,550	2,010	360×240
CUHK Avenue [383]	30,652	15,328	15,324	640×360
ShanghaiTech [388]	317,398	274,515	42,883	856×480
StreetScene [501]	203,292	56,847	146,445	1280×720

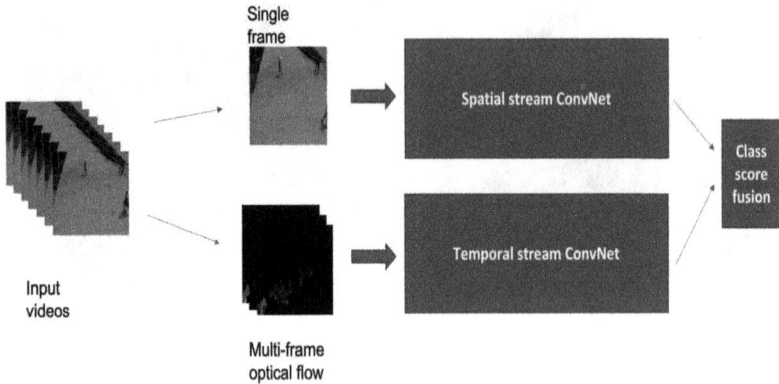

Figure 5.5   Data flow in the Two-Streams model. The input to the spatial and temporal architectures includes single frames (grayscale or RGB) and optical flow frames. The fusion of both output streams assigns the anomaly score.

5.3.1   OPTICAL FLOW AS A STREAM

Optical flow frames encode the difference between two frames, the frame t and the frame $t + 1$. The difference between those two frames is encoded using the differential between the pixel intensity on each image. As a result, optical flow obtains a helpful representation of temporal features. The assumptions for the calculation of the optical flow include the constant brightness assumption and the slight motion of the object from one frame to the following.

Simonyan and Zisserman demonstrate that optical flow helps describe temporal features using supervised learning [571]. They presented a model with two streams of Convolutional Neural Networks (CNN): one of the CNNs is trained using regular RGB scale frames, and the second CNN is trained several times using different temporal extraction techniques. A simplified version is observed in Figure 5.5. The temporal extraction techniques included single-frame optical flow (optical flow stacking with 5 and 10 frames), trajectory stacking (with 10 frames), and bi-directional optical flow (stacking with 10 frames). Their best results were obtained using optical flow and bi-directional optical flow.

The results of the two streams from the CNNs are combined by late fusion or class score fusion. Late fusion integrates the predictions of the two models to generate a final prediction improving overall accuracy. In [571], two late fusion methods were tested: averaging and using a multi-class linear SVM. The training of both CNNs is the same as proposed by Krizhevsky [318]. That is, frames are randomly sampled from the classes, and the optimization technique is the mini-batch stochastic gradient descent with momentum set to 0.9. The model of Simonyan and Zisserman outperformed the results of state-of-the-art papers at the time.

In particular, the Two-Stream Convolutional Networks model is applied to action recognition tasks. Nevertheless, the ideas presented in their paper are useful for video analysis in general and have been used in video surveillance anomaly detection tasks. A vital part of the two-stream technique is the technique for temporal feature extraction. The optical flow was the best option for pre-processing the temporal frames.

Other deep learning techniques have also been developed to process optical flow. For instance, FlowNet [158] obtains optical flow frames using convolutional neural networks with two consecutive images as input. FlowNet is composed of contracting convolutional and pooling layers and a refinement module. The contracting module learns how to extract features needed to generate the optical flow frames with a reduced size. Two versions of the FlowNet were assessed, the FlowNetSimple and FlowNetCorr. FlowNetSimple is a standard CNN that receives both consecutive frames stacked. FlowNetCorr has two separate streams that receive each one of the consecutive frames. Each frame is processed separately, and features are extracted and combined at a higher level. The combination is achieved using a *correlation layer*; both feature maps are correlated with a multiplicative patch comparison. The correlation, c, between two patches with the first map centered at $\mathbf{x}_1$ and the second map centered at $\mathbf{x}_2$ is defined by Equation 5.1. In the equation, $\mathbf{f_1}$ and $\mathbf{f_2}$ are multi-channel feature maps.

$$c\left(\mathbf{x}_1,\mathbf{x}_2\right) = \sum_{\mathbf{o}\in[-k,k]\times[-k,k]} \langle\mathbf{f}_1\left(\mathbf{x}_1+\mathbf{o}\right),\mathbf{f}_2\left(\mathbf{x}_2+\mathbf{o}\right)\rangle \qquad (5.1)$$

Both FlowNetSimple and FlowNetCorr versions use the refinement module. The refinement module improves the results of the contracting layers and recovers the original image resolution. The refinement module has upconvolutional, a type of up-sampling layers, and a convolutional layer. The features maps are upconvoluted and concatenated with the corresponding feature map from the contracting network part. The concatenation refines the results already achieved by the contracting network. The optical output flow image is four times smaller than the original input image. Finally, Endpoint Error (EPE) is used as training loss. EPE is a standard metric for optical flow estimation techniques that measures the Euclidean distance between the predicted flow vector and the ground truth flow vector at each pixel. The optimization problem is solved using the Adam optimizer.

FlowNet 2.0 has also been proposed [263]. FlowNet 2.0 was designed to improve the results of the original FlowNet to be more competitive against iterative methods. Therefore, FlowNet 2.0 combines a group of FlowNets for different displacement sizes. The complete architecture has two separate streams that each receive two consecutive frames. The first stream is for large displacement, and it contains three networks. The reason to stack three networks is to have iterative refinement. The first network is a FlowNetC, and it receives as input the consecutive images, image I_1 and image I_2. The FlowNetC outputs an optical flow estimate $w_i = (u_i, v_i)^\top$, i is the index of the stacked network. The next network is FlowNetS and it receives as input the previous network output, both images, a warped image, and its brightness error. The second image is warped operationally using flow $w_i = (u_i, v_i)^\top$ and bilinear interpolation as observed in Equation 5.2.

$$\tilde{I}_{2,i}(x,y) = I_2\left(x+u_i,y+v_i\right) \qquad (5.2)$$

Table 5.2

Comparison Average Endpoint Error (AEE) per frame of FlowNet approaches using different benchmarks.

Model	Sintel clean	Sintel final	KITTI 2012	Middlebury	Runtime ms
	Train/Test	Train/Test	Train/Test	Train/Test	GPU
FlowNetS	4.50/6.96	5.45/7.52	8.26/-	1.09/-	18
FlowNetC	4.31/6.85	5.87/8.51	9.35/-	1.15/-	32
FlowNet2	2.02/3.96	3.14/6.02	4.09/-	0.35/0.52	123

The brightness error is calculated using Equation 5.3.

$$e_i = \left\| \tilde{I}_{2,i} - I_1 \right\| \tag{5.3}$$

Bilinear interpolation allows the computation of the derivates of the warping operation, therefore, it is possible to train the stacked networks end-to-end. A second FlowNetS follows the first one. FlowNetS expects the same five inputs as the first network. The difference is that these inputs are calculated using the output of the first FlowNetS. The small displacement stream has a single network, FlowNet-SD, a modified version of FlowNet-S. The modifications were made at the beginning of the network by making the kernels smaller and adding convolutions between upconvolutions to alleviate the noise for small displacements. Finally, a fusion network combines the output of both streams. The input to the fusion network is the flow, flow magnitude, and brightness error of each of the streams. The output of the fusion network would be a flow estimate. As a result, FlowNet 2.0 improves the optical flow estimation for small displacements.

Table 5.2 shows a comparison between FlowNet and FlowNet 2.0. It uses Average Endpoint Error (AEE), the average EPE used as a loss. Four datasets are compared: Sintel clean, Sintel final, KITTI 2012, and Middlebury. Sintel datasets obtain their ground truth from rendered artificial scenes. Sintel clean data has no added blur or atmospheric effects as Sintel final has. KITTI datasets are obtained from real-world scenarios using a camera and a 3D laser scanner. The motion that it contains is a very specific type and distant objects' motion is not captured. Middlebury is the dataset that contains the smallest displacement and it is also the smallest of the three datasets. The AEE values are obtained from FlowNet 2.0 paper [263].

Table 5.2 shows that there are no significant differences between FlowNetS and FlowNetC. The improvement for the FlowNet 2 is significant as it is having an AEE of more than 50% lower than previous techniques. Nevertheless, the system is more than 6 times slower than previous systems. Deep-learning techniques for optical flow extraction are not used as a base for video anomaly detection cases because iterative methods are still obtaining better results.

5.3.2 AUTOENCODERS

Autoencoders are considered unsupervised learning techniques aiming to learn a compressed representation of the input data and, based on the compact representation, reconstruct the input data as accurately as possible. An autoencoder is usually trained with unlabeled data without explicit labels or categories. Additionally, it is possible to use autoencoders in a semi-supervised learning setting by incorporating labeled data into the training process. Thus, autoencoders are promising semi-supervised learning approaches for the anomaly detection task in video surveillance. This way, autoencoders are trained only with normal data, while anomalies are unknown to the trained autoencoder. Therefore, in the testing stage, the encoder fails to generate the correct representation in the case of an anomaly; the decoder can not reconstruct the original data from the generated representation. As a result, the reconstruction loss increases, indicating an anomaly.

Autoencoders have been the base pipeline for several settings for video anomaly detection. In addition, convolutional layers help improve the feature extraction part. In general, autoencoders have two neural networks in their structure: an encoder and a decoder (see Figure 5.6). The encoder learns to generate a new representation for the data in a latent space while the decoder reconstructs the original data from the representation obtained from the encoder (in the latent space). Both networks are trained using a reconstruction loss function. Reconstruction loss is defined by the difference between the encoder's input and the decoder's output. In Equation 5.4, the encoder transforms the input data $\mathbf{x}$ using the weights $\mathbf{W}$, biases $\mathbf{b}$, and an activation

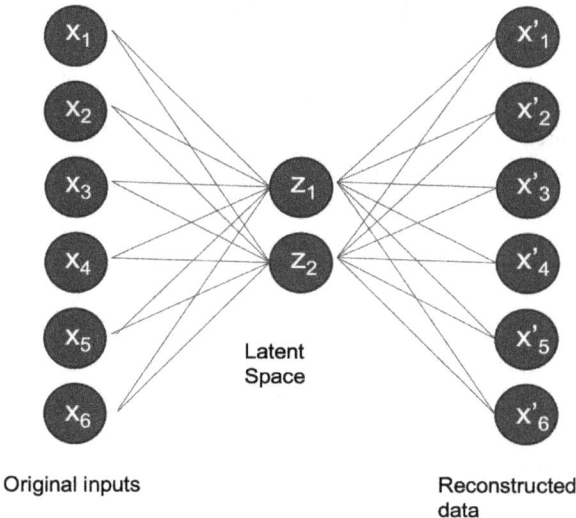

Original inputs   Latent Space   Reconstructed data

Figure 5.6 A generic autoencoder structure: x_i symbolizes the original data and x_i' the reconstructed data. The latent representation is given by z_i.

function σ to generate the latent representation $\mathbf{z}$.

$$\mathbf{z} = \sigma(\mathbf{Wx} + \mathbf{b}) \tag{5.4}$$

In Equation 5.5, the decoder transforms the latent representation $\mathbf{z}$ into reconstructed data $\mathbf{x}'$ using another group of weights $\mathbf{W}'$, biases $\mathbf{b}'$, and another activation function σ'.

$$\mathbf{x}' = \sigma'\left(\mathbf{W}'\mathbf{z} + \mathbf{b}'\right) \tag{5.5}$$

The loss is defined by Equation 5.6, which is the difference between the original and reconstructed data.

$$\mathscr{L} = \left\|\mathbf{x} - \mathbf{x}'\right\|^2 = \left\|\mathbf{x} - \sigma'\left(\mathbf{W}'(\sigma(\mathbf{Wx} + \mathbf{b})) + \mathbf{b}'\right)\right\|^2 \tag{5.6}$$

One potential model for anomaly detection using autoencoders is the model Appearance, and Motion DeepNet (AMDN) [682]. Its authors proposed three information streams with different temporal, spatial, and spatiotemporal inputs. Each stream uses a Stacked Denoising Autoencoder (SDAE) to detect anomalies. The inputs to the model are grayscale images, optical flow maps, and a fusion of those at the pixel level. Each network focuses on extracting a different feature type from the video data. The first network focuses on spatial feature extraction. The second network focuses on temporal feature extraction. The third network focuses on spatiotemporal feature extraction, a combination of spatial and temporal features. Denoising Autoencoders (DAEs) receive a corrupted x vector and output a corrected version of the x vector. An SDAE uses several DAEs together. The encoders are stacked together at the beginning of the model. Each of the encoders is smaller than the previous one. As a result, a bottleneck is generated as the output of the stacked encoders. The bottleneck is used as input to stacked decoders that have a symmetrical form to the stacked encoders. The training of the autoencoders is divided into pretraining and fine-tuning. The pretraining is for a single SDAE at the time and layer-wise using Equation 5.7. The data is represented by $\mathbf{x}_i$ and the reconstructed data by $\hat{\mathbf{x}}_i$. The learned parameters are represented by $\mathbf{W}$ and $\mathbf{W}'$ for the weights and $\mathbf{b}$ and $\mathbf{b}'$ for biases. The Frobenius norm is denoted by $\|\cdot\|_F$. The user defines λ parameter which defines the importance of any of the two parts.

$$\min_{\mathbf{W},\mathbf{W}',\mathbf{b},\mathbf{b}'} \sum_{i=1}^{N} \left\|\mathbf{x}_i - \hat{\mathbf{x}}_i\right\|_2^2 + \lambda \left(\left\|\mathbf{W}\right\|_F^2 + \left\|\mathbf{W}'\right\|_F^2\right) \tag{5.7}$$

The fine-tuning training is applied to a single SDAE at the time, and all layers are considered a single model. The objective function for the fine-tuning is shown in Equation 5.8. The first half of the equation is the reconstruction loss, $\mathbf{x}_i^k$ is the original data and $\hat{\mathbf{x}}_i^k$ is the data reconstructed by the SDAE. The training set is given by $\mathscr{T}^k = \{\mathbf{x}_i^k\}_{i=1}^{N^k}$ with N^k training samples. Given the training set, the SDAE learns the $\mathbf{W}$ and $\mathbf{W}'$ parameters. The user defines λ_F parameter. The Frobenius norm is denoted by $\|\cdot\|_F$.

$$J\left(\mathscr{T}^k\right) = \sum_{i}^{N^k} \left\|\mathbf{x}_i^k - \hat{\mathbf{x}}_i^k\right\|_2^2 + \lambda_F \sum_{i=1}^{L} \left(\left\|\mathbf{W}_i^k\right\|_F^2 + \left\|\mathbf{W}_i'^k\right\|_F^2\right), \tag{5.8}$$

Three One-Class SVMs are used to classify the latent representation of the autoencoders. One-Class SVMs learn to generate a topological space that covers a particular class, therefore, it has been useful for anomaly detection. In this case, the One-Class SVMs are used to classify the data from the autoencoders' latent representation in an unsupervised manner. The data used to input the autoencoders combines normal and abnormal data. Anomalies are the ones that lie outside the generated topological space for the normal data. Given a set of training samples $\mathscr{S} = \{\mathbf{s}_i^k\}_{i=1}^{N^k}$, the One-Class SVMs underlying problem can be defined by Equation 5.9. The learned weight vector is denoted by $\mathbf{w}$, the offset by ρ, and the feature projection function by $\blacksquare(\cdot)$. The user defines the parameter $v \in (0,1]$ and defines the expected fraction of outliers outside the normal topological space.

$$\min_{\mathbf{w},\rho} \quad \tfrac{1}{2}\|\mathbf{w}\|^2 + \tfrac{1}{vN^k}\sum_{i=1}^{N^k}\xi_i - \rho$$

$$\text{s.t.} \qquad \mathbf{w}^T\Phi\left(\mathbf{s}_i^k\right) \geq \rho - \xi_i, \xi_i \geq 0. \tag{5.9}$$

Then, using the results from the three One-Class SVMs, a decision-level late fusion is applied to obtain the final classification of abnormal or not. Late fusion is an optimization problem that learns the importance and weights of each result. To learn the weights $\alpha = \left[\alpha^A, \alpha^M, \alpha^J\right]$, the optimization problem in Equation 5.10 has to be solved.

$$\min_{\mathbf{W_s}^k,\alpha^k} \sum_k \alpha^k \operatorname{tr}\left(\mathbf{W_s}^k\mathbf{S}^k\left(\mathbf{W_s}^k\mathbf{S}^k\right)^T\right) + \lambda_s\|\alpha\|_2^2$$

$$\text{s.t.} \quad \alpha^k > 0, \quad \sum_k \alpha^k = 1 \tag{5.10}$$

The training set samples are represented by the matrix $\mathbf{S}^k = \left[\mathbf{s}_1^k, \dots, \mathbf{s}_{N^k}^k\right]$. $\mathbf{W_s}^k$ are the weigths that map the k-th feature $\mathbf{s}_i^k$ into a new subspace. The k-th feature covariance is represented by $\mathbf{W_s}^k\mathbf{S}^k\left(\mathbf{W_s}^k\mathbf{S}^k\right)^T$. A regularizer is added using the term $\|\alpha\|_2^2$ and user-defined λ_s.

As Figure 5.7 shows, patches were obtained from the original images. For the spatial part, three different patch sizes (15×15, 18×18, 20×20) were extracted using a sliding window. Then, from the more than 50 million obtained patches, 10 million randomly selected patches were warped to 15×15 for training data. For the temporal part, patch size was fixed to 15×15, and 6 million patches were used as training data. In the test phase, the sliding window is a fixed size of 15×15. Also, it is possible to pre-process the image with dynamic background subtraction to improve computing efficiency in the test part.

The AMDN method was assessed with UCSD Ped 1 and 2 datasets outperforming some of the state-of-the-art papers of the time, like Detection at 150 FPS [383], or Social Force [411]. AMDN was tested for the frame and pixel level of the Ped 1 subset. Xu *et al.* did not test the AMDN with more complex data sets like Avenue.

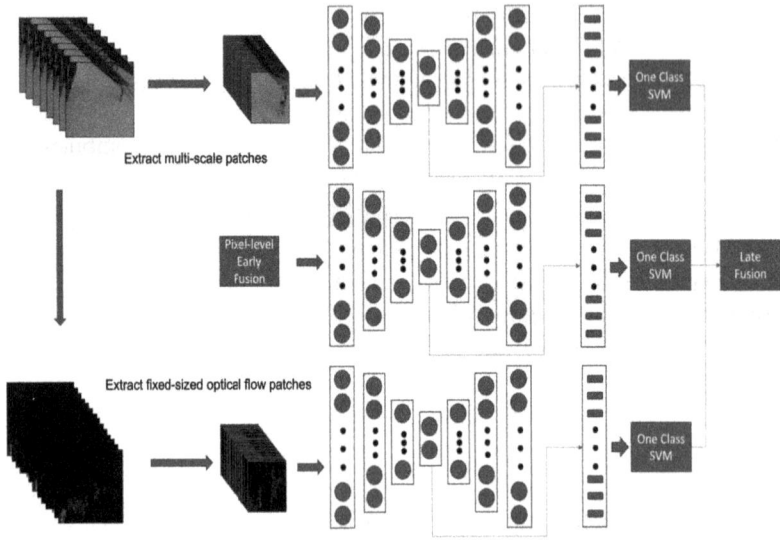

Figure 5.7 Data flow in the AMDN model, which has three streams and applies late fusion at the end to detect anomalies.

In 2016, other autoencoder architectures were applied for video surveillance anomaly detection tasks. The HOG + HOF Autoencoder was presented by Hasan *et al.* [228]. The authors used two autoencoders for their model. The first is a deep autoencoder that uses a temporal cuboid of Histogram of Oriented Gradients (HOG) and Histogram of Optical Flows (HOF) as input. HOG extracts the spatial features, and HOF the temporal features. Because the HOG + HOF input values go from 1 to 0, sigmoid or hyperbolic tangent is used as an activation function. The objective function of the deep autoencoder is defined in Equation 5.11. The reconstruction loss is the Euclidean loss of the original input $\mathbf{x}_i$ and the reconstructed input $f_W(\mathbf{x}_i)$. The optimal $\mathbf{W}$ vector of weights has to be found. An L_2 regularization term is added $\|W\|_2^2$. The γ is added to balance the loss and regularization. The HOG + HOF Autoencoder uses AdaGrad technique, stochastic gradient descent with an adaptive subgradient method, for the objective function.

$$\hat{f}_W = \arg\min_W \frac{1}{2N} \sum_i \|\mathbf{x}_i - f_W(\mathbf{x}_i)\|_2^2 + \gamma\|W\|_2^2, \qquad (5.11)$$

Nevertheless, HOG + HOF descriptors are not enough to learn the normal temporal pattern of videos. Therefore, the second autoencoder uses convolutional and upconvolutional layers to improve spatiotemporal feature extraction. Convolutional layers can learn to extract low spatiotemporal features that can be more useful in some cases than the manually tuned HOG + HOF descriptors. Upconvolutional layers can learn to generate dense data from sparse data and reconstruct the original data. Convolutional layers are alternated with max pooling layers to avoid translation variance and decrease the size of the output of the last layer. Unpooling layers reconstruct the original data modified by the pooling layers. The second autoencoder optimization equation is Equation 5.12. The equation is similar to Equation 5.11 but $\mathbf{X_i}$ is i^{th} cuboid.

$$\hat{f}_W = \arg\min_W \frac{1}{2N} \sum_i \|\mathbf{X}_i - f_W(\mathbf{X}_i)\|_2^2 + \gamma\|W\|_2^2, \tag{5.12}$$

Finally, the HOG + HOF model does not use the anomaly of a frame but the normality. That is, the pixels' intensity value reconstruction error is calculated with Equation 5.13. The $I(x,y,t)$ term represents the pixel's intensity in location (x,y) and frame t. The f_W represents the trained convolutional model. Then, a regularity score is calculated using Equation 5.14. All the pixels reconstruction error is summed $e(t) = \sum_{(x,y)} e(x,y,t)$. The regularity score is defined by $s(t)$

$$e(x,y,t) = \|I(x,y,t) - f_W(I(x,y,t))\|_2, \tag{5.13}$$

$$s(t) = 1 - \frac{e(t) - \min_t e(t)}{\max_t e(t)}, \tag{5.14}$$

Figure 5.8 shows the training procedure for the HOG + HOF method. Hasan *et al.* assessed their HOG + HOF method in Avenue, Subway, and UCSD datasets. The results were close to AMDN and, in other cases, worst than the state-of-the-art papers of the time. However, the model is less complex than AMDN because of the use of convolutions and pre-processed features.

These two techniques, AMDN and HOG + HOF Autoencoders were the first to apply deep learning to solve the video surveillance anomaly detection problem. Those authors considered using spatiotemporal features relevant by applying more than a single model to improve results.

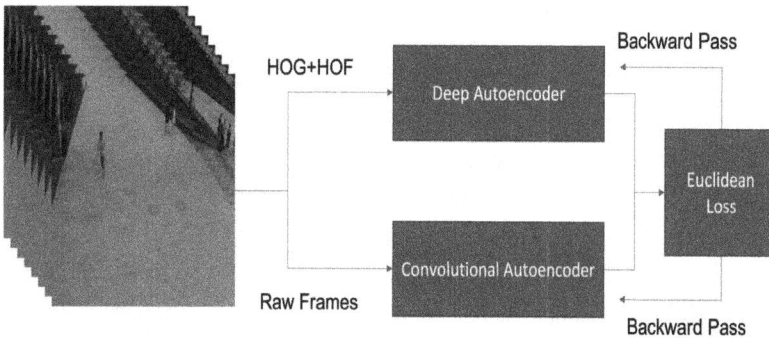

Figure 5.8 Visualization of the training pipeline proposed by Hasan *et al.* It uses the Deep Autoencoder or Convolutional Autoencoder [228].

There are still techniques that have been applied for video surveillance anomaly detection that rely on something other than optical flow. For instance, the Chong and Tay's spatiotemporal autoencoder [115] used a temporal encoder for temporal feature extraction instead of optical flow. The encoding part of the model contains two convolutional layers, a spatial encoder and a temporal encoder. The temporal encoder includes three convolutional Long Short-Term Memory (ConvLSTM) layers to extract the temporal features from the feature maps obtained from the spatial encoder. LSTM layers are part of the Recurrent Neural Networks, which get feedback from previous data by weighting the short-term and long-term data. The decoding part is symmetrical to the encoding part, but the decoding part includes upconvolutional layers instead of convolutional layers. Adding convolutional layers to the LSTM allows using fewer weights. Equations 5.15 to 5.20 define the ConvLSTM. The weights to be learned for each equation are represented by $\mathbf{W}$ and their biases by b. The input vector is denoted by x_t. There are two different states, cell state C_t and hidden state h_t, both at time t. $\otimes$ denotes the Hadamard product. Equation 5.15 is the layer that allows forgetting. Equations 5.16 and 5.17 are the ones used to add new information. Equation 5.18 combines the new with old information. Equations 5.19 and 5.20 outputs the learned.

$$f_t = \sigma \left(W_f * [h_{t-1}, x_t, C_{t-1}] + b_f \right) \tag{5.15}$$

$$i_t = \sigma \left(W_i * [h_{t-1}, x_t, C_{t-1}] + b_i \right) \tag{5.16}$$

$$\hat{C}_t = \tanh \left(W_C * [h_{t-1}, x_t] + b_C \right) \tag{5.17}$$

$$C_t = f_t \otimes C_{t-1} + i_t \otimes \hat{C}_t \tag{5.18}$$

$$o_t = \sigma \left(W_o * [h_{t-1}, x_t, C_{t-1}] + b_o \right) \tag{5.19}$$

$$h_t = o_t \otimes \tanh \left(C_t \right) \tag{5.20}$$

The before-mentioned regularity score is used as it was applied by Hasan *et al.* in Equation 5.14. During the training, the objective is to minimize the reconstruction error concerning the input. Adam optimizer is applied to solve the optimization problem. Data is augmented by skipping frames using different stride sizes. The first data group is formed by using all the frames from the dataset. The second group uses one frame and skips one from the original dataset. The third group uses one frame and skips two from the original dataset. As a result, the temporal feature extraction is improved. Each of the datasets has changed the temporality of the dataset slightly from the other. The datasets used to test the model were Subway, UCSD, and Avenue. There were improvements in Ped1, Avenue, and Subway Exit against other methods like Social Force [411] or a regular Convolutional Autoencoder. Nevertheless, a regular convolutional autoencoder outperformed the method in Ped 2 and Subway Entrance. This could mean that the temporal feature extraction has improved, but the spatial feature extraction has also worsened. Also, LSTM models are prone to overfitting, and their training might take a lot of computational resources.

5.3.3 VARIATIONAL AUTOENCODERS

Variational autoencoders (VAE) have also been applied to video surveillance anomaly detection tasks. A VAE aims to learn to encode and decode data while learning a probability distribution of the data. A representative example of VAEs for anomaly detection is the Gaussian Mixture Fully Convolutional Variational Autoencoder (GMFC-VAE) [169]. Fan *et al.* use two GMFC-VAEs, one for the temporal features and one for the spatial features. The temporal features extraction stream uses dynamic flow instead of optical flow [658]. Dynamic flow is the combination of sequential optical flow frames. Dynamic flow extracts longer-term temporal information than the one extracted from standard optical flow frames. Dynamic flow frames are obtained through an optimization problem that can be solved using linear ranking machines such as RankSVM [579]. The grayscale and dynamic flow frames are given as patches instead of the complete frame. The patches are obtained by sliding a window through the frames. The GMFC-VAE is trained using only normal data as in the standard anomaly autoencoder case, but variational autoencoders define their latent space with a Gaussian distribution. VAEs can be used to generate data that is similar to the distribution of the original training data. Figure 5.9 shows a VAE sample structure.

Their loss function is defined by Equation 5.21. To make the decoder $p_\theta(\mathbf{x} \mid \mathbf{z})$ learn to reconstruct $\mathbf{x}$, the expected log-likelihood of $\mathbf{x}$ is added in the first term. In the second term, the Kullback-Leibler divergence is calculated. It measures the difference between two probability distributions, and the goal is to learn a distribution with the encoder and the prior distribution of the latent representation $\mathbf{z}$.

$$L(\theta, \phi, \mathbf{x}) = E_{\mathbf{z} \sim q_\phi(\mathbf{z}|\mathbf{x})} \left[\log p_\theta(\mathbf{x} \mid \mathbf{z}) \right] - D_{KL} \left(q_\phi(\mathbf{z} \mid \mathbf{x}) \| p(\mathbf{z}) \right) \qquad (5.21)$$

Gaussian distribution makes the VAEs a good choice for detecting anomalies. Nevertheless, anomalies are not clustered in a single Gaussian centroid. Therefore, the GMFC-VAE model uses a Gaussian Mixture (GM) model instead. The GM-VAE loss is redefined as observed in Equation 5.22. The difference is in the Kullback-Leibler divergence as it is applied to the Mixture of Gaussians prior $p(\mathbf{z}, c)$ and variational posterior $q(\mathbf{z}, c|\mathbf{x})$.

$$L = -E_{\mathbf{z} \sim q(\mathbf{z},c|\mathbf{x})} \left[\log p(\mathbf{x} \mid \mathbf{z}) \right] + D_{KL}(q(\mathbf{z}, c \mid \mathbf{x}) \| p(\mathbf{z}, c)) \qquad (5.22)$$

Anomalies are not detected using the output of the two GMFC-VAEs. The joint probabilities of the GM model are used instead. They generate an energy-based score that is used as an anomaly score. The energy method is presented in Equation 5.23.

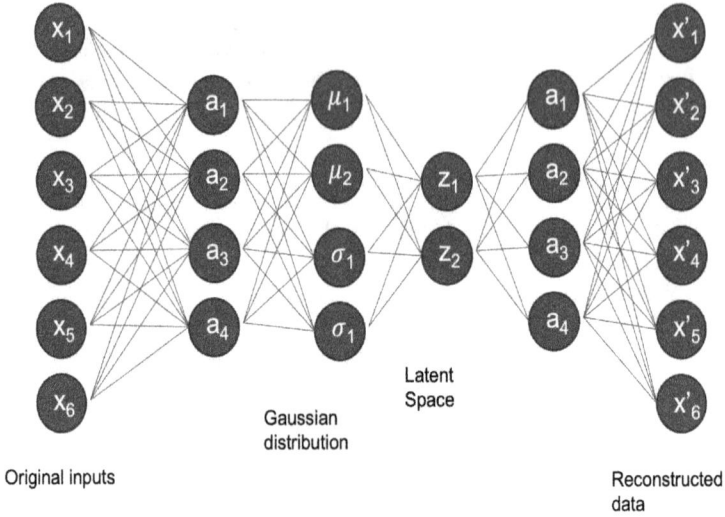

Figure 5.9 A generic variational autoencoder structure. The original data is represented by x_i and reconstructed data is represented by x'_i. The main difference is the use of a Gaussian distribution and its defining elements, mean μ_i and standard deviation σ_i, that shape the latent space representation z_i.

It combines both streams, spatial and temporal. The Gaussian components are represented by c'.

$$E(\mathbf{z}) = -\log\left(\sum_{c'=1}^{K} p\left(\mathbf{z} \mid c'\right) p\left(c'\right)\right) \qquad (5.23)$$

Anomalies are the higher scores. The complete score is calculated using both appearance and emotion scores. The scores are balanced with α and β.

$$E_{\text{overall}} = \alpha E_{\text{appearance}} + \beta E_{\text{motion}} \qquad (5.24)$$

The assessment of the GMFC-VAE model was using UCSD and Avenue. Results improved from models like AMDN or the HOG+HOF mentioned before. Nevertheless, in the Avenue dataset, results are far from perfect yet. Another problem that could arise for this model is that training VAEs raises some difficulties for the correct hyperparameter selection. Also, the size of the model is considerable as it uses two complete VAE networks.

5.3.4 GANS

Generative adversarial networks (GANs) are complex models integrating two networks to generate data: generative and discriminator networks. GANs are trained

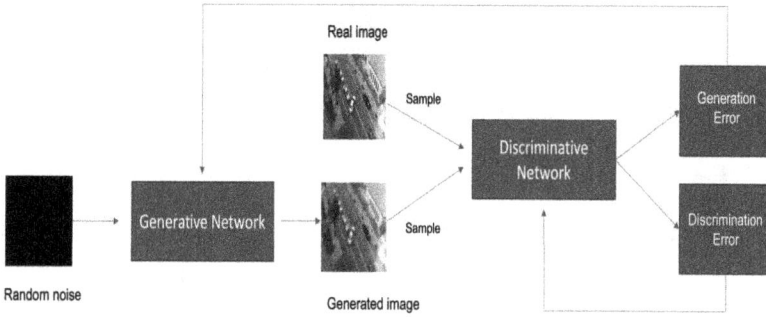

Figure 5.10 A generic GAN structure. The structure is defined by two networks, the generative and the discriminative. Each of them is updated using their respective error which is calculated in the discriminative network.

with both networks competing. The generative network generates new data based on the training data. GAN's discriminative network learns to detect the data that the GAN's generator has been generated. The generator has to improve the generated output to avoid being caught by the discriminator. The discriminator has to adapt to the generator improvements. As a result of the competition, both networks' accuracy is improved. In sum, GANs learn how to emulate the behavior of the input data to generate similar data. GAN loss equation can be observed in Equation 5.25 proposed by Goodfellow *et al* [212]. Both models are trained simultaneously. The first part trains the $D(\mathbf{x})$ to represent the probability that $\mathbf{x}$ is a real data sample and not generated. The second part trains the generator $G(\mathbf{z})$ with the help of the discriminator. Figure 5.10 shows a sample structure of a GAN.

$$L_{GAN} = \mathbb{E}_{\boldsymbol{x} \sim p_{\text{data}}(\boldsymbol{x})}[\log D(\boldsymbol{x})] + \mathbb{E}_{\boldsymbol{z} \sim p_{\boldsymbol{z}}(\boldsymbol{z})}[\log(1 - D(G(\boldsymbol{z})))] \qquad (5.25)$$

More recently, GANs have been applied to video surveillance anomaly detection tasks learning the behavior of normal inputs. For instance, Ravanbakhsh *et al.* propose to detect anomalies using a two-stream GAN model [509]; therefore, four networks. One of the GANs is for spatial feature extraction. The other GAN is for temporal feature extraction. The spatial stream receives optical flow frames as input and generates raw frames, the $N^{F \to O}$ case. The temporal stream receives raw frames as input and generates optical frames, the $N^{O \to F}$ case. Both GANs are trained using only normal data using a crossed-channel approach. Equations 5.26 and 5.27 show the losses used to train the generative network G and discriminative network D, respectively. In the $N^{F \to O}$ case, training set contains frames and optical flow images $\mathscr{X} = \{(F_t, O_t)\}_{t=1,\dots,N}$. Therefore, in L_{L1}, $x = F_t$ and $x = O_t$. Discriminators learn the normal distribution of the training data using optical flow and raw frames. Therefore, for testing, only discriminators are employed. The results of each discriminator are accumulated to determine if the frame is anomalous.

$$\mathscr{L}_{L1}(x, y) = \|y - G(x, z)\|_1, \qquad (5.26)$$

$$\mathscr{L}_{cGAN}(G, D) = \mathbb{E}_{(x,y) \in \mathscr{X}}[\log D(x, y)] + \mathbb{E}_{x \in \{F_t\}, z \in \mathscr{X}}[\log(1 - D(x, G(x, z)))] \quad (5.27)$$

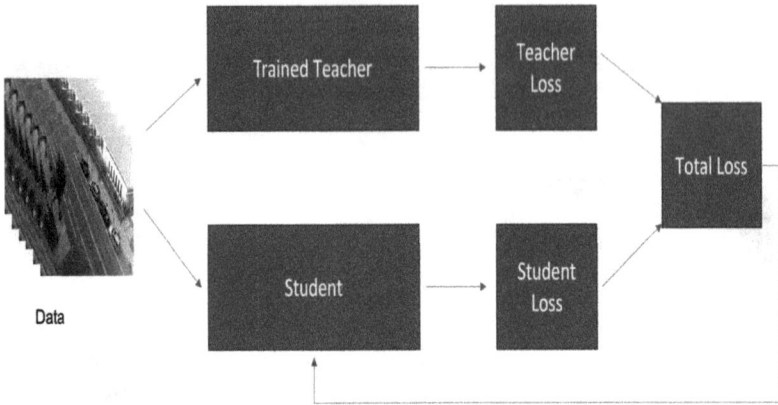

Figure 5.11 Knowledge Distillation sample training structure. The trained teacher is fixed for training the student network. Once trained, only the student network is used.

In general, using two streams has obtained better results at the cost of increasing computating requirements. The two-stream GAN model outperformed AMDN and regular convolutional autoencoder using the UCSD dataset for the assessment. Nevertheless, training the model using four networks is costly. Also, it has yet to be tested for more difficult datasets for comparison purposes.

5.3.5   TRAINING WITH KNOWLEDGE DISTILLATION

Knowledge distillation has been used to transfer knowledge from a teacher network to a student network. The student network is generally smaller than the teacher network. The student network learns to replicate teacher network output using fewer neurons. When labels are known for the dataset, the knowledge distillation can be done by generating a weighted addition of two objective functions as shown in Equation 5.28. The sample structure of knowledge distillation can be observed in Figure 5.11.

$$L_d(\theta) = L_s(\theta) + \lambda L_t(\theta) \tag{5.28}$$

Knowledge distillation is also useful for transferring knowledge between networks with different but related inputs. Stroud *et al.* implemented a knowledge distillation technique for the action recognition task named the D3D network [586]. D3D network distills the knowledge from a temporal network into a spatial network. The spatial network uses RGB scale frames as input. The temporal network uses optical flow frames as input. The temporal stream becomes the teacher network, and the spatial stream becomes the student network. The idea is that the spatial stream learns some of the temporal stream behavior and learns to generate the temporal representation. A custom loss is presented in Equation 5.29. The model includes the action classification loss and the distillation loss.

$$L(\theta) = L_a(\theta) + \lambda L_d(\theta) \tag{5.29}$$

Distillation loss penalizes the model each time the spatial network results differ from the temporal network results; this can be observed in Equation 5.30. The equation includes two models, f_s is the student spatial stream, and f_t is the teacher temporal stream. Student network f_s includes learning parameter θ, but the teacher network f_s is fixed. After the distillation, the spatial network obtains spatial and temporal features using only RGB scale frames as input.

$$L_d(\theta) = \frac{1}{N} \sum_{i=0}^{N-1} \left(f_s \left(x^{(i)}; \theta \right) - f_t \left(x^{(i)} \right) \right)^2, \tag{5.30}$$

Video anomaly detection techniques also benefit from distillation learning. One of those techniques is anomaly score distillation which applies knowledge distillation techniques similar to the D3D [135]. The main difference is that D3D is a classification technique, and anomaly score distillation is a reconstruction technique. The anomaly score is calculated using Equation 5.31. The anomaly score increases if the spatial stream reconstruction loss L_s differs from the temporal stream reconstruction loss L_t. The anomaly score is added to the model loss of Equation 5.32. The anomaly score and the spatial reconstruction loss are added to obtain the complete loss. The variable x_i represents the video data and θ the learning parameters.

$$s(\theta, x^{(i)}) = (L_t(x^{(i)}) - L_s(x^{(i)}; \theta))^2 \tag{5.31}$$

$$L(\theta) = \sum_{i=0}^{N-1} (L_s(x^{(i)}; \theta) + s(x^{(i)}; \theta)) \tag{5.32}$$

The other assessed technique in [135] is Joint Spatiotemporal training which combines the loss of spatial and temporal networks. The combined loss is used for backpropagation in both networks. θ_1 and θ_2 represent the learning parameters.

$$L(\theta_1, \theta_2) = L_s(\theta_1) + (L_s(\theta_1) * L_t(\theta_2)) \tag{5.33}$$

The models were assessed with four datasets Avenue, UCSD, ShaghaiTech Campus, and StreetScene. The four datasets were merged into a single dataset to have more data and improve model results. Anomaly score and joint spatiotemporal training obtained results close to the state of the art but using a smaller single network model. Nevertheless, the results for the temporal feature extraction datasets were not the best in their cases and had room for improvement.

5.4 COMPARISON OF RELEVANT TECHNIQUES FOR ANOMALY DETECTION

The previous section described representative models for anomaly detection using supervised and semi-supervised learning. Therefore, this section shows the measured performance of those methods using standard benchmarks. Method's authors employed different combinations of datasets described in section 5.2 to validate their results. Thus, we can directly compare those methods using standard AUC-ROC and EER metrics published in their papers.

Area Under the Curve – Receiver Operating Characteristic (AUC-ROC) and Equal Error Rate (EER) have been used as evaluation metrics for video surveillance anomaly detection works. Both metrics are useful for binary classification tasks, and video surveillance anomaly detection is a binary classification between anomalous and normal. The AUC-ROC is a better metric than accuracy for binary classification as it considers the trade-off between true-positive rate (TPR) and false-positive rate (FPR). Therefore, the TPR and FPR are calculated for various thresholds and plotted to calculate the AUC-ROC. FPR is on the plot's x-axis, and TPR is on the y-axis of the plot. The area under this plot is the AUC-ROC. EER is the optimal point in the ROC graph.

Table 5.3 lists the AUC-ROC values obtained by the different models in the anomaly detection task. Table 5.4 lists the EER values obtained by the same models under the standard benchmarks.

The AMDN is the first model listed in Table 5.3 and Table 5.4. AMDN model outperforms in AUC-ROC and EER metrics compared to the two following and newer methods HOG+HOF CAE and ConvLSTM. There is a tradeoff between those

Table 5.3

AUC-ROC results as reported in the different models' publications using standard datasets for assessment.

Models	Subway Entr.	Subway Exit	UCSD Ped1	UCSD Ped2	Avenue	Shanghai Tech	Street Scene
AMDN [682]	-	-	92.1	90.8	-	-	-
HOG+HOF CAE [228]	94.3	80.7	81.0	90.0	70.2	-	-
ConvLSTM [115]	84.7	94.0	89.9	87.4	80.3	-	-
GMFC-VAE [169]	-	-	94.9	92.2	83.4	-	-
Adversarial Discriminator [509]	-	-	96.8	95.5	-	-	-
Anomaly Score [135]	-	-	71.1	87.3	83.3	87.1	57.3

Table 5.4

AUC-ROC results as reported in the different models' publications using standard datasets for assessment.

Models	Subway Entr.	Subway Exit	UCSD Ped1	UCSD Ped2	Avenue	Shanghai Tech	Street Scene
AMDN [682]	-	-	16.0	17.0	-	-	-
HOG+HOF CAE [228]	26.0	9.9	27.9	21.7	25.1	-	-
ConvLSTM [115]	23.7	9.5	12.5	12.0	20.7	-	-
GMFC-VAE [169]	-	-	11.3	12.6	22.7	-	-
Adversarial Discriminator [509]	-	-	7.0	11.0	-	-	-
Anomaly Score [135]	-	-	31.2	14.1	21.8	18.1	42.7

models' efficiency and efficacy; in this case, the AMDN model has three streams and fully connected autoencoders and uses optical flow frames for one of its streams. HOG + HOF CAE and ConvLSTM present simpler, less computationally intensive models. HOG + HOF CAE uses convolutional layers and feature extractors, reducing the computational cost. In counterpart, using convolutional layers for spatial feature extraction improves the results for UCSD Ped2 because of the high quantity of spatial anomalies in that dataset. Regarding the ConvLSTM model, the best results are using the datasets with the highest quantity of temporal anomalies, UCSD Ped1 and Avenue dataset. The good performance of ConvLSTM is achieved by applying an LSTM network which aids in extracting temporal features. Accordingly, using datasets with more spatial anomalies results in slightly worse performance; for instance, using UCSD Ped2 for the assessment.

In the next half of both tables, there is the comparison of two generative models: a VAE and a GAN network. Both generative models have better results than previous models because of their capability to handle more complex data such as video. In the case of the GMFC-VAE, the improvement is related to a more complex representation given by the Gaussian Mixture model in the latent space. The cross-channel training of the adversarial discriminator helps to find a more complex normal space to classify anomalies. Nevertheless, as observed, both techniques were assessed in more complex datasets like ShanghaiTech and StreetScene. Additionally, the generative models are harder to train because of several challenges, such as high computational costs and hard parameterization. Finally, the results of the assessment of the anomaly score model are observed in the tables. The anomaly score results have room for improvement for the UCSD Ped dataset; however, their performance is much better in the other datasets. In this case, the advantages of using the anomaly score model are that it introduces a new approach to training the model and has a lower computational complexity because it uses a single stream for inference. A direct observation from these comparison tables is the necessity of improving results in UCSD Ped1 without sacrificing spatial feature extraction results, that is, losing accuracy in UCSD Ped2. Some interesting challenges for the future of anomaly detection models in video surveillance are related to improving temporal feature extraction while having more efficient models that explore understanding the dynamics and behavior of objects in the video sequences.

5.5 CONCLUSION

This chapter has reviewed the most outstanding supervised and semi-supervised learning methods for video surveillance anomaly detection in recent years. Although methods have improved over the years, more challenging and realistic datasets for assessment have also emerged. For instance, in earlier years, datasets were more limited to single scenes or only pedestrian paths. In contrast, more realistic and less constrained datasets have recently appeared, requiring more sophisticated models for their solution. Moreover, researchers have focused on improving the accuracy of the models and refining their understanding over the years.

As observed throughout the chapter, the two-Streams approach has been crucial in implementing several current methods for the anomaly detection task. However, using a second stream comes at the expense of larger and more computationally complex models. For example, the AMDN model utilizes three streams of fully connected layers achieving better accuracy than the HOG+HOF CAE or the ConvLSTM models. Nevertheless, the AMDN's most significant disadvantage is its complexity, which has been reduced in the newer models of HOG+HOF CAE and ConvLSTM. The two-stream approach also inspired two generative models based on VAE and GAN. Both of these models, namely GMFC-VAE and the Adversarial Discriminator, obtained high accuracy values, and successfully tackled the UCSD Pedestrian dataset. As expected, higher precision comes with increased computational complexity. The adversarial discriminator employs four networks during training, whereas GMFC-VAE can be tricky to train. Finally, the anomaly score was a more efficient but less accurate model in this selected list.

The research area of video anomaly detection has been dedicated to improving the accuracy and efficiency of the models while enhancing our understanding of potential solutions. Some models described in this chapter are preferable for accurate detections when complexity is not a concern, such as the adversarial discriminator model. On the other hand, models like anomaly score are useful when less complex models are required. However, as observed in the results of the more realistic ShanghaiTech and StreetScene datasets, a better understanding of temporal feature extraction is still necessary to develop more accurate models that do not necessarily increase in complexity. Therefore, there is ample room for improvement in video surveillance anomaly detection models and their understanding, paving the way for future research.

6 Deep Learning to Classify Pulmonary Infectious Diseases

Dora-Luz Flores, Ricardo Perea-Jacobo, Miguel Angel Guerrero-Chevannier
Universidad Autónoma de Baja California

Rafael Gonzalez-Landaeta
Universidad Autónoma de Ciudad Juarez, Universidad Autónoma de Baja California

Raquel Muniz-Salazar
Universidad Autónoma de Baja California

CONTENTS

6.1 INTRODUCTION

Pulmonary infectious diseases, such as COVID-19 and tuberculosis (TB), are a significant cause of morbidity and mortality worldwide. Timely and accurate diagnosis is critical to ensure appropriate treatment and prevent disease transmission. Deep learning, a subset of machine learning, has shown promise in various medical

DOI: 10.1201/9781003385615-6

applications, including image analysis and classification. In recent years, there has been increasing interest in using deep learning techniques to classify pulmonary infectious diseases using chest radiographs (CRX) or computed tomography (CT) scans. By training deep learning algorithms on large datasets of annotated medical images, these models can learn to accurately classify different types of pulmonary infections, allowing for more efficient and accurate diagnosis. This approach has the potential to improve patient outcomes, reduce healthcare care costs, and improve public health efforts by facilitating early detection and treatment of lung infections.

Deep learning models can be trained on large data sets of annotated medical images, allowing them to identify patterns and features that are characteristic of different types of lung infections. For example, a deep-learning model can learn to recognize the presence of infiltrates, nodules, or cavities in CRX or CT scans that are indicative of specific types of pulmonary infections. Furthermore, deep learning models can continuously learn and improve their diagnostic accuracy over time as they are exposed to more data. This means that as more medical images are available for training, deep-learning models can become even more accurate in their diagnosis of pulmonary infections.

6.2 COUGH MONITORING AND ANALYSIS

Coughing is considered a defense mechanism by which the body expels secretions or any blockage that restricts the passage of air in the upper airways. The causes of coughing can be environmental (dust, smoke), bacteria, viruses, or some chronic or acute health condition [412]. In several respiratory diseases, cough is one of the main symptoms, and depending on the type of cough, it is possible to obtain a clinical diagnosis. The cough has different characteristics that can give information about the severity of a certain disease, including disease identification. Among the main attributes of cough are intensity, frequency, duration, and pattern [489]. In patients with TB, for example, the pattern of coughing depends on the amount of M. tuberculosis present in the lungs [733]. Also, it has been possible to diagnose COVID-19 in asymptomatic patients using Artificial Intelligence (AI) on the record of a forced cough [327].

There are some criteria to classify cough. According to the resistance, which indicates the duration of the cough, it is classified as acute (less than 3 weeks), subacute (3-8 weeks), and chronic (more than 8 weeks). According to its sound, which is due to the sudden expulsion of air, the cough is classified as dry or wet. In patients with TB and chronic obstructive pulmonary disease (COPD), the majority experience a chronic, wet cough, while in patients with COVID-19, or with exacerbation of asthma, acute and dry cough predominates [262]. It can then be seen that the cough produces characteristic sounds that help identify some respiratory diseases.

The cough is compounded by two sounds and an intermediate stage between them. The first sound provides information about the peripheral airways at the level of the tracheal bifurcation, the second sound provides information about the larynx, and the intermediate zone reflects the processes in the trachea [313]. The duration and intensity of each of these stages will depend on the respiratory health condition of the subject, as shown in Figure 6.1.

Condition	Cough patterns	Duration
Normal airways		0.79 seconds
Narrowed airways (obstruction)		0.85 seconds
Widened airways (obstruction)		0.82 seconds
Scared lungs (restriction)		1.41 seconds
Fluid filled lungs (restriction)		1.82 seconds

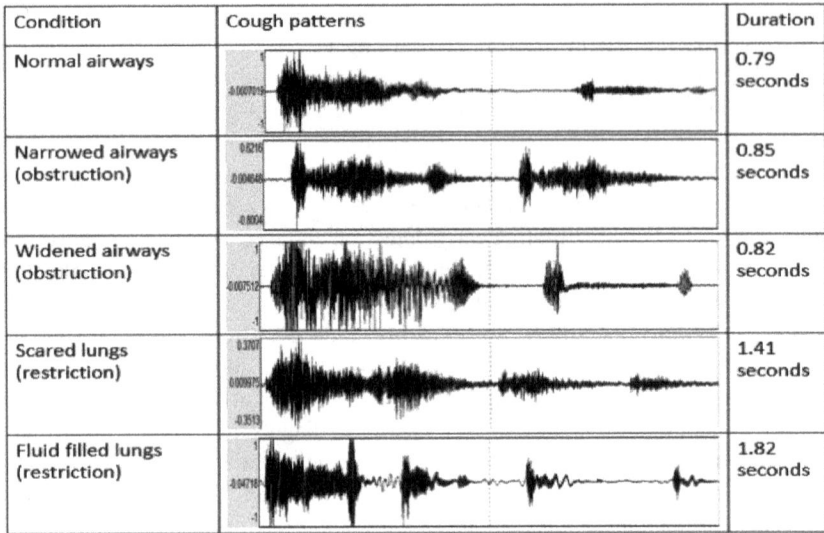

Figure 6.1 Pattern of cough sounds. Reprinted from (Rudraraju et al., 2020) with permission from Elsevier.

There are medical devices that detect cough in hospital environments, such as spirometers, and pneumotachographs, among others. However, these devices require special skills to operate and use procedures that are performed in controlled environments, giving short-term information, which makes it difficult to observe the evolution of the disease or the effectiveness of treatment. Carrying out continuous or periodic measurements on an outpatient basis, and in real-time, would allow coughing to be recorded for prolonged periods, helping to assess the progress of the disease. In this sense, there are ambulatory cough monitoring systems that allow 24-h records to be obtained. Among them are the Leicester monitor [408] and the Cayetano monitor [490]. These devices have a microphone and an audio recorder, which stores the data of the cough sounds so that they can be analyzed later. In the case of the Leicester monitor, the data is stored in .mpeg format, while in the Cayetano monitor the data is stored in .mp3 format.

Some proposals detect cough in non-hospital conditions for longer than 24 hours using the wearable paradigm. Because coughing causes a turbulent flow of air in the airways, the vibrations that this flow produces in the upper airways can be measured. In this sense, some proposals use piezoelectric or piezoresistive flexible films, which are attached to the subject's throat and thorax to measure the vibrations produced by coughing [31, 373]. Although it is a proposal allowing real-time measurements, its long-term use can cause discomfort in the subject because an object must be attached to the skin at the throat or thorax level. There are proposals that detect the sounds of coughing more comfortably. One of the best-known systems is the LifeShirt^{TM} (Vivometrics) [577], which is a jacket with multiple sensors that allow the detection of several cardiac and respiratory parameters. This system can detect cough sounds, thanks to the addition of a microphone placed near the throat. LifeShirt^{TM} has been used to detect cough in COPD outpatients for 24 hours with a sensitivity and specificity of 78.1% and 99.6%, respectively [133]. Other simpler approaches use microphones incorporated into wearable electronic stethoscopes [373] or use microphones that are integrated into smartphones [641]. Smartphones have been used to detect biomarkers related to COVID-19, and lower respiratory tract infections [65]. Also, to discriminate TB coughs from COVID-19 and healthy coughs [464]. However, the drawback of using microphones is that cough sounds are superimposed on environmental sounds, which often require complex algorithms to reliably extract cough characteristics.

From the signal point of view, cough sounds have frequency components that can vary according to the state of health of the respiratory tract. In healthy subjects, cough sounds range from 300 Hz to 500 Hz; however, these frequencies may increase up to 1200 Hz in subjects with bronchitis [313]. Cough sounds can be analyzed in the time domain using their amplitude-time characteristics, or in the frequency domain. The spectrogram of cough sounds has also been used to extract their characteristics using AI algorithms [733].

There are several characteristics of cough sounds that provide information about various respiratory diseases, but these characteristics are not audible to the human ear. In that sense, various AI-based algorithms have been developed that analyze these characteristics. Among the most widely used spectral characteristics are the Mel-Frequency Cepstral Coefficients (MFCC) [28, 69, 116, 412], Mel-Scaled Spectrogram, Tonal Centroid, Chromagram, Spectral Contrast [116], and Log Spectral Energies [69]. Regarding the classifiers, there are different points of view about their performance. In TB patients, the rapid increase in signal energy has been used to be able to differentiate the cough from the voice signal [620]. The algorithm detected cough and non-cough events using classifiers based on machine learning (ML) algorithms. Multilayer perceptrons (MLP), machine support vectors (SVM), and minimum sequential optimization (SMO) were compared. They chose SMO for its simplicity, obtaining a sensitivity of 81% and false positives of 3.3/hour. The algorithm was able to detect a reduction in cough events in 28 patients with drug-sensitive TB. Some works have compared the Log Spectral Energies with MFCC in short-term recordings of cough in subjects with TB, to then apply classifiers through statistical models using linear regression, Hidden Markov Models (HMM), and Decision Trees [69]. In this study, cough could be distinguished between TB and healthy subjects with an accuracy of 80% and a specificity of 95%. MFCCs discard information that is useful for classifying sounds, so their accuracy was only 63% and their specificity was 80%. Linear regression has been used as a classifier to differentiate cough from non-cough in various respiratory diseases, including COVID-19 and TB, and to differentiate the cough from other sounds present in the environment. Artificial neural networks (ANN) and Random Forests (RF) have been also used as classifiers in cough sounds detected by spirometry [28]. In patients with COVID-19, it has been possible to diagnose the disease from the sounds of coughing in symptomatic and asymptomatic subjects. Chowdhury et al. (2022) used different ML classifiers to distinguish patients with COVID-19. The classifiers based on Extra-Trees, HGBoost, and RF showed the best performance, obtaining accuracies of the order of 87%

6.3 ELECTROCARDIOGRAM MONITORING AND ANALYSIS

The electrocardiogram (ECG) is the representation of the bioelectric potentials of cardiac cells. It is composed of several waves (P, Q, R, S, T) that describe the depolarization (electrical activity before contraction) and repolarization (recovery after depolarization) of the atria and ventricles. In summary, the P-wave represents the depolarization of the atria, the QRS complex represents the depolarization of the ventricles and repolarization of the atria, and the T-wave represents the repolarization of the ventricles [116]. In addition, some intervals also provide information about the bioelectrical functioning of the heart. These intervals represent the time that elapses between two ECG waves. Said intervals are, RR, PQ, and QT. The RR interval provides information about the heart rate since it is the time that elapses between one beat and another. The PQ interval represents the time between the onset of atrial depolarization and the onset of ventricular depolarization. The QT interval is the time it takes for the ventricle to start to contract and to finish relaxing and is measured from the beginning of the Q wave to the end of the T wave [116]. The amplitude and duration of each of these waves, and the duration of the intervals provide information about the heart condition of a subject, therefore, any alteration in the ECG is an indication of a cardiac problem.

ECG detection for diagnostic purposes is carried out using 12 leads that are widely known in the literature [666]. These leads are I, II, III, aVR, aVL, aVF, and V1-V6. To do this, 10 electrodes are attached to the surface of the skin at different locations on the body. The 12-lead ECG is measured in clinical settings and is performed by qualified personnel; however, under these conditions, patterns or abnormalities in the ECG that can appear during daily activities are often not detected. In this sense, there are Holter systems that record the ECG continuously for 24–48 hours on an outpatient basis. However, they also require various electrodes that cause discomfort and need a specialist to place them on the subject. The importance of ECG detection during activities of daily living has driven the development of portable and wearable systems that detect long-term ECG comfortably and simply, and without the need for specific knowledge to use them. There are systems based on smartwatches [199], which detect lead I when used in the standard way, however, it has also been possible to detect leads II and V2 using a different configuration [580]. There are also systems based on mobile applications, such as the KardiaMobile® (ALIVECOR), which can obtain 6-lead medical grade records of the ECG (https://www.kardia.com/). Other wearable technologies, but less conventional, are embedded in objects such as glasses, bands, and patches [199, 540]. Figure 6.2 shows various wearable and portable systems capable of detecting multiparameter, including ECG in daily life.

There is evidence that some respiratory diseases produce ECG abnormalities. In subjects with COPD, the most common abnormalities are a rightward P wave axis ($\geq 70°$) and a rightward QRS axis ($\geq 90°$), as well as transient atrial and ventricular arrhythmias [516]. The presence of P pulmonale (a peaked P wave in lead II)

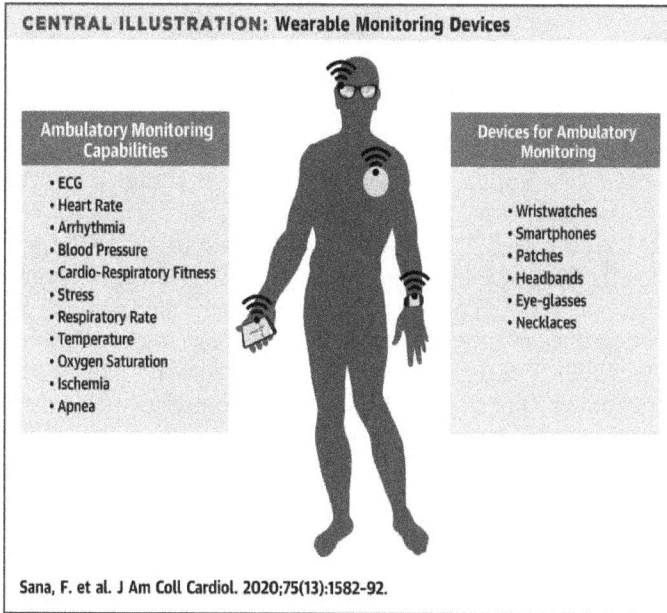

Figure 6.2 Depiction of wearable and smartphone-based solutions for continuous ambulatory cardiac monitoring, along with a summary of potential use cases. Reprinted from Sana et al. (2020) with permission from Elsevier.

is indicative of the severity of COPD even without the need for spirometry. COPD does not cause changes in the duration of the P wave, nor in the amplitude of the QRS complex [16]. On the other hand, there is evidence that P wave dispersion is a good predictor of COPD exacerbations. In the case of subjects with TB, although the disease itself does not cause significant changes in the electrocardiogram, certain drugs used for the treatment of TB cause a prolongation of the corrected QT interval (QTc) [227]. The QTc is used since the QT interval changes with the heart rate, so, this effect must be corrected. For this, several formulae have been proposed, namely, Bazett, Fridericia, Framingham, Hodges, and Rautaharju [636]. A prolonged QTc (450 ms in men, 460 ms in women) is an indicator of risk for Torsade de Pointe [202], which is a rare ventricular tachycardia and is also a potential indicator of sudden death. Therefore, long-term ECG monitoring is considered good practice in subjects receiving treatment for multidrug-resistant tuberculosis. Some ECG abnormalities have also been found in subjects with COVID-19. These include T-wave inversions [373, 410] and sinus tachycardia [379]. Prolongation of the QTc interval has also been found, but it is mainly caused by the different pharmacological treatments [410]. Some studies reveal alterations in the ST segment in subjects with COVID-19, specifically an elevation of this segment, which is due to coronary artery disease or heart muscle disease (myocarditis) [373]. ST segment deviations are also considered a predictive indicator of mortality risk in subjects with COVID-19 [330].

From all the ECG alterations caused by respiratory diseases, one of the most studied is the prolongation of the QTc interval due to the implications it has on the subject's life. Accurately estimating the QTc is not easy since it is necessary to identify the point where the end of the T-wave coincides with the isoelectric line. In situations where there are motion artifacts, this can be a difficult task. Commonly, the QTc is measured manually by a trained personal using printed or digital ECG recordings. QTc should be measured using leads II, V5, or V6, and should be estimated from the average of at least 3 beats [206]. Artificial intelligence for the QTc from the ECG recorded using the KardiaMobile® device (ALIVECOR) has been proposed. Using deep neural networks (DNN), it was possible to estimate the QTc with a specificity of 94.4%, making it an effective method to estimate congenital or acquired prolongation of the QTc [202]. The algorithm was trained, tested, and validated on more than 1.6 million 12-lead ECG records from 538,200 subjects. These types of tools can be helpful for subjects with TB and COVID-19 who are under drug treatment. There are other proposals that use the single-lead ECG, which is the one normally detected by smartwatches. Maille et al. (2021) used a deep convolutional neural network, composed of 11 U-net architectures with 11 convolutional layers and six residual blocks. The results showed a great similarity with the QTc measured manually in ECG recordings in leads II, V5 and V6, where a difference of 50 ms was obtained in 98.4% of the 85 patients who participated in the study.

There is a wide variety of artificial intelligence algorithms that have been used to detect abnormalities in the ECG recordings, especially arrhythmias. According to Saini and Gupta, the most popular are ANN, Fuzzy Systems, Neuro-fuzzy Systems, Genetic-Fuzzy Systems, Probabilistic Neural Networks (PNN), CNN, Support Vector Machines (SVMs), and Linear Discriminants (DNs) [537]. However, the most used in the extraction and classification of ECG characteristics are the CNNs due to their self-learning capacity.

Among the characteristics of ECG recordings, the most studied are the following: time-domain, frequency-domain, time-frequency domain, statistics features, and non-linear features. Commonly, these features are extracted from the raw ECG data, however, the generated ECG image has been used instead of the raw data to apply neural networks. The advantage of using images of ECG traces is that they can be captured by smartphones. Using heat maps helps the algorithm to focus on a certain area of the image to make decisions, which helps the clinician to find abnormalities in the ECG that may be indicative of COVID, TB, and other cardiac diseases [223,500]. Figure 6.3 shows the image of an abnormal ECG trace of a subject with COVID-19, where those areas where the abnormalities are highlighted, and that is where the algorithm should focus on making decisions.

6.4 CHEST RADIOGRAPHY AND DEEP LEARNING

There are numerous diagnostic methods for TB, each with advantages and limitations. Chest radiography (CXR) is one of the most used diagnostic tools to detect pulmonary TB [443]. We will discuss the benefits and disadvantages of using CXR for pulmonary TB diagnosis compared to other diagnostic techniques and how the combination of CXR with AI addresses its limitations.

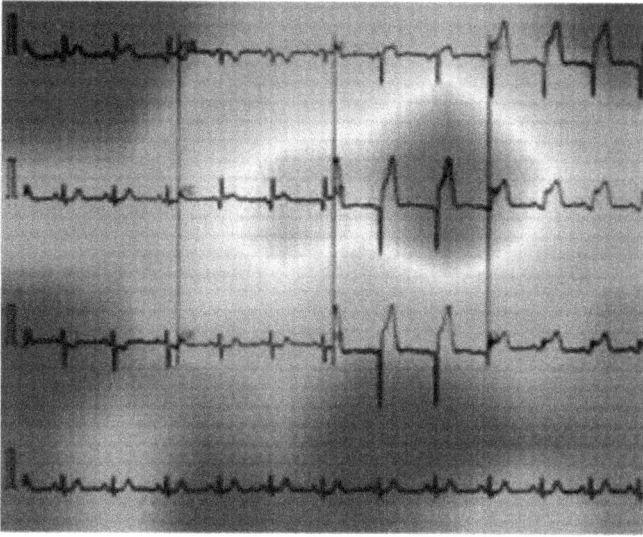

Figure 6.3 Score-CAM visualization of abnormal (COVID-19) ECG. Taken from Rahman et al. (2022) with the permission of Springer Nature BV.

CXR is a relatively inexpensive, non-invasive, and widely available diagnostic tool that can provide valuable information on the presence and extent of TB-related lung abnormalities. It is beneficial for detecting characteristic lung parenchymal changes and the presence of cavitations indicative of active TB disease [322].

CXR has a reported sensitivity of 92% [72]; it can detect TB cases in early stages, even before characteristic symptoms occur, allowing to find individuals at high risk of developing the disease, such as close contact with patients with active disease or people from TB-endemic regions (2016). CXR can be used as a screening method. It can be applied in large groups to find individuals initially considered healthy, leading to rapid treatment and reduced disease transmission (2016). On the other hand, in recent years, there have been significant improvements in the portability of the equipment for taking CXR; currently, there is FDA-approved equipment with a weight of less than 4 kg, which makes them accessible to remote areas where there are no other diagnostic tests Table 6.1.

Although CXR is a useful diagnostic tool for pulmonary TB, it has some limitations that should be considered. CXR reports low levels of specificity; in practice, the CXR is examined and interpreted by a medical radiologist, so the process is subjective to the medical staff's experience. Different diseases show similar radiological patterns to pulmonary TB, leading to a high false-positive rate in people with other lung diseases or conditions, such as COPD or lung cancer [633]. One of the significant problems with CXR is the need for more trained radiologists available in low-resource areas, in conjunction with the absence of other diagnostic techniques, which play a significant role in the prevalence and spread of the disease [496].

Table 6.1

Characteristics of commercially available ultraportable X-ray equipment.

Manufacturer	Fujifilm	Delft
Model	FDR Xair	Light
Price(USD)	$ 47,000	$ 66,750
X-ray generator weight (kg)	3.5	7
Weight of complete system (kg)	29.4	33.2
Sockets per battery charge	100	200

AI has the potential to revolutionize medical care by improving diagnostic accuracy, efficiency, and access to care. In recent years, there has been growing interest in using AI to develop computer-aided diagnosis (CAD) systems to detect pulmonary TB [311]. CAD systems use machine learning algorithms to analyze medical images, such as chest X-rays or CT scans, and provide diagnostic results automatically [311].

AI-based CAD systems have several potential benefits for TB diagnosis. First, they can provide accurate and consistent diagnoses without a radiologist or trained medical professional, which is very helpful in resource-limited settings where access to trained medical professionals may be limited [493]. Second, AI-based CAD systems can improve diagnostic efficiency by reducing the time required for image interpretation and diagnosis, increasing the throughput of diagnostic services, reducing waiting times, and improving patient outcomes. Third, AI-based CAD systems can help reduce diagnostic errors and improve the accuracy of TB diagnosis, leading to better patient outcomes [493]. On the other hand, AI-based CAD systems have the potential to be used for active screening of people at high risk for TB, identifying early TB cases, leading to prompt treatment and reduced disease transmission [266]. AI-based CAD systems can also monitor disease progression and response to treatment, providing valuable information for patient management and care [321].

One of the main problems of commercially available CAD systems for pulmonary TB diagnosis is their high cost; an InferRead DR software license costs USD 5,552, including installation service and support, while a CAD4TB license costs USD 28,475, including installation service and support for three years [495]. The high costs of commercial systems make it difficult for them to be acquired and implemented in high-burden countries, so developing new CAD systems to support TB diagnosis is a viable option. An example is the MIA-TB RX UABC system [218], developed by the Autonomous University of Baja California in collaboration with the Tijuana Tuberculosis Clinic and Laboratory, which serves patients in northwestern Mexico. This CAD system was developed by implementing pre-trained convolutional neural networks using CXR radiographic images of patients collected over two years. It can classify radiographs into three categories: TB, normal, or other pathology. The system obtained a sensitivity value of 1.0 and an accuracy of 0.92 for pulmonary TB Table 6.2

Table 6.2

Characteristics of the software currently available on the market.

Company	Delft Imaging Systems	Infervision	JLK	Lunit	Qure.ai	RadiSen
Country	The Netherlands	Beijing, China	Seoul, Republic of Korea	Seoul, Republic of Korea	Mumbai, India	Seoul, Republic of Korea
Product	CAD4TB	InferRead DR Chest	JLD-02K	Lunit INSIGHT CXR	qXR	AXIR
Version	7	1.0	1.0	3.1.0.0	3.0	1.1.2.2
Intended Age Group (years)	4+	16+	10+	6+	6+	16+
Chest X-ray image format input	DICOM, PNG, JPEG	DICOM, PNG, JPEG	DICOM, PNG, JPEG	DICOM	DICOM, PNG, JPEG	DICOM
Chest X-ray type input	PA	AP/PA	AP/PA	AP/PA	AP/PA	PA
Output	Abnormality score for TB, Heat map, Binary classification "TB" or "not TB"					
Product Development Method	Supervised deep learning (CNN, RNN) plus manual feature engineering	Supervised deep learning (CNN, RNN)	Supervised deep learning (CNN, DBNs)	Supervised deep learning (CNN)	Deep learning to analyse chest X-ray scans.	Supervised deep learning (CNN)

In conclusion, CXR is a useful diagnostic tool for pulmonary TB, particularly in resource-limited settings where other diagnostic techniques may not be available or affordable. CXR can provide valuable information on the presence and extent of TB-related lung abnormalities and can be used to detect TB in high-risk individuals. However, CXR has some limitations, such as the possibility of false-positive results when the interpreting physician confuses other pulmonary pathologies. AI-based CAD systems have the potential to revolutionize TB diagnosis by reducing the limitations of CXR alone, thereby providing accurate, efficient, and consistent diagnostic results.

6.5 NEXT-GENERATION SEQUENCING IN THE DIAGNOSIS OF INFECTIOUS LUNG DISEASES

Next-generation sequencing (NGS) is a high-throughput sequencing technique that has revolutionized the field of genomics and transcriptomics. This technology involves the fragmentation of DNA or RNA into small pieces that are simultaneously amplified and sequenced into thousands or even millions of fragments. The next-generation sequencing process generates a large amount of data that are processed by bioinformatics algorithms to reconstruct the complete genome or transcriptome [37, 567].

NGS has become a fundamental tool in the study of genetic diversity and evolution of microorganisms, allowing the identification of mutations and genetic variations associated with diseases and the elucidation of complex metabolic pathways and gene regulatory networks [512, 585].

The ability to sequence large amounts of genetic material in a short period of time is one of the main advantages of next-generation sequencing compared to traditional sequencing techniques. Furthermore, NGS has enabled an exponential increase in the identification of genetic causes in rare diseases and heterogeneous disorders. NGS is used in research and clinical settings, accelerating diagnosis, and reducing costs. Despite the enthusiasm, there are limitations in coverage and accuracy and challenges in the interpretation of variants and ethical issues. There is a need to define quality and control standards in NGS to further improve its application for the benefit of patients [375].

In the study of bacterial genomics, next-generation sequencing has allowed the characterization of the genetic diversity of *Mycobacterium tuberculosis* populations, which has facilitated the identification of strains associated with resistance to anti-tuberculosis drugs. In addition, NGS has improved the detection and diagnosis of tuberculosis by identifying mutations and genetic variations associated with disease, characterizing the genetic diversity of populations of microorganisms, and elucidating complex metabolic pathways and gene regulatory networks [458, 541, 649].

6.5.1 MOST WIDELY USED SEQUENCING TECHNOLOGIES

NGS technologies have become valuable tools in the diagnosis of infectious lung diseases. Bioinformatics processes in NGS significantly influence disease management and patient care. Lack of standardization leads to variability in bioinformatics procedures, generating inaccurate results that affect patient care. Therefore, in recent years, several guidelines have been proposed to homogenize processes and standardize knowledge of bioinformatics data management [307, 530]. Among the most widely used technologies are Illumina, Ion Torrent, PacBio, and Oxford Nanopore. All these technologies are used in the diagnosis of infectious lung diseases to identify respiratory pathogens, detect mutations associated with antibiotic resistance and characterize the genetic diversity of pathogens. The main features, advantages and disadvantages, and specific applications of each of these technologies are described in Table 6.3.

NGS technologies are valuable tools in the diagnosis of infectious lung diseases. Each of these technologies has its own advantages and disadvantages, and their specific application depends on the type of analysis required. They are constantly changing and being updated, so these descriptions of their advantages and disadvantages can quickly become obsolete. It is necessary to constantly update their performance and recommended implementations. The choice of the appropriate technology will depend on factors such as the complexity of the genome, the length of sequences required, and the available budget.

6.5.2 SEQUENCING DATA PROCESSING AND ANALYSIS

NGS data analysis is a complex process that requires the use of advanced bioinformatics tools for processing, alignment, and identification of genetic variants. The WHO recently published a technical guide for workflow in NGS of *Mycobacterium tuberculosis* . It is mainly structured in four steps DNA extraction and quality control, DNA library preparation, Sequencing, Data Analysis [458]. Since the first three steps vary depending on access to the sequencing platform, available budget, available sample quality, and the desired quality of the output files, we proceed with general comments on the process of analyzing the raw sequencing data. The analysis process begins with obtaining raw sequencing data, which undergo a series of steps before they can be analyzed.

The first step in sequencing data processing is the removal of low-quality sequences and filtering of adapter and contaminant sequences. This is done using quality control programs such as FastQC or Trim Galore, which can identify and remove low-quality sequences and filter out adapter and contaminant sequences [33, 319]. Once low-quality sequences have been removed and adapter and contaminant sequences have been filtered out, sequences are aligned to a genomic or transcriptomic reference using alignment programs such as Bowtie or BWA [144]. This allows the identification of genetic variants and characterization of the genetic diversity of pathogens.

There are several alignment programs available (Bowtie, BWA, HISAT2, TopHat, and other). Each of these programs has its own features and advantages, and the choice of the appropriate alignment program will depend on the type of sequencing data and the purpose of the analysis.

The next step is the identification of genetic variant using bioinformatics tools such as GATK, FreeBayes or SAMtools, or pipelines as MTBseq, which allow the identification of SNPs (Single Nucleotide Polymorphisms), indels and other types of genetic variants. These tools can also be used to identify mutations associated with antibiotic resistance. This program may differ in their ability to detect different types of variants, such as SNPs, indels, and other types of structural variants. These programs use different approaches for the identification of SNPs, such as the comparison of read sequences with a reference genome or the detection of base changes in the alignment of reads [138, 144, 308, 344, 409, 634].

Indels (insertions and deletions) are variants that involve the insertion or deletion of one or more nucleotides in a DNA sequence. The identification of indels can be more difficult than the identification of SNPs, due to the greater complexity of the

Table 6.3

Next-Generation Sequencing technologies

Platform	Key features	Advantages	Disadvantages	Ref.
Illumina	Sequencing technology based on DNA strand synthesis. It uses DNA cluster amplification to generate large amounts of short, high quality DNA sequences. Sequencing products are read simultaneously on millions of DNA fragments using a fluorescent camera system.	Is a high-throughput sequencing technology with high accuracy and reproducibility. It can generate large amounts of data in a short period of time at relatively low cost. The read length is suitable for genetic variant analysis and mutation detection in pathogens.	The read length is limited and does not allow sequencing of longer genomic regions. In addition, there may be problems in sequence alignment due to the high homology between some genomic regions.	[426]
Ion GeneStudio S5 System and the Ion Torrent Genexus System.	A sequencing technology based on the detection of protons released during DNA strand synthesis. It uses DNA cluster amplification to generate high-quality short DNA sequences. Sequencing products are read sequentially using a pH sensor system.	Fast, accurate and easy-to-use sequencing technology. It allows sequencing of short DNA fragments and can detect genetic variants and mutations in pathogens.	Sequencing quality may be affected by the presence of homopolymers, and the number of read errors may increase in repetitive regions.	[328]
PacBio	Sequencing technology based on the detection of fluorescent light generated during DNA strand synthesis. It uses real-time DNA amplification to generate long DNA sequences with high quality.	High-quality sequencing technology that can generate long DNA sequences with high quality, allowing the resolution of complex genomic regions and the identification of genetic variants. It is also capable of detecting epigenetic modifications.	Expensive sequencing technology that requires a large amount of data to generate reliable results. In addition, the sequencing error rate can be high in regions with high homology.	[4, 497]
Oxford Nanopore	Sequencing technology based on the detection of electrical currents generated by the interaction between DNA and pores in a membrane. It uses nanopores to directly sequence DNA molecules without amplification	Portable, real-time sequencing technology that can generate long DNA sequences. It allows the detection of epigenetic modifications and the identification of genetic variants in real time.	Sequencing error rate can be high, especially in repetitive or high-homology regions. In addition, sequencing quality can be affected by the presence of contaminants and genome complexity.	[385]

regions containing indels and the variability in their size. Programs such as GATK and VarScan are commonly used for the identification of indels from sequencing data. These programs use different methods for indel identification, such as comparison of read sequences with a reference genome, identification of misaligned read positions or detection of base read imbalances [306, 409].

In addition to SNPs and indels, there are other types of genetic variants that can be detected by next-generation sequencing data analysis programs. For example, structural variants, which include inversions, translocations, and duplications, can be detected by programs such as Delly, Lumpy, Manta, Visor among others [66, 108, 332, 508]. However, adequate detection of sequence copy number changes remains a challenging problem. Recently published machine learning approaches suggested that it is more effective than standard methods at accurately detecting sequence copy number changes in lower quality or coverage next-generation sequencing data, and is equally powerful in high-coverage data, including the identification of novel CNVs in genomes previously analyzed for CNVs using long-read data [244].

6.5.3 APPLICATIONS OF NEXT-GENERATION SEQUENCING IN THE DIAGNOSIS OF INFECTIOUS PULMONARY DISEASES

Identification of differential gene expression and analysis of genetic diversity are two important approaches in *Mycobacterium tuberculosis* research. In Mtb, identification of differential gene expression is often performed using next-generation RNA sequencing (RNA-Seq) technologies. RNA-Seq data can be analyzed techniques, such as differential cell lysis, probe-based ribosomal depletion, and genome-wide metabolic network analysis, scientists can investigate the regulatory networks and gene expression patterns of Mtb and its host during infection to identify genes that are over- or under-expressed compared to control conditions. This approach has been used to investigate the molecular mechanisms underlying M. tuberculosis virulence, drug resistance, and host immune response [131, 188, 390, 420, 734].

Analysis of the genetic diversity of M. tuberculosis involves the analysis of genetic variation within and between populations of the bacterium. This can be accomplished by analyzing the DNA sequence of multiple strains of M. tuberculosis. Next-generation sequencing data are often used to generate complete or partial genomes of M. tuberculosis strains. These genomes can be compared to identify mutations and genetic variations that occur in different strains. Genetic diversity analysis is used to understand the epidemiology of tuberculosis, including the spread of drug-resistant strains, the identification of new emerging strains, and the evolutionary history of the bacterium [274, 392, 402, 449, 452, 541, 542, 546, 653].

6.5.4 INTEGRATION OF DEEP LEARNING AND NEXT-GENERATION SEQUENCING FOR TB DIAGNOSIS

NGS enables rapid and accurate identification of Mtb bacteria in sputum samples and other body fluids. While deep learning enables rapid analysis of large genomic datasets to identify patterns of genetic variation and classify Mtb strains. The use of these techniques could allow a faster and more accurate diagnosis of TB, which could help reduce the spread of the disease and improve treatment outcomes. In addition, it could be a useful tool for monitoring disease progression and identifying potential drug-resistant strains [274].

Recently, WHO has published and endorsed the first catalog of resistance-associated genetic variants based on more than 38,000 MTBC isolates to predict clinically relevant resistance phenotypes from genetic data. This mutation catalog provides a common, standardized reference for the interpretation of resistance to all first-line drugs (RIF, INH, ethambutol, and pyrazinamide) and also to second-line group A drugs (levofloxacin , moxifloxacin , bedaquiline, and linezolid), group B (clofazimine), and group C (delamanid, amikacin , streptomycin, ethionamide, and prothionamide [457].

Barely a year later, the Comprehensive Resistance Prediction for Tuberculosis: an International Consortium (CRyPTIC) has published a compendium of data from 12,289 global clinical isolates of *Mycobacterium tuberculosis* , all of which have been subjected to whole genome sequencing and measured for their minimum inhibitory concentrations against 13 antituberculosis drugs processed uniformly in 23 countries in a single assay. This is the largest matched phenotypic and genotypic dataset of Mtb to date. The compendium contains 6,814 isolates resistant to at least one drug, including 2,129 samples that fully meet the clinical definitions of rifampicin-resistant (RR), multidrug-resistant (MDR), pre-extensively drug-resistant (pre-XDR) or extensively drug-resistant (XDR). This combination of an extensive catalog and open availability of an immense amount of data provides an ideal framework for the development of artificial intelligence implementations, especially deep learning [128, 129].

Several investigations have addressed the processing of genome datasets with machine learning techniques [39, 146, 287, 315, 694]. However, most with smaller scale sets than those recently published. With the recent increase in the availability of massive data, a trend to explore deep learning-based solutions can be observed. Moving from the predominance of machine learning classification models (Support Vector Machine, Random Forest, and ensemble models, etc.) to solutions based on deep learning, (convolutional networks, LSTM, GRU, ANN, etc.) [106, 269, 273, 534].

These technologies can also be used to develop new diagnostic approaches, such as early detection of TB. However, the integration of these technologies also presents challenges. Adequate computational infrastructure is needed to process large amounts of sequencing and phenotypic data. In addition, it is important to validate the results in clinical studies to ensure their accuracy and reliability.

6.6 LIMITATIONS

First, deep learning models rely on large amounts of high-quality annotated data for training. However, such data may not always be readily available, particularly in low-resource settings. Additionally, data quality can vary, which may affect the accuracy of the model's diagnosis.

Secondly, deep learning models can be limited by the scope of the data they are trained on. If a model is trained on a specific set of medical images, it may not be able to accurately diagnose pulmonary infections that present differently or have atypical features.

Third, deep-learning models are often considered "black boxes" because they are highly complex and difficult to interpret. This can make it difficult to understand how the model arrived at its diagnosis, which can be problematic for healthcare professionals who need to justify their diagnoses to patients or other healthcare providers.

Lastly, there may be ethical and legal concerns related to the use of deep learning for medical diagnosis. For example, if a deep learning model produces a false-positive or false-negative diagnosis, this may lead to unnecessary treatments or missed diagnoses, respectively, which can have serious consequences for patients.

6.7 CONCLUSIONS

Timely and accurate diagnosis is essential in treating and preventing the spread of pulmonary infectious diseases like COVID-19, COPD, and TB, which are significant causes of illness and death globally. Deep learning, a machine learning type, has proven useful in medical applications, particularly in analyzing and classifying biosignals and medical images. More recently, there has been growing interest in using deep learning techniques to diagnose pulmonary infections by analyzing cough signals, the ECG, chest radiographs, or computed tomography scans. By training deep learning algorithms on large medical datasets, these models can learn to accurately classify different types of pulmonary infections, leading to more efficient and accurate diagnosis. This improves patient outcomes by enabling early detection of lung infections and enhancing the effectiveness of treatments. Machine learning can also be used to predict drug resistance by summarizing the predictive ability of various factors. This can aid in clinical decision-making and detect single nucleotide polymorphisms as whole genome sequencing data increases.

7 Memristor-Based Ring Oscillators as Alternative for Reliable Physical Unclonable Functions

Joseph Herbert Mitchell-Moreno
INAOE, Department of Electronics, México

Guillermo Espinosa Flores-Verdad
INAOE, Department of Electronics, México

Arturo Sarmiento-Reyes
INAOE, Department of Electronics, México

CONTENTS

7.1 INTRODUCTION

The first decades of the 21st century have demonstrated that the conception of the world as it was previously known has changed, becoming governed mainly by digital technology. The number of insecure environments in which data is transmitted has opened a gap that represents a space of opportunity for third parties to carry out attacks on intellectual property "IP" and private information, resulting in detrimental

DOI: 10.1201/9781003385615-7

economical and social impacts for the target sectors. Embedded systems, mobiles, and computing, in general, have been approaching toward everyday tasks, e.g., device authentication, secure communication, or confidential information storage. For example, in financial transactions using smartphones, the mobile device acts as a token for authentication purposes and stores the user's secret information; however, smartphones typically remain in insecure environments, so attackers can take private information. Currently, the best solution in order to avoid exposing the device token is by using non-volatile memory "NVM", erasable programmable read-only memory "EEPROM" or Static Random Access Memory "SRAM" [564] backed by a battery to store the secret keys. Nevertheless, these are costly techniques in terms of design area, economic cost, and energy-consuming for networks with limited resources such as the Internet of Things, in addition to the fact that some conventional encryption techniques have been infringed before [618]. Also, it has been proved that by tamper attacks of invasive and non-invasive nature, such as glitch attacks, micro-probing, laser cutting, and power analysis [309] is possible to infringe the layer of security of conventional memories. Therefore, protecting the secret keys against such attacks require active anti-tamper mechanisms [240], which means even more increases in energy and economic expenditure.

The needs of society have evolved, and it has become notorious that many tasks such as communications, business, security, financial transactions, and identification are more efficient when done with technology. With the arrival of the digital era, networks such as IoT require very high-security environments. Nowadays, a substantial vast amount of authentication methods require software execution, which can be risky since an attacker with high computational capacity may overcome that security barrier and finally obtain confidential data, such as passwords or secret keys. Considering the aforementioned scenario, it is identified a need related to developing new technologies to guarantee the security of intellectual property, including new physical security primitives. So far, the advances have mainly focused on software-based schemes leaving the physical security at the expense of the attacker's ability [214]. The first approach to the concept of hardware security "HS" by using physical unclonable functions "PUFs" dates back to 2002 when physical one-way functions were introduced [469], leading the foundation for PUFs as a promising alternative to provide innovative security primitives based on physical schemes. The concept is based on using physical phenomena during manufacturing to provide identity to each chip. To exemplify this idea, slight differences exist in nature, serving as unique identifiers for individuals, e.g., human fingerprints and zebra stripes. Analogously in silicon, a correct design of a PUF can take advantage of the "birth process" of integrated circuits "ICs" by stamping a birthmark in the form of a digital code used as a silicon identifier, as depicted in Figure 7.1.

Although the IC manufacturing process provides an opportunity for innovation in PUF structures, some authors have bet on conventional CMOS technology as the source of HS frameworks motivated by the constant size reduction of channel lengths (up to 7nm to date). In contrast, others have chosen to focus on the use of emergent

Figure 7.1   The manufacturing properties of silicon could be used to identify ICs, an example of similar identifiers is the genetic code that creates the pattern of lines of the zebras.

nanotechnologies such as resonant tunnel diodes, spin-transfer torque "STT", carbon nanotube field effect transistors "CNFET" or memristors [194], [12], [34] which present a high native degree of randomness in their manufacture process. This work considers memristors as a device whose range of inherent process variations can be leveraged for the creation of PUF structures. In this proposal, memristors are used to develop PUFs based on ROs. The variability of the oscillators is due to the process variation of the memristors. In addition, a specially tailored memristor model is used to control the memristance variation with internal symbolic parameters of the device.

7.2   PHYSICAL UNCLONABLE FUNCTIONS

In nature, biometrics features are used to find notorious differences among individuals. Biometric traits are impossible to duplicate since many complex factors affect their final physical structure. Expanding this concept to physics, there are physical entities with similar properties that ensure that an individual's identity cannot be known a priori and not even replicated.

7.2.1   CONCEPT

A PUF is an entity embodied in a physical structure that uses the intrinsic complexity and irreproducibility of physical systems to generate secret information. Its response is due to the properties of the system itself, which are directly related to the internal structure of the PUF, [363]. In PUFs an output is provided by the hardware only when powered instead of storing any information. A challenge (input) and its associated response are known as challenge-response pair (CRP). The set of all possible CRPs in the PUF instance defines the identity of the PUF itself, which is also unique and unclonable. PUF designs focus on exploiting the process variations of manufacturing technologies that are not variation re-creatable or controllable, leading to instances capable of creating high entropy cryptographic keys [200].

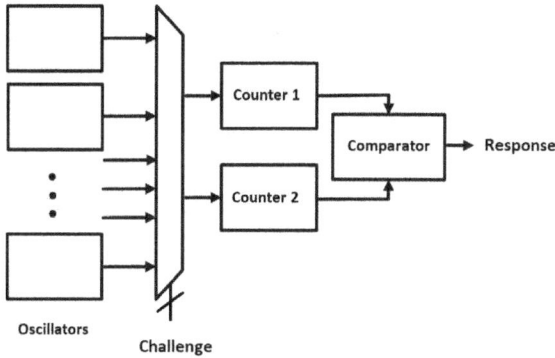

Figure 7.2 Ring oscillator PUF.

7.2.2 RING OSCILLATOR PUF

A ring oscillator "RO" consists of an odd number higher than 1 of inverter stages that fulfill the Barkhausen criteria [67]. Each chained inverter contributes to the delay of the signal; adding more inverters increases the total delay and decreases the oscillator's frequency. In addition, a higher number of stages produces more robustness to process variations; more stages imply a compensation of the overall variations [539].

An electronic PUF design based on delay loops is implemented in this work, this scheme was first introduced by Suh and Devadas [590], (RO PUF). It consists of a bench of ROs subjected to the same design process. Their outputs are connected to an N to 2 multiplexer controlled by the challenge, which addresses two ROs output signals to a pair of counters and finally to a comparator that detects the higher frequency oscillator in order to generate an output bit, as seen in Figure 7.2. The operation of the RO PUF is based on identifying frequency differences in multiple oscillators.

7.2.3 PERFORMANCE METRICS

The response of interest in a RO PUF, instead of being an analog voltage or current, is rather a bitstring that comes from several comparisons among oscillators. In order to find a quantitative performance indicator of quality, many different researchers have proposed metrics; this work only applies the following three measures widely used on HS applications reported in [396].

Uniformity:
Measures the proportion of 1 and 0 bits of a response bitstring. Uniformity of an n-bit PUF is defined as its Hamming weight percentage.

$$(Uniformity)_i = \frac{1}{n} \sum_{l=1}^{n} r_{i,l} * 100 \qquad (7.1)$$

Where $r_{i,l}$ is the l-th binary bit of an n-bit response for a chip i.

Uniqueness:
Represents the ability of a PUF to be distinguished from a group of chips of the same type. The Hamming distance between k devices is used to evaluate uniqueness. If two different instances P_i and P_j produce R_i and R_j responses, respectively, to a C challenge.

$$(Uniqueness) = \frac{2}{k(k-1)} \sum_{i=1}^{(k-1)} \sum_{j=i+1}^{k} \frac{HD(R_i, R_j)}{n} * 100 \qquad (7.2)$$

Bit-aliasing:
It allows to know whether different chips may produce identical PUF responses which is an undesirable outcome. The bit-aliasing of the l-th bit of the PUF is calculated as the percentage of Hamming weight for l-th bit of PUF across k devices.

$$(Bit\text{-}aliasing)_l = \frac{1}{k} \sum_{i=1}^{k} r_{i,l} * 100 \qquad (7.3)$$

7.3 APPLYING MEMRISTOR IN SECURITY PRIMITIVES

PUFs have grown with time as a novel alternative for security primitives and, over the years, have become more promising. Emerging technologies that scale to the nanometer range produce inherently high process variation, so technologies such as memristor, CNFET, PCM, nanoscale diodes, or STT-MRAM become viable options for implementing security in ICs [194]. Figure 7.3 depicts an overview of some of the emerging technologies currently in use for developing PUFs.

The attention of this work is focused on memristor-based technologies since they exhibit high levels of randomness on their parameters (thickness, cross-sectional area, or doping profile) as a consequence of scaling down up to nano scale [396]. In addition, important properties, such as cycle-to-cycle variations, bidirectionality, non-volatility, low energy consumption, small footprint, and the possibility of being implemented on nanocrossbar architectures [193], make memristors suitable for HS primitives.

7.3.1 MEMRISTOR MODEL

Leon O. Chua expounded, in his seminal paper [117], the theoretical concepts of the fourth fundamental circuit element (the memristor) that establishes a direct relationship between the flux linkage (ϕ) and the electric charge (q). Nevertheless, it was until 2008 that an actual memristor was fabricated at HP Laboratories, the device complied with the whole theory from almost 40 years before. From that moment, the efforts of researchers have been focused on developing appropriate models and various applications of the novel device.

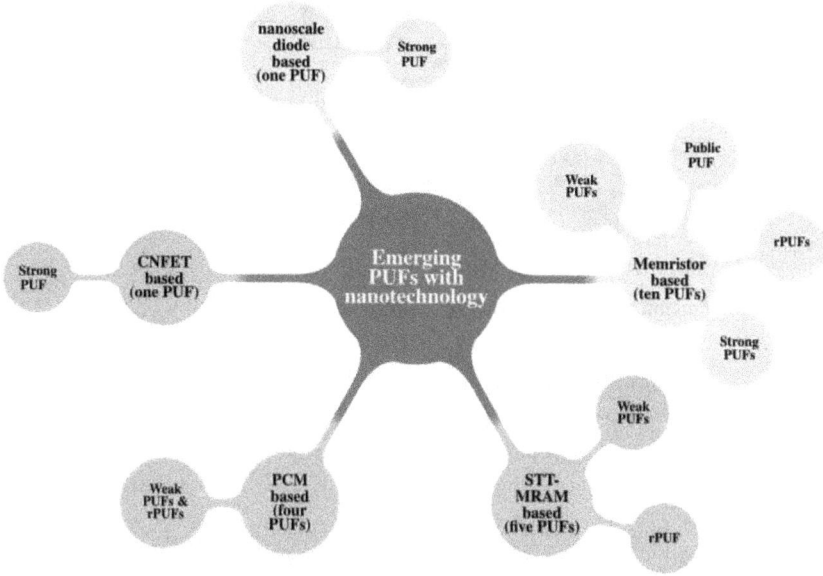

Figure 7.3 Emerging PUFs with nanotechnology. A general classification according to the technology employed [194].

The mechanism governing the device's physical behavior has been recast in the drift equation. If a linear drift mechanism is assumed, then the relationship that defines the dynamics of a state variable $x(t)$ that represents the position of the interface that separates the doped and undoped regions of the device (Figure 7.4) concerning the current $i(t)$ is given as:

$$\frac{dx(t)}{dt} = \eta \frac{\mu_v R_{on}}{\Delta^2} i(t) \tag{7.4}$$

Where μ_v is the mobility of the charges in the doped region and η indicates the direction of displacement of $x(t)$, R_{on} is the resistance value in *ON* state, Δ is the length of the device. In fact, η defines whether the movement of the interface compresses ($\eta = -1$) or widens ($\eta = 1$) the doped region under a positive polarization. The general procedure is to solve the differential equation above for $x(t)$. Then, the expression for the memristance can be obtained by the coupled resistance relationship:

$$M(t) = R_{on}x(t) + R_{off}(1 - x(t)) \tag{7.5}$$

Rewriting the state variable $x(t)$ as a function of Δ the relationship $x(t) = \frac{x}{\Delta}$ is defined, also R_{off} as the OFF state resistance of the device. These concepts form the coupled resistor equivalent [587], as depicted in Figure 7.4.

The memristance model of (7.5) exhibits some drawbacks mentioned in [9], such as the incapacity to explain the behavior in the boundaries of the device and the lack of a physical explanation of the formation of conductive filaments when the device is

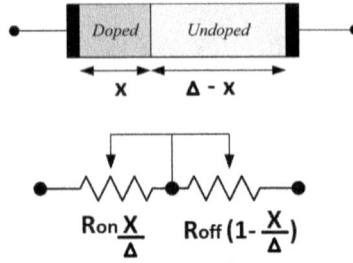

Figure 7.4   HP memristor: coupled resistor equivalent.

switched from a conductive state to the another. For these aforementioned reasons, it is necessary to include a function that allows to correctly model. Furthermore, since (7.4) does not deal with nonlinear aspects that are present in the dynamics of the device, it becomes necessary to introduce the nonlinear drift differential equation as:

$$\frac{dx(t)}{dt} = \eta \frac{\mu_v R_{on}}{\Delta^2} i(t) f_w \tag{7.6}$$

Where f_w is a window function that adds the nonlinear dependence. Several window functions have been reported in the literature [276] [61].

After defining $\kappa = \frac{\mu_v R_{on}}{\Delta^2}$, a slightly different version of (7.6) can be obtained, as seen in (7.7):

$$\frac{dx(t)}{dt} = \eta \kappa i(t) f_w(x(t)) \tag{7.7}$$

A charge-controlled model has been developed in [545], here the nonlinear drift equation is expressed as a function of the charge q in order to obtain a symbolic solution $x(q)$ and generate the expression of the memristor from it, using joglekar's window function and taking into account that the current can be expressed as $i(t) = dq/dt$ the nonlinear drift equation is expressed as (7.8):

$$\frac{dx(q)}{dq} = \eta \kappa (1 - (2x(q) - 1)^{2z}) \tag{7.8}$$

Parameter z controls the linearity level of the window function, and $x(q)$ represents the state variable as an electric charge (q) function.

Homotopy-perturbation method (HPM) HPM is a methodology to solve ordinary/partial and linear/non-linear differential equations. The method adds a perturbation parameter $p \in [0, 1]$ to the ODE and separates it into linear and nonlinear part. As a result, the HPM function is given as:

$$H(v, p) = (1 - p)[L(v) - L(u_0)] + p[L(x) - N(x) - f(r)] = 0 \tag{7.9}$$

Where $L(x)$ and $N(x)$ are the linear and non-linear parts, respectively, of the original ODE. Besides u_0 is the initial condition which satisfies boundary conditions and $f(r)$ is a known analytic function.

The general (order n) solution for a HPM is described as:

$$v = \sum_{n=0}^{\infty} v_n p^n \tag{7.10}$$

The order of the homotopic solution depends on the amount of terms (n) taken from (7.10) to solve (7.9). For $p = 1$, the homotopic formulation becomes the formulation for the original ODE:

$$H(v,1) = L(x) - N(x) - f(r) = 0 \tag{7.11}$$

The approximate solution to the original ODE is given as:

$$x = \lim_{p \to 1} \sum_{n=0}^{\infty} v_n p^n = v_0 + v_1 + v_2 + \dots + v_n \tag{7.12}$$

When HPM is applied on (7.8), the homotopy formulation is expressed as:

$$H(x(q), p) = (1-p) \left[\frac{dx(q)}{dq} - C_1 \eta \kappa x(q) \right]$$
$$+ p \left[\frac{dx(q)}{dq} - \eta \kappa (1 - (2x(q) - 1)^{2z}) \right] = 0 \tag{7.13}$$

Here C_1 is a coefficient required to complete the linear part of the ODE.

If (7.13) is solved using $z = 1$, $p = 1$ and $n = 1$, an equation controlled by the electric charge (q) that represents the memristance is obtained:

$$M(q) = \begin{cases} \eta_+ \begin{cases} \dfrac{(R_{off} - R_{on})(X_o - 1)[(X_o - 2)}{e^{4\kappa q} - (X_o - 1)e^{8\kappa q}] + R_{on}} & q \leq 0 \\[2ex] \dfrac{-(R_{off} - R_{on})X_o[X_o e^{-8\kappa q} -}{(X_o + 1)e^{-4\kappa q}] + R_{off}} & q > 0 \end{cases} \\[6ex] \eta_- \begin{cases} \dfrac{-(R_{off} - R_{on})X_o[X_o e^{8\kappa q} -}{(X_o + 1)e^{4\kappa q}] + R_{off}} & q \leq 0 \\[2ex] \dfrac{(R_{off} - R_{on})(X_o - 1)[(X_o - 2)}{e^{-4\kappa q} - (X_o - 1)e^{-8\kappa q}] + R_{on}} & q > 0 \end{cases} \end{cases} \tag{7.14}$$

Where X_o is the initial condition of the state variable. Due to the massive size of (7.14), mathematical operators θ and Λ are defined according to Figure 7.5. The θ operator has not a physical meaning, it is used to adjust the sign of the drift η to reproduce (7.14), and its value is [1,0]. On the other hand, Λ operator describes the sign of the electric charge q, yielding to $\Lambda = 1$ when $q \leq 0$ and $\Lambda = -1$ if $q > 0$.

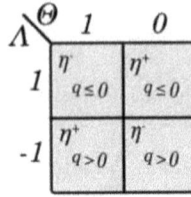

Figure 7.5 Operators used on memristance model.

Applying operators Λ and θ to (7.14), the equation of the memristance as a function of q is found as (7.15).

$$M(q) = [\theta] * [(R_{off} - R_{on})(X_o - 1)[(X_o - 2)e^{\Lambda 4\kappa q} - (X_o - 1)e^{\Lambda 8\kappa q}] + R_{on}]$$
$$+ [\theta - 1] * [(R_{off} - R_{on})X_o[X_o e^{\Lambda 8\kappa q} - (X_o + 1)e^{\Lambda 4\kappa q}] + R_{off}] \qquad (7.15)$$

An important limit value of the memristance is R_{init} which can be regarded as the slope of the hysteresis cycle at t=0, it as well can be used to determine the memristance value when operating at high frequencies It is expressed as:

$$R_{init} = X_o R_{on} + (1 - X_o)R_{off} \qquad (7.16)$$

The compact model expressed in (7.15) considers the basic physical equivalent of the memristor as shown in Figure 7.4. However, it does not cover physical phenomena such as filament forming during the resistive switching mechanism as reported in other compact models such as [728]. However, the proposed model results are adequate for RO-PUF applications because the critical point consists in the use of the memristor in high-frequency, when it behaves as a linear resistor whose final value is randomly established by X_o, R_{off} and R_{on}.

7.3.2 MEMRISTIVE RING OSCILLATOR PUF

ROs with memristive load are proposed to be used instead of regular resistors to take advantage of their reduced area footprint, compatibility with CMOS manufacture [381], and their inherent randomness. Memristors exhibit high percentages of process variability due to manufacturing steps that produce unwanted uncertainties wafer to wafer and lot to lot [575]. These variations are mainly attributed to chemical vapor deposition "CVD" process used to deposit the layers of TiO_2, the spatial distribution of oxygen vacancies and mismatch in sizing [23]. On the other hand, it has been proved that memristors are affected by non-idealities effects leading to device-to-device changes in $\frac{R_{off}}{R_{on}}$ ratio after the programming phase [316]. The memristance of each oscillator may depend mainly on process variations, mismatch, and environmental factors such as temperature.

ROs are implemented by cascading common source stages, see Figure 7.6. Transistors sizing is $\frac{W_n}{L_n} = \frac{5.4um}{0.18um}$, capacitor values are established by parasite capacitance of the technology, and load capacitors and coupling elements are used to modify the position of the poles of the circuit to alter the frequency oscillation and provide randomness. The nominal parameters used in the PUF for load and coupling memristors are listed in Table 7.1. X_o parameter in load devices is set randomly in the

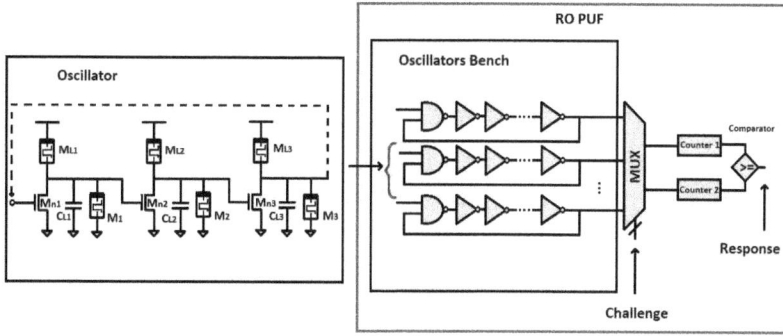

Figure 7.6   RO PUF with memristive load.

range of $[0.5 \pm 0.025]$ to reproduce the initial position of the device interface when manufactured [87], values of high and low state resistance are set as $20k\Omega$ and 100Ω respectively, while Δ and μ_v are typical values for TiO_2 memristors. Coupling devices have parameters $R_{off}=23k\Omega$ and $R_{on}=200\Omega$. The choice of these parameters is fundamental, because, based on these values, it depends on complying with the Barkhausen oscillation criterion to maintain the oscillatory behavior in the system. The minimum electrical resistance value for coupling and load memristors is $7.5k\Omega$ and $1.5k\Omega$, respectively, to avoid losing oscillatory behavior. A nominal value of X_o must be established considering that this oscillation is not compromised due to process and environmental variability.

The oscillation frequency of a regular RO is theoretically calculated using (7.17):

$$f_{osc} = \frac{1}{\pi N t_d} \tag{7.17}$$

Here N is the number of inverter stages and t_d is the time delay produced by a single stage.

Table 7.1

Nominal parameters for load and coulping memristors.

Device	X_o	η	$\Delta\,[m]$	$R_{on}\,[\Omega]$	$R_{off}\,[\Omega]$	$\mu_v\,[\frac{m^2}{Vs}]$
Load	-	+1	10n	100	20k	10f
Coupling	-	+1	10n	200	23k	10f

7.4 RESULTS

This section is devoted to presenting the main results of the HS scheme. On one side, the results related to the performance of the ROs are treated. On the other side, the results of the PUF and its performance metrics for two mapping architectures.

7.4.1 ANALYSIS OF RING OSCILLATOR PERFORMANCE

The main goal of the analysis of the performance of the RO is focused to determine how the oscillation frequency behaves with respect to environmental and internal parameters of the circuit.

The memristive RO-PUF was implemented using 3-stage ROs with memristors as dynamic loads with coupling memristors between stages. Simulations were carried out with Synopsys Hspice software using 180nm UMC technology. The response was obtained for variations in temperature of -30, 27, and 120 Celsius considering process variations in three different corners provided by the technology manufacturer (SS, TT, FF). A bench of memristors with different effective memristance is used (assuming a normal distribution with a dispersion of 5 % on X_o parameter from a Monte-Carlo analysis) to select the load of each inverter randomly. 3 memristors are grouped as a so-called "memristive array." Each stage is loaded with a random memristive array to perform simulation.

A full-factorial experiment with three levels $(+, *, -)$ and three factors (process, memristive array, temperature) is designed, as seen in Table 7.3, the nomenclature is given in Table 7.2. 27 treatments are considered with the purpose of analyzing the effect of these factors on the output frequency. Results are shown in Figure 7.7. It is noted that the most potent effect for all treatments is due to temperature, changing the frequency by a maximum of 0.78 GHz. No changes are observed due to the memristive array, and regarding the process, a noticeable variability that is represented by the spaces between the planes of about 0.25 GHz between the TT and FF corner is observed.

A frequency analysis is performed to estimate the dispersion in the fundamental frequency of the output signals. SS TT FF corners were simulated assuming changes in temperature as shown in Table 7.2, memristive array is not considered since it did not show to be a critical factor. From 9 curves obtained the variability in the

Table 7.2

Nomenclature used.

Nomenclature	-	*	+
Process	TT	SS	FF
Memristive array	1	2	3
Temperature (C)	-30	27	120

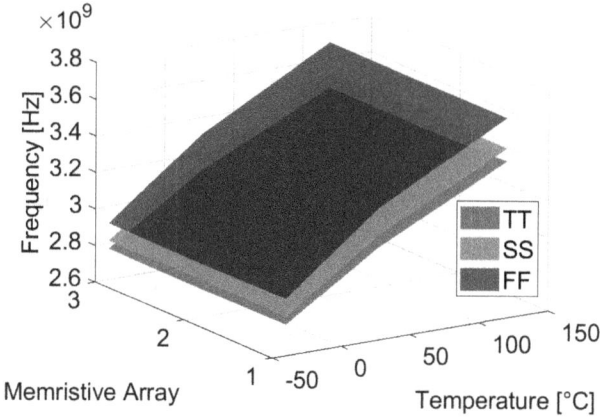

Figure 7.7   Effect of the 3 factors on the frequency of a RO.

frequency is observed (Figure 7.8), to achieve a higher dispersion in the fundamental frequency, it is proposed to design for higher frequencies; it is pointed out that this would increase the effect of parasitic capacitances in the system which could benefit from the randomness of the PUF. To study the effect of process and temperature, 144 samples were simulated (48 samples per corner). From Figure 7.9 on x-axis, it is labeled HT, MT, and LT which means high (120 C), medium (27 C), and low (-30 C) temperatures, respectively. For each corner, the temperature is varied 48 times randomly within a range of ± 15 Celsius. Since the variation on the process-axis does not show any notable trend, it is concluded that temperature changes mitigate the variation due to the corner process.

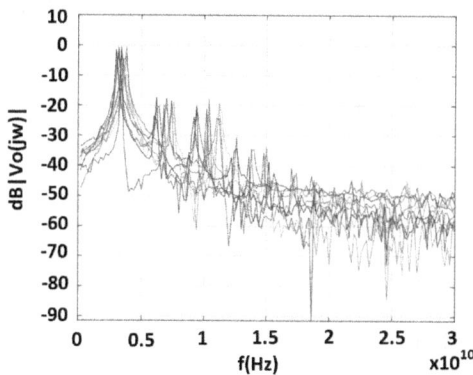

Figure 7.8   FFT analysis of the output signal carried out in the full-factorial experiment.

Table 7.3

Complete factorial design used for the experiment.

Treatment	Temp	Array	Process	Freq
1	-	-	-	2.7683E+09
2	*	-	-	3.1066E+09
3	+	-	-	3.4560E+09
4	-	*	-	2.7643E+09
5	*	*	-	3.1121E+09
6	+	*	-	3.4464E+09
7	-	+	-	2.7692E+09
8	*	+	-	3.1115E+09
9	+	+	-	3.4473E+09
10	-	-	*	2.8130E+09
11	*	-	*	3.1748E+09
12	+	-	*	3.5285E+09
13	-	*	*	2.8198E+09
14	*	*	*	3.1832E+09
15	+	*	*	3.5316E+09
16	-	+	*	2.8134E+09
17	*	+	*	3.1764E+09
18	+	+	*	3.5268E+09
19	-	-	+	2.9101E+09
20	*	-	+	3.3184E+09
21	+	-	+	3.6950E+09
22	-	*	+	2.9016E+09
23	*	*	+	3.3064E+09
24	+	*	+	3.6907E+09
25	-	+	+	2.9084E+09
26	*	+	+	3.3089E+09
27	+	+	+	3.6893E+09

7.4.2 ANALYSIS OF PUF RESULTS

The analysis of the performance of the PUF has been oriented to determine the most important metrics that are used to evaluate its quality.

Two memristive RO PUFs with 32 oscillators each were simulated with a different mapping algorithm. By pairwise and by combinations of all possible pairs, only the most likely cases of corner and temperature are considered in both cases. Memristors were randomly taken from a generic bench where the parameter X_o was varied according to the same distribution as done in section 4.1 (meaning that variations in

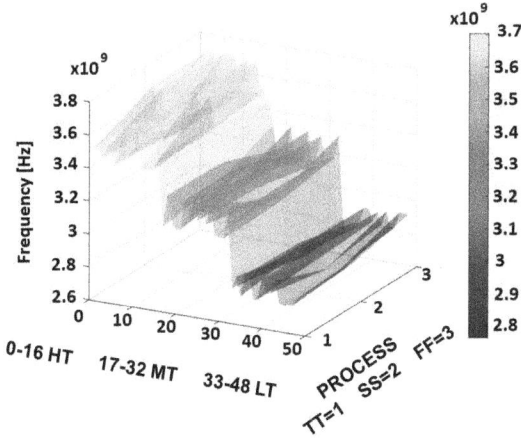

Figure 7.9 The experiment results in a 3D plot show up the dependence of a sample oscillator with respect to the process.

the memristance are result of manufacturing depicted by X_o parameter). Variations in parameters such as R_{on} or R_{off} from (7.16) would change the output frequencies producing similar results. Finally, a comparison between both cases is performed.

Mapping by pair-wise $\frac{n}{2}$

An analysis was made using 32 oscillators, by pairing pair-wises without reuse. That is, a challenge addressing oscillators A and B to generate a bit does not reconsider any of these for any other output bit. The challenge in this case has the same number of bits as the response, and it is given by $\frac{n}{2} = k$ where n is the number of oscillators; since the system does not reuse hardware it keeps the entropy as high as possible (entropy of $\frac{n}{2} = k$). In order to generate more reliable bits, a bigger area in silicon is needed compared to other mapping algorithms. When using this mapping scheme, it is evident that the generated keys are small, although due to the lack of correlation between each PUF execution, these are highly secure bits. In this scheme, the multiplexer selects RO_1 paired RO_2 and it generates a bit, then RO_3 paired RO_4 generates another bit, the process is repeated until finishing the last pair.It is highlighted that the more hardware on silicon, the more bits.

2864 PUF instances were analyzed, and each challenge was provided randomly so that the pairing is always random as well, only focusing on nonrepeating any oscillator for a single instance. These simulations show deterministic results. A normal curve for each one of the metrics allows highlighting that in the vast majority of PUF instances the performance will approach a Gaussian distribution Figure 7.10, indicating a very low probability of extracting keys through hardware attacks.

From Figure 7.10 a), it is seen that the Hamming distance surrounds 8 bits for a 16-bit-string, meaning that there are no notable differences in the vast majority of PUF instances, with standard deviation $\sigma = 1.7944$. From b), it can be seen that the response is not skewed toward either side, the Gaussian bell centered at

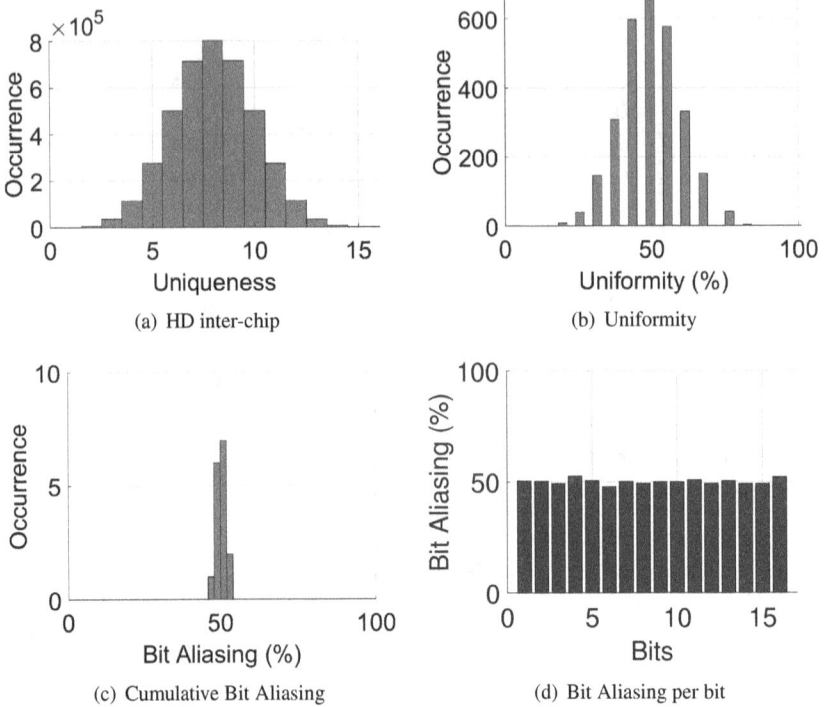

(a) HD inter-chip

(b) Uniformity

(c) Cumulative Bit Aliasing

(d) Bit Aliasing per bit

Figure 7.10 PUF metrics for pairs algorithm, 2864 samples were taken into account for this simulation, the output bitstring was generated by pairing random oscillators without reusing.

50 indicates that PUF instances produce the same quantity of 1 and 0, $\sigma = 10.4905$, c) and d) both provide information about the bit-aliasing, cumulatively in c) and d) For a vast majority of PUF instances, there is a measurement close to the ideal value with $\sigma = 1.1615$. In addition, Figure 7.11 shows the footprint provided for a single PUF instance subjected to 100 challenges randomly selected. The vertical columns

Figure 7.11 PUF Footprint obtained by performing simulation, it is representing the identity of a single PUF instance under 100 random challenges (white=logical HIGH, black=logical LOW).

contain the bits of a single output bitstring (white=1, black=0), Figure 7.11 represents the entire identity of the PUF for the 100 challenges.

Mapping by all possible combinations $\binom{n}{2}$
In this case of study, the same bench of $n = 32$ oscillators as section 4.2.1 is used, the combinatorial $\binom{n}{2} = \frac{(n)(n-1)}{2} = k$ generates 496 output bits, the challenge of the system must address the multiplexer to all the possible combinations of oscillators. This algorithm requires a multiplexer logic considering all possible combinations of oscillators, RO_1 paired RO_2 (RO_1-RO_2), then RO_1-RO_3... RO_1-RO_{32}, RO_2-RO_3, RO_2-RO_4... RO_2-RO_{32} and in this way complete the 496 possible pairs.

496 bits are gotten, and only ($log_2 n!$ with $n = 32$) 117 of 496 are reliable bits that contribute to the system's entropy. Since oscillators are not used only once, they generate correlated bits. Third parties might find a bridge to perform successful attacks because it is noticed that whether $f_a > f_b$ and $f_b > f_c$ then the frequency $f_a > f_c$, this correlation is unwanted.

The results for this case are shown in Figure 7.12, 100 PUF instances are simulated, presenting Gaussian bells with $\sigma = 33.5790$, $\sigma = 5.7810$ and $\sigma = 4.9701$ respectively. In this case, more bits were generated. This outcome might be useful for low-risk environments applications that require high number of keys. Is it also worth mentioning that it is verified that both logic algorithms lead to reliable PUFs which can be used according to the designer's specifications, pointing out an important trade-off between entropy, area, and power consumption. For this case, a single PUF instance is considered, 100 random challenges are applied and the resulting footprint is displayed in Figure 7.13.

Comparison
From several works compared in Table 7.4, it is observed that memristors based PUFs presents exceptional results, as well as conventional implementations. In order to validate the true unpredictability in PUF keys, the auto-correlation function (ACF)

Table 7.4
Comparison of different RO-PUF systems.

Uniformity	Uniqueness	Bit-Aliasing	Memristor	Work
51.43	48.57	51.43	Yes	[381]
47.31	-	50.72	No	[395]
51.20	49.40	53.37	Yes	[103]
49.66	50.17	-	Yes	[195]
50.56	94.07	50.56	No	[396]
50.17-51.20	45.15-39.79	50.17-50.14	No	[441]
-	47.20	50.20	No	[198]
50.04	**49.99**	**50.04**	**Yes**	$\binom{n}{2}$
50.19	**50.01**	**50.19**	**Yes**	$n/2$

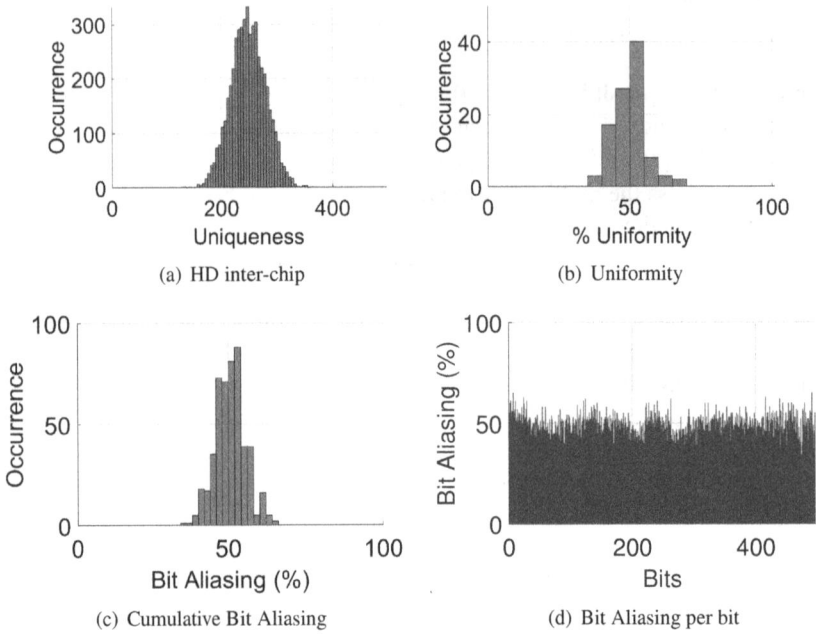

(a) HD inter-chip

(b) Uniformity

(c) Cumulative Bit Aliasing

(d) Bit Aliasing per bit

Figure 7.12 PUF metrics for combinations algorithm, 100 samples were simulated, the output bitstring have $\binom{n}{2} = \frac{(n)(n-1)}{2} = k$ number of bits, n is the number of oscillators.

is shown in Figure 7.14. Statistical values indicate a better level of correlation in $\frac{n}{2}$ PUF, which is expected since its mapping algorithm does not reuse any oscillator. These results also prove the randomness of the PUF bits and their resilience under

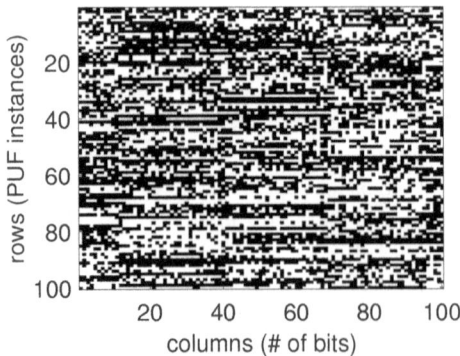

Figure 7.13 Footprint generated through the memristive RO-PUF for 100 random challenges (white=logical HIGH, black=logical LOW).

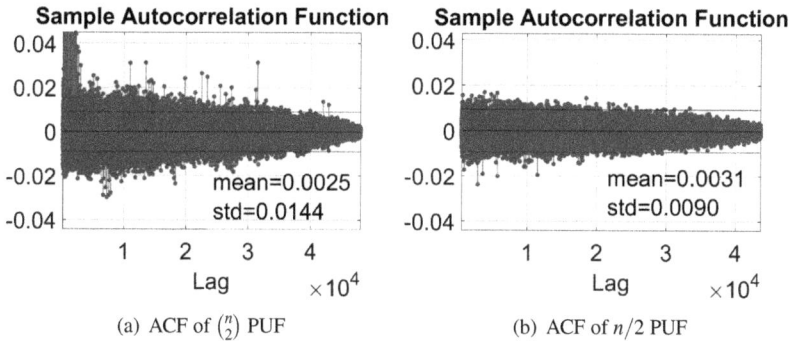

(a) ACF of $\binom{n}{2}$ PUF   (b) ACF of $n/2$ PUF

Figure 7.14 Autocorrelation functions for both implementations.

auto-correlation attacks. With no improvements in the capacity of CRPs, a RO works as a weak PUF; however, some works of researchers have focused their efforts on increasing the reliable CRPs generated from this configuration PUF [148], so that its performance can be improved for more security applications.

Temperature was considered; nevertheless, the lack of a memristive model that includes the parameter T means that the memristance only considers effects with temperature in the same way as a resistor would do on SPICE. Authors are still aware of the need of a memristive model including temperature parameters in order to have more complete characterization of the PUF. Environmental variations such as aging and polarization voltages are not taken into account throughout this work.

7.5 CONCLUSIONS

The feasibility of using functional analytical models of memristors in HS schemes was demonstrated. The models were tested on a memristive RO PUF system that showed excellent figures regarding the most commonly used metrics. Special emphasis has been given to the analysis of the effects of the variations of memristor parameters on the performance metrics and the overall behavior of the HS system. In addition, the effect of the temperature on the frequency of the RO is highlighted.

Since the model of the memristor has been recast in an analytical expression, it is possible to establish the functioning of the memristive PUF with basis on the memristor parameters, in particular X_o, R_{on} and R_{off}.

The variability in the process is analyzed and quantified, assumptions such as variations of temperature, process, and load were made to observe the affectations in the output frequency with the aim of taking design guidelines that consider environmental factors in order to avoid bit flipping.

The mapping of the challenge into a PUF directly affects its characteristics and applicability. Two different maps were tested to use the same type of PUF. The benefits and drawbacks for each case were highlighted, and the randomness and entropy of the responses for both cases were also verified.

Part II

Machine Learning for Unmanned Systems

8 Past and Future Data to Train an Artificial Pilot for Autonomous Drone Racing

Leticia Oyuki Rojas-Perez, Alejandro Gutierrez-Giles, Jose Martinez-Carranza
Computer Science Department, Instituto Nacional de Astrofísica, Óptica y Electrónica (INAOE), Puebla 72840, Mexico.

CONTENTS

8.1 INTRODUCTION

Autonomous drone racing (ADR) presents the challenge of developing an artificial pilot capable of autonomously flying on a racetrack [176, 391, 427]. This research area has led to improvements in algorithms for obstacle detection, trajectory generation, high-speed control methods, and autonomous indoor drone navigation. However, the development of an artificial pilot does not end with simply surpassing a human pilot. The scientific community continues to strive for the development of an artificial pilot that can adapt to new situations, much like human pilots.

Currently, ADR solutions are based on controlling high-speed autonomous racing drones. To achieve this, positioning systems such as VICON and optimisation methods based on time and trajectory have been implemented [177]. These have achieved speeds comparable to those of human pilots. However, this solution is not capable of adapting to changes in the racetrack, such as moving gates from their original position, as the approach is based on the drone's position rather than the position of the gate. In contrast, human pilots guide the drone in the direction of the following gates and are able to identify areas on the track that allow them to increase or decrease their speed as well as position the drone to head towards the next target [483, 484]. Additionally, human pilots do not necessarily seek optimal flight but instead, familiarise themselves with the track to improve their overall performance.

On the other hand, there are methods for controlling autonomous racing drones based on deep learning (DL), which involve training convolutional neural networks (CNNs) and using computer vision to improve drone performance and adaptability. For instance, DeepPilot [523] is a neural pilot based on a CNN that processes images from the drone's camera and generates real-time flight control commands to enable autonomous navigation of the drone on an unfamiliar racetrack. The flight commands provided by DeepPilot align the drone with the centre of the gate to cross it. However, the flight speed is not very fast. Nevertheless, when combined with object detection, this neural pilot can identify areas of interest that allow the drone to complete the track up to 7 times faster [525] using a waypoint controller. The preparation stage of human pilots inspired this strategy before a competition and has no need for prior information about the racetrack, such as its dimensions, number of gates, or the positions and orientations of the gates.

Although [525] identifies points of interest during the preparation stage, it does not learn from this new information. In this work, we present a methodology based on [525] for DeepPilot to improve its performance and learn flight commands from controllers such as the line follower and the Model Predictive Controller (MPC). First, we design basic sections, such as straight lines, zig-zag, up and down, to train two new DeepPilot models. The first trained with flight commands provided by the line follower, and the second model used the MPC flight commands. Our method has been evaluated on the RotorS simulator implemented for Gazebo [184]. DeepPilot has been trained with runway sections so that it is able to navigate on any runway without prior knowledge of it. We have compared it with the performance of a human pilot and the new DeepPilot models on the same track with very similar results.

This chapter has been organised as follows to describe our approach: Section 14.2 discusses the related work; Section 8.3 describes in more detail our approach; Section 14.3 presents our experimental framework; and finally, Section 14.5 outlines our conclusions and future work.

8.2   RELATED WORK

In recent years, ADR has become increasingly important in the scientific community as it presents fundamental challenges for aerial robotics. These challenges include a real-time association of perception and action, adapting to dynamic environments, dealing with inaccurate sensors and actuators, as well as state estimation in environments with low texture or motion blur. Since 2016, competitions such as the IEEE

IROS Autonomous Drone Racing (ADR) [427, 428], the AlphaPilot [176], and Microsoft Game of Drones [391] have encouraged the development of artificial pilots capable of autonomously executing tasks that were previously only achievable by human pilots.

Most ADR strategies are based on four processes: perception, localization, trajectory generation and control, which are executed simultaneously. In [521], the authors analyse the methods and strategies used in ADR competitions and highlight the critical role of data processing, as the data received from the sensors is used to interpret the environment. In ADR, the perception module allows the drone to interpret the environment to detect obstacles or targets and to estimate the vehicle's position and orientation on the racetrack. Actually, different machine learning techniques, such as deep and reinforcement learning, have been explored to improve the drones' ability to learn and adapt to changing situations. For example, deep learning networks to detect the gate [84, 278, 279], classification networks to determine the drone's movements [427], CNN and Multilayer perceptron to estimate direction and velocity [285, 286], the relative position [286], CNN to obtain relative position with respect to the gate [126, 127], CNN to identify the blind spot [125, 525] and even CNN to predict the flight commands [485, 518, 519, 523].

Concerning the vehicle position and orientation estimation, the most common vision-based localization systems in robotics are Simultaneous Localization and Mapping (SLAM) [439] and Visual Odometry (VO) [140, 149]. Nevertheless, these systems are computationally expensive, resulting in a position estimate of less than 15 Hz in embedded systems [177, 405, 439]. However, the methods mentioned earlier for pose estimation accumulate errors. Also, these methods are not appropriate for performing an agile fight due to is necessary to obtain the pose estimation at a higher frequency. Alternatively, low-cost methods are implemented on embedded systems, such as [427] where authors implemented IMU-based position estimation methods using the Kalman Filter [353, 355]. Another solution to obtain a higher frequency position is using neural networks combined with the Extended Kalman Filter. For example, authors in [522] proposed a convolutional neural network which associates a 2D image with a 3D pose to relocate in an environment. Although they only work in a known environment, this approach obtains the drone's pose at a frequency of 65 Hz in an intelligent camera with an average error of 0.25 m in translation and 2.0Â° in the heading w.r.t. ORB-SLAM, in contrast to motion capture systems, which provide positions up to 500Hz without error accumulation.

This last module is crucial for the trajectory planner, as it requires knowledge of the drone's current state (position and orientation) and starts and destination point information indicated in GPS coordinates or 3D points. It also requires knowledge of movement constraints, such as maximum speed, acceleration, and turning limits. The methods for trajectory planning used in ADR are: 1) sampling-based, 2) graph-based, and 3) optimisation-based. Sampling-based trajectory generation allows the generation of optimal trajectories in terms of time and distance. This method is proper when specific constraints such as speed limits or arrival time requirements exist. However, they are computationally intensive and can generate non-smooth trajectories, unlike graph-based trajectories, which can generate smooth and controllable trajectories. Nevertheless, in the context of ADR, they are not suitable since a race track involves environments with dynamic obstacles, curves, and turns. For this reason, optimisation-based trajectory generation methods, such as time-optimal planning [177], time-optimal flight [526], and minimum-time trajectory [581] are commonly used. These methods generate smooth and controllable trajectories suitable for environments with dynamic obstacles and trajectories with many curves and turns. However, they must be pre-computed as they are computationally expensive. Since these methods seek to find an optimal trajectory that satisfies certain constraints, such as distance travelled, flight time [177], speed and maximum accelerations [413,435], among others.

Finally, the control module must be able to drive the drone in real-time. In ADR, the controllers implemented are feedback-based, observer-based and state estimation based. Feedback-based controllers such as Proportional Integral Derivative Controller (PI) [125, 127, 427, 524] or visual servoing [277–279, 427] use visual or position feedback from the drone to adjust its behaviour. In contrast, observer-based controllers use algorithms such as Kalman Filter to estimate the drone's position and velocity from sensor data such as Inertial Measurement Units (IMU) [351, 352, 354, 427]. Finally, state-estimation-based controllers such as the Model Predictive Controller (MPC) use an explicit dynamic system model to compute control commands to minimise tracking time or error [177,526]. It also allows the prediction of the future states of the drone and provides information about the stability properties of the system, suitable for agile manoeuvres, unlike and visual servoing, which do not have information about future states.

Nevertheless, these four modules must be executed simultaneously, so it is necessary to have the appropriate hardware to process the information to execute agile flight at high speeds. Therefore, in recent years, learning-based methods have been proposed to replace the planner and controller by using only the perception module since deep learning networks are able to identify patterns of characteristics in images or states, which can be associated with control commands [485,518,519,523]. However, these methods require a data collection process that is long, laborious, and error-prone, as it is necessary to collect training data that provides relevant information for the network to learn a specific task.

Motivated by this, we propose a methodology to train a neural pilot, DeepPilot, by using flight commands from two different flight controllers: one based on feedback (PI) and the other on state estimation (MPC). Our methodology involves learning using key racetrack sections that are crucial for completing an unfamiliar race track.

8.3 METHODOLOGY

This section outlines the modules required to implement our approach. First, we provide a description of the DeepPilot neural pilot. Next, we describe the waypoint discovery method and the flight controllers employed to navigate a racetrack. These modules will serve as the basis for generating two new datasets that will enable DeepPilot to enhance its performance.

Table 8.1
Ranges values used in the original dataset of DeepPilot.

Flight Command	Roll (S_ϕ)	Pitch (S_θ)	Yaw (S_ψ)	Altitude (S_h)
Max	0.9	1.0	0.1	0.1
Min	0.1	0.1	0.05	0.05

8.3.1 DEEPPILOT

DeepPilot [523] is a convolutional neural network (CNN) that processes images from a drone's camera and generates flight control commands in real time to enable autonomous navigation of the drone on the racetrack. The CNN is trained on a dataset that includes images from a single gate and the corresponding control commands. During training, the network learns to identify patterns in the images and associate them with the appropriate control commands. Once the network is trained, it processes the drone's camera images in real time and generates flight control commands based on captured images. These commands adjust the drone's speed, direction, and orientation to enable autonomous navigation of the racetrack.

The DeepPilot network architecture comprises four branches to obtain the flight signals in parallel. Each branch consists of four convolutional layers and three inception modules to extract feature characteristics. Finally, one fully connected layer is connected with one regressor to obtain the flight signal value. Next, DeepPilot associates a set of images with a flight signal to produce translational and rotational motion as a tuple $(S_\phi, S_\theta, S_\psi, S_h)$, which corresponds to the signal values in roll, pitch, yaw and altitude. Finally, the tuple produced by DeepPilot is sent to the drone's internal control to navigate autonomously through the racetrack.

The DeepPilot model learns seven basic movements: right, left, up, down, right rotation, left rotation and forward displacement with respect to a gate. For this, the authors used a dataset of 10,334 mosaic images associated with a flight value as the label. The label ranges are summarised in Table 8.1. Also, DeepPilot required three specialised models: the first to learn the roll and pitch values, the second to learn yaw values, and the third to learn altitude values. Additionally, authors implement noise filters on the output values to smooth out the spikes in the flight signals, which produce oscillatory behaviour and jerkiness.

8.3.2 WAYPOINTS DISCOVER

Drawing inspiration from how human pilots familiarise themselves with unknown racetracks, [525] proposes a two-step approach for autonomous drone racing that does not require prior knowledge of the track's position, orientation, height, or number of gates. The first step involves racetrack exploration, where an artificial pilot (DeepPilot) and an object detector are used to obtain a set of waypoints

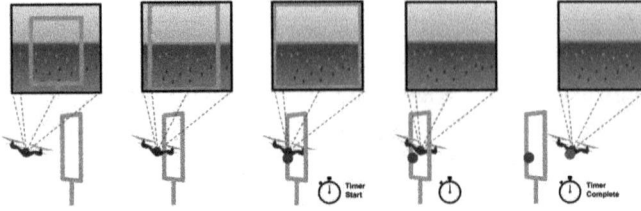

Figure 8.1   The system automatically sets the entry point to the blind spot zone (blue dot) when the drone stops seeing the gate. Once the time is up, the system places the exit from the blind zone (red dot). The timer prevents placing the dot in the centre of the gate.

corresponding to the entrance and exit (blind spot zone) of each gate on the race-track. The authors use the Single Shot Detector 7 (SSD7) to identify an orange gate and implement a strategy based on a state machine to discover the blind spot zone. The state machine consists of four states; the first state identifies when the drone stops seeing the gate in front of it to establish the blind spot zone entry. The second state prevents the system from adding a blind spot zone entry while the drone is moving towards the gate. The third state sets a timer; once this time elapses, the system adds the blind spot zone exit to the waypoint list. Finally, the fourth state prevents the system from duplicating the exit points of the blind spot zone.

The second step consists of using a flight controller to perform a much faster flight, which uses the waypoints discovered in the first step and the global position obtained from the Gazebo simulator. The flight controller is a PI that translates and orientates the drone to the next point, where the maximum value in pitch is 1.

8.3.3   FLIGHT CONTROLLERS

This section provides a brief description of the flight controllers used to navigate a racetrack. Specifically, the Proportional-Integral (PI) controller and the MPC model were employed to create new datasets for training a new DeepPilot model. These controllers are used to enhance the model's performance in terms of navigation and time efficiency while operating on the racetrack.

The PI controller is widely used for its simplicity, versatility, and effectiveness in various applications. In the context of ADR, the PI controller enables the drone to respond quickly and accurately to real-time input and output signals, allowing for prompt adjustments to the drone's motion in response to the operating environment.

To complete the racetrack as quickly as possible, we require the drone's state (position and orientation) and a set of waypoints. We then implemented a proportional-integral controller for roll and height and a proportional controller for yaw. The pitch was set to the maximum control signal of 1 to move the drone forward as quickly as possible. To help the drone gradually reduce speed, the controller switches to a proportional controller, reducing the speed from 1 to 0.5 when the distance between the drone and the entry waypoint is less than 3 metres. This reduces the inertia of movement, allowing the drone to turn towards the exit waypoint. This is especially useful on the curved sections of the circuit, where the turn involves a significant angle change in yaw. This controller uses the drone's current position obtained by the Gazebo simulator to calculate the errors between the drone's current position and the reference waypoint.

Assuming that the drone flies on a horizontal plane, we operate with vectors obtained from the translation $\mathbf{t}$ and rotation matrix $\mathbf{R}$ estimated with the Gazebo simulator. Using a unit vector $\mathbf{v} = [1,0,0]$, a heading vector is set as $\mathbf{h} = \mathbf{R}\mathbf{v}$. A departing waypoint $\mathbf{w}_s$ and the next waypoint $\mathbf{w}_g$ are used to define the direction vector $\mathbf{d} = \mathbf{w}_g - \mathbf{w}_s$, with its corresponding rotation matrix representation $\mathbf{R}_d = Rot(\mathbf{d})$, where $Rot(\cdot)$ is a function that calculates such matrix. Finally, we compute the drone's position relative to $\mathbf{w}_s$:

$$\mathbf{r} = \mathbf{R}_d^{\top}(\mathbf{t} - \mathbf{w}_s) \tag{8.1}$$

The control signals are calculated as follows:

$$s_\theta = \begin{cases} 1: & (\|\mathbf{d}\| - r_x) > 3 \\ 0.5 + 0.5\frac{(\|\mathbf{d}\| - r_x)}{3}: & (\|\mathbf{d}\| - r_x) \le 3 \end{cases} \tag{8.2}$$

$$s_\phi = K_{p_\phi}(-r_y) + K_{i_\phi}\int(-r_y)dt \tag{8.3}$$

$$\mathbf{n} = \mathbf{d} \times \mathbf{h} \tag{8.4}$$

$$s_\psi = K_{p_\psi}sign(\mathbf{n})acos\left(\frac{\mathbf{d}\cdot\mathbf{h}}{\|\mathbf{d}\|\|\mathbf{h}\|}\right) \tag{8.5}$$

$$s_h = K_{p_h}(w_{gz} - r_z) + K_{i_h}\int(w_{gz} - r_z)dt \tag{8.6}$$

where *sign* is defined as:

$$sign(\mathbf{n}) = \begin{cases} 1: & n_z \ge 0 \\ -1: & n_z < 0 \end{cases} \tag{8.7}$$

and $\mathbf{w}_g = [w_{gx}, w_{gy}, w_{gz}]$, $\mathbf{r} = [r_x, r_y, r_z]$, $\mathbf{n} = [n_x, n_y, n_z]$.

Our controller received the pose estimation, published by the Gazebo simulator at $1000\ Hz$, and the set of enter/exit blind spot waypoints. Finally, the gains $K_{p_\phi}, K_{i_\phi}, K_{p_\psi}, K_{p_h}, K_{i_h}$ were tuned empirically aiming to avoid oscillations or excessive flight speed that would make the drone hit a gate.

The second flight controller utilised, MPC is a well-established technique in the control community. It has its foundations in optimal control theory. Roughly speaking, an MPC controller performs an optimisation procedure over a finite horizon at a given time step. In the next time step, the horizon, which has the same time length, is moved forwards in time and the optimisation is carried again along with the additional aid of the new feedback information obtained at this time step. Since the optimisation horizon is of a fixed length but moving along the time axis, it is commonly called *receding horizon*.

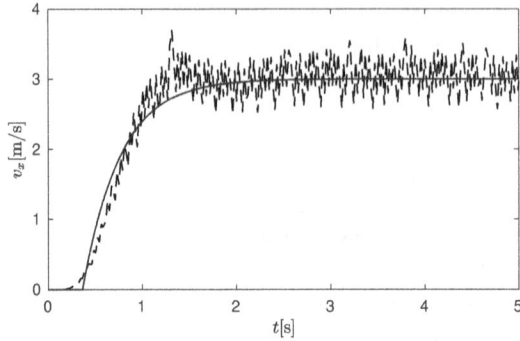

Figure 8.2 Step response of the input $u_x = 1$ for model identification: velocity in the x direction (- - -) and model adjustment (—).

Although having a receding horizon instead a fixed or an infinite one is the main difference between MPC and the standard Linear Quadratic Regulator (LQR), there are some other important differences. One of the most important of such differences is that MPC is able to explicitly handle input and state constraints, both *hard* and *soft* ones. The incorporation of these constraints can be then incorporated to the Optimisation Control Problem (OCP). The MPC can also handle nonlinear dynamic models and more general cost functions in contrast with the LQR controller which only handles linear systems and quadratic cost functions.

To implement the MPC in the context of ADR, the first step is to obtain a dynamic model of the system. Since the goal is to obtain the least processing time required to solve the MPC problem, in this work we consider a very simplified model of the quadrotor employed, i.e. a linear and decoupled one. The inputs of this model are the roll, pitch, and yaw angles, and the vertical thrust. The outputs are defined as the linear positions in all three Cartesian directions and the rotation around the vertical axis. For the model identification, it is better to plot the step responses with the velocities as outputs instead of positions, and then use an integrator to obtain the positions. For this step, a first-order derivative filter was implemented to obtain the velocities from position measurements. The velocity in the x coordinate for a unit step in the input u_x is shown in Figure 8.2

Since the dynamics is assumed to be decoupled, one can obtain a transfer function from the inputs to the velocities, in each direction, in the form

$$Y_i(s) = \frac{K_i}{\tau_i s + 1} e^{-T_{di}s} U_i(s), \tag{8.8}$$

where s is the Laplace variable, $Y_i = \{v_x, v_y, v_z, \omega_z\}$, K_i are the step-response gains in steady state, τ_i are the time constants of each coordinate, and T_{di} are the time-delays. To obtain the transfer function from the inputs to the Cartesian positions, one can just multiply the above transfer function by an integrator, i.e. $1/s$.

We want to obtain a dynamic model that permits the online solution of the OCP problem with the least computational load. Firstly, it was noticed that the rotation around the vertical axis, i.e. the yaw coordinate, is not critical for the ADR problem. Therefore it can be left out of the online optimisation carried out in the MPC framework. Instead, this coordinate will be controlled by a simple PI. Secondly, the time delays in (8.8) can be neglected to obtain a linear time-invariant model of the form

$$\dot{\mathbf{x}} = \mathbf{A}\mathbf{x} + \mathbf{B}\mathbf{u} \qquad (8.9)$$

$$\mathbf{y} = \mathbf{C}\mathbf{x}, \qquad (8.10)$$

where $\mathbf{x} \in \mathbb{R}^6$ is the state vector containing both the Cartesian positions and velocities of the drone with respect to a fixed coordinate frame, $\mathbf{u} \in \mathbb{R}^3$ is the vector of inputs, which are the roll and pitch angles, and the thrust. Notice that these inputs are normalised, i.e. $|u_i| \leq 1, i = 1, 2, 3$, $\mathbf{y} \in \mathbb{R}^3$ is the vector of outputs, which are the Cartesian positions, and $\mathbf{A}, \mathbf{B}, \mathbf{C}$ are matrices of appropriate dimensions obtained from (8.8). This model is used both for the MPC implementation and for a Kalman filter, which in turn is employed to obtain the unmeasured states, i.e. the drone velocities.

A third-order polynomial is proposed to generate the desired trajectory for each Cartesian coordinate. It is well known that these kind of polynomials can generate optimal trajectories for simple linear mechanical systems [569, Sec. 4.2]. In our case, the polynomials are given as a functions of time. A third-order polynomial trajectory is created for each pair of gates. The coefficients of such polynomials can be uniquely determined by defining the initial and final positions and velocities. The positions are given simply by the gate-entrance waypoints described at Section 8.3.2. In turn the velocities are defined as a vector in the Cartesian space, whose magnitude is equal to the desired speeds at the start and at the end of the trajectory, and its direction is computed as the normal vector from the gate-entrance to the gate-exit waypoints. In these terms, the resulting trajectories are functions of the desired time t_f between waypoints. If in addition one imposes the constraint $\|\mathbf{v}(t_0)\| = \|\mathbf{v}(t_f)\| = \|\mathbf{v}((t_f - t_0)/2)\| = v_{max}$, where v_{max} is the maximum attainable speed of the drone, then one obtains a minimum-time trajectory. Notice that this minimum-time trajectory is theoretical and it would be reachable only if all the dynamic forces are neglected. In practice, we can tune this value to have a percentage of the maximum speed, e.g. $0.8\,v_{max}$.

To implement the MPC, the following OCP is defined:

$$\min_{\mathbf{u}} \int_{t_0}^{t_0 + t_h} \mathbf{x}^{\mathrm{T}}\mathbf{Q}\mathbf{x} + \mathbf{u}^{\mathrm{T}}\mathbf{R}\mathbf{u}\, dt + \mathbf{x}(t_h)^{\mathrm{T}}\mathbf{Q}_h\mathbf{x}(t_h)$$

subject to

$$\dot{\mathbf{x}} = \mathbf{A}\mathbf{x} + \mathbf{B}\mathbf{u}$$
$$-1 \leq u_i \leq 1, i = 1, 2, 3,$$

where t_h is the *receding horizon* time, $\mathbf{Q} \in \mathbb{R}^{6\times 6}$ and $\mathbf{R} \in \mathbb{R}^{3\times 3}$ are weighting matrices accounting for the costs of the states and the inputs, respectively, and $\mathbf{Q}_h \in \mathbb{R}^{6\times 6}$ is the weighting matrix of the state at the end of the receding horizon.

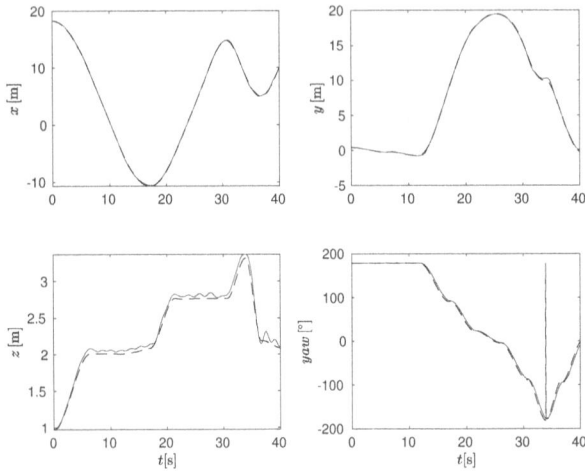

Figure 8.3 Trajectory tracking in Cartesian coordinates of MPC in autonomous mode: references (- - -), measured (—).

The above OCP can be solved with the aid of the ACADO Toolkit libraries in C++[1], along with the qpOASES library. The employed method was *single shooting* with Gauss Newton approximation of the Hessian matrix and a fourth-order Runge-Kutta numerical integrator. An example of trajectories generated with the polynomials for several gates and the corresponding tracking with the MPC in an autonomous-mode operation is displayed in Figure 8.3.

8.4 EXPERIMENTS

In this section, we provide a brief description of our hardware system and communication architecture, as well as the dataset collection process that utilises flight commands from the PI controller and MPC. We outline the DeepPilot training process and introduce the evaluation race tracks, along with the experimental constraints.

8.4.1 SYSTEM OVERVIEW

We performed our experimental framework on an Alienware R5 laptop. This laptop has a Core i7 processor, 32GB of RAM, and an NVIDIA GTX 1070 graphics card. It also runs on the Ubuntu 20.04 LTS and the Robot Operating System (ROS) Noetic Ninjemys version. To run the DeepPilot network, we utilised CUDA 11.6, Cudnn 8.1 libraries, and TensorFlow and Keras 2.8.0 frameworks.

Our communication architecture uses the Robot Operating System (ROS) as a communication channel to integrate the modules in an aerial system. We utilised three different configurations for each of the following processes: 1) exploration, 2)

---

[1] Available online: https://acado.github.io/

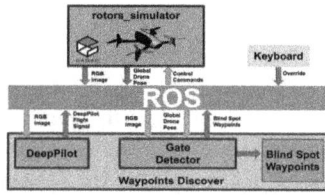

(a) Exploration Process

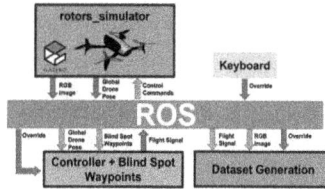

(b) Navigation and Dataset Collection

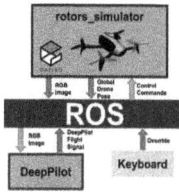

(c) Neural Pilot

Figure 8.4 A Robotic Operating System-based communication system was used. Figure (a) illustrates the communication during exploration to detect entry and exit points of the blind zone. Figure (b) displays the configurations for generating two training datasets using a PI controller and MPC. Figure (c) shows the evaluation methodology configuration.

navigation and dataset generation, and 3) Neural Pilot. To facilitate the simulations, we leveraged RotorS [184], a node that simulates the Parrot Bebop2 Drone in a virtual environment on Gazebo 11. With RotorS, we were able to control the drone's movements and receive a live video stream from the onboard camera of the Bebop2 at 30 fps. Additionally, we implemented a secondary node for a manual operation that allowed us to control the drone using a keyboard and manually initiate or cancel autonomous flights.

Figure 8.4(a) depicts the first configuration for autonomous navigation on the racetrack using a neural pilot called DeepPilot, combined with a gated detector to identify a set of waypoints. DeepPilot employs visual temporal information, which is achieved by capturing the video stream and generating a mosaic image composed of six frames, updated every five frames, to serve as input for the DeepPilot Network. This network provides four flight signals $(S_\phi, S_\theta, S_\psi, S_h)$ at 14 fps. This configuration enables the drone to detect the entrance and exit of the blind spot zone of any unknown racetrack without prior information, such as the racetrack's dimensions or the number and positions of the gates.

In Figure 8.4(b), we present the second configuration used in this work. We utilise the waypoints discovered in the exploration process to enable faster flight on the racetrack. Additionally, we implemented a dataset generation node that collects a mosaic image updated every five frames and the flight command signal provided by the controller to train a new model of DeepPilot. This configuration uses the same communication setup to collect data from the PI controller and MPC. Finally, in Figure 8.4(c), we illustrate the configuration used to execute the new models of DeepPilot, which are trained using a dataset collected from the PI controller and MPC flight command signals, along with a mosaic image updated every five frames.

8.4.2 DEEPPILOT TRAINING

In contrast to the original DeepPilot dataset, we designed a set of key training sections to teach the new DeepPilot model the specific movements required to guide a drone along a racetrack with respect to a group of three gates. To accomplish this, we utilised the flight commands provided by the PI controller and MPC. As shown in Figure 8.5, eight key sections were created, covering an area of 40m × 6m. To obtain a variety of data, we replicated each section three times to provide examples with gates of different heights (2m, 2.5m, and 3m).

We designed tracks by interleaving gate heights to provide a variety of challenges for the DeepPilot model. For example, Figure. 8.5(f) shows a straight section with

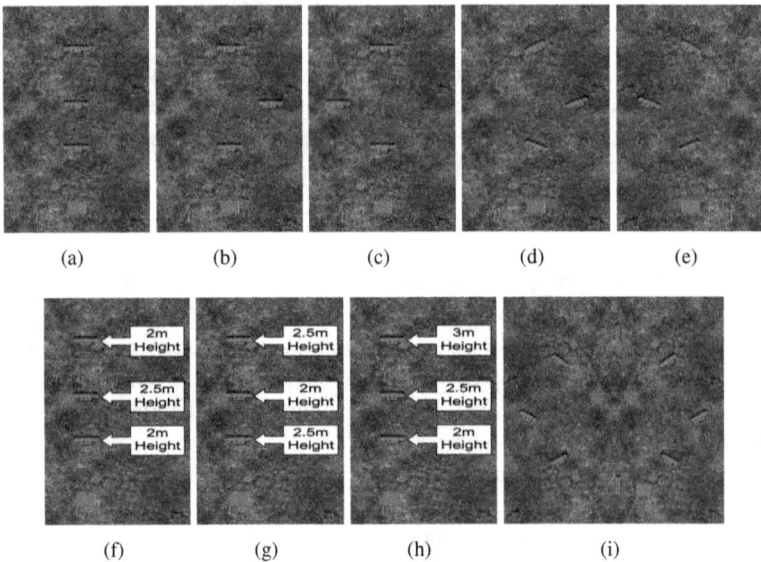

Figure 8.5 Track sections for training DeepPilot models: (a) straight segments with gates at 2, 2.5, and 3m; (b) right zig-zag segments; (c) left zig-zag segments; (d) right curve segments; (e) left curve segments; (f-h) straight segments with elevation changes; and (i) circle example.

Table 8.2

Data distribution that composes the PI dataset to improve the performance of DeepPilot. The dataset comprises 614 images from straight lines, 1127 from zig-zag sections, 1132 from curves, 1040 from cricle and 603 from elevation changes. Each image is associated with a flight signal provided by the PI controller.

		Racetrack Section						
		Zig-Zag		Curve		Circle		
Heigth	Line	Right	Left	Right	Left	Right	Left	UP & Down
2m	207	189	183	191	193	512	528	207
2.5m	202	185	190	189	192	-	-	201
3m	205	195	185	177	190	-	-	195
Total of Images per section	614	569	558	557	575	512	528	603

three gates, where gates 1 and 3 are 2m high, and the middle gate is 2.5m high. The next section, depicted in Figure. 8.5(g), also has three gates, but this time gates 1 and 3 are 2.5m high. The last section consists of a straight segment where gate height increases gradually; gate 1 is 2m high, gate 2 is 2.5m high, and the final gate is 3m high, as shown in Figure. 8.5(h). Additionally, we included a circular track to obtain different yaw values, as seen in Figure. 8.5(i). This track covers an area of 20m × 20m.

To avoid drone collisions, we conducted ten runs of an exploration process, see Figure 8.4(a), to determine the entry and exit points of the blind zone. The average of these waypoints was used as a reference during the navigation and dataset generation process, as depicted in Figure. 8.4(b). We then captured two new datasets, the first involved associating the image mosaic with a flight signal provided by the PI controller, and the second involved associating the image mosaic with a flight signal provided by the MPC. Table 8.2 shows the number of images collected from each track segment for the PI dataset, which contains 4,516 images from 18 racetrack segments and circle examples. Similarly, the MPC dataset contains 5,885 images from the same segments, as shown in Table 8.3.

Finally, Table 8.4 compares the flight commands used in the original DeepPilot dataset against the flight command values of the two new datasets, MPC and PI. In this comparison, it can be observed that the new datasets require fewer mosaic images than the original dataset, and the range of flight commands is smaller. Specifically, the range in roll goes from 0.2 to 0, in pitch from 1 to 0, in yaw from 1 to 0, and in altitude from 0.3 to 0. This distribution suggests that the waypoint controller primarily controls pitch and yaw, whereas height and roll do not require constant change. This behaviour is similar to a human pilot's, where the control signals prioritise yaw and pitch. In contrast, the MPC dataset prioritises roll and pitch, which helps keep the drone close to the trajectory.

Table 8.3

Data distribution that composes the MPC dataset to improve the performance of DeepPilot. The dataset comprises 868 images from straight lines, 1484 from zig-zag sections, 1464 from curves, 1226 from circle and 603 from elevation changes. Each image is associated with a flight signal provided by the MPC.

		Racetrack Section						
		Zig-Zag		Curve		Circle		
Heigth	Line	Right	Left	Right	Left	Right	Left	UP & Down
2m	266	252	249	243	246	612	614	269
2.5m	303	233	254	241	247	-	-	285
3m	299	248	248	246	241	-	-	289
Total of Images per section	868	733	751	730	734	612	614	843

Table 8.4

Distribution of ground truth flight command (Flight CMD) values associated as labels to the images in the original dataset used in [523] to train DeepPilot and the two new datasets (PI and MPC) recorded during a navigation and dataset generation process.

Dataset	Total of Mosaic Images	Flight CMD	Roll (S_ϕ)	Pitch (S_θ)	Yaw (S_ψ)	Altitude (S_h)
Original	10,334	mean	0.028	**0.232**	0.0	0.0023
		std	0.296	**0.357**	0.036	0.141
		max	± 0.9	$+1$	± 0.1	± 0.1
		min	± 0.1	± 0.1	± 0.05	± 0.05
PI	4,516	mean	0.004	**0.918**	0.006	0.044
		std	0.058	0.138	**0.259**	0.131
		max	± 0.2	$+1$	± 0.99	± 0.3
		min	0.0	0.0	0.0	0.0
MPC	5,885	mean	0.0182	**0.668**	0.0008	0.0316
		std	0.2287	**0.2882**	0.0280	0.1969
		max	± 1	$+1$	± 1	± 1
		min	0.0	0.0002	0.0	± 0.00008

After collecting the datasets, we used the DeepPilot Network available on GitHub[2]. During training, the network learns to identify image patterns and associate them with the corresponding control commands. Once the network is trained,

---

[2]https://github.com/QuetzalCpp/DeepPilot/tree/Noetic

it is used in real-time to process the drone's camera images and generate flight control commands for the drone based on the images it is capturing in real-time. These control commands are used to adjust the drone's speed, direction, and orientation to navigate the racetrack autonomously. We trained the PI and MPC DeepPilot models for 100 epochs using the following parameters: the batch size of 60, Adam optimiser, learning rate of 0.001, epsilon of 1e-08, and clip value of 1.5. Finally, the loss function used for each branch is shown in equation 8.11, where S_ϕ, S_θ, S_ψ, and S_h corresponds to the flight command values for each image (I) recorded when piloting the drone using the PI controller and MPC, and $\hat{S}_\phi, \hat{S}_\theta, \hat{S}_\psi$, and $\hat{S}_h$ are the flight command predicted by the model. The loss function is evaluated four times, once for each control command: S_ϕ, S_θ, S_ψ, and S_h. Thus, DeepPilot predicts values for $S_\phi, S_\theta, S_\psi, S_h$, and each variable falls within the $[-1, 1]$ range.

$$loss(I) = \alpha||\hat{S}_\phi - S_\phi||_2 + \alpha||\hat{S}_\theta - S_\theta||_2 + \alpha||\hat{S}_\psi - S_\psi||_2 + \alpha||\hat{S}_h - S_h||_2 \quad (8.11)$$

We also multiply alpha to each term of the loss function to control the influence of each variable on the total loss function. Assigning a value of 0.3 to each term of the loss function indicates that each variable similarly influences the total loss function. Moreover, this technique modifies the loss function by adding a penalty term favouring smaller network weights and avoiding overfitting [211]. As a result, the new DeepPilot models require no more than 100 epochs and no more than 6,000 training images to obtain a satisfactory result. This is a significant improvement compared to the original DeepPilot, which required three specialised models, 500 training epochs, and 10,334 images.

8.4.3   EVALUATION RACETRACK DESCRIPTION

We have used two racetracks to evaluate our approach. In Figure 8.6, we show the characteristics of the gates used in the racetracks, and each gate has a square frame measuring $1m \times 1m \times 0.05m$ with three different heights, 2m, 2.5m and 3m. Depending on the complexity of the racetrack, the gates vary in height. For example, the first racetrack consists of five gates, each measuring two metres in height and positioned at varying angles to form a curved shape, as depicted in Figure 8.7, within an area of

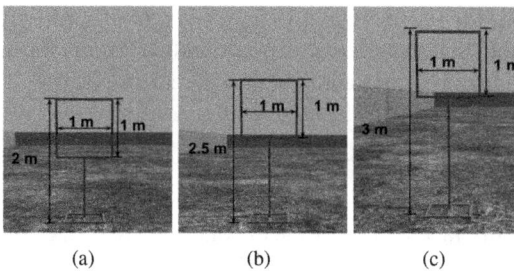

Figure 8.6   The dimensions of the gates. Each gate has a square frame of $1m \times 1m \times 0.05m$.

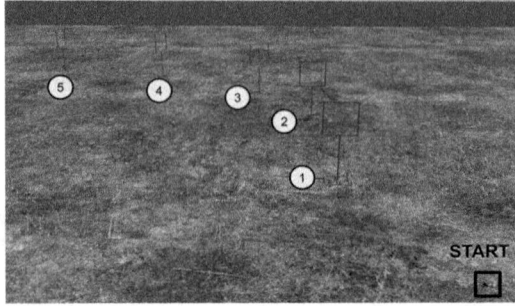

Figure 8.7 The figure shows the initial racetrack test with five gates placed at two meters, varying orientations. The track measures $25m \times 35m$, with gate spacing of $7.2m$ to $9.03m$. Each gate has a $1m \times 1m \times 0.05m$ square frame. Highlighted gates indicate the crossing order.

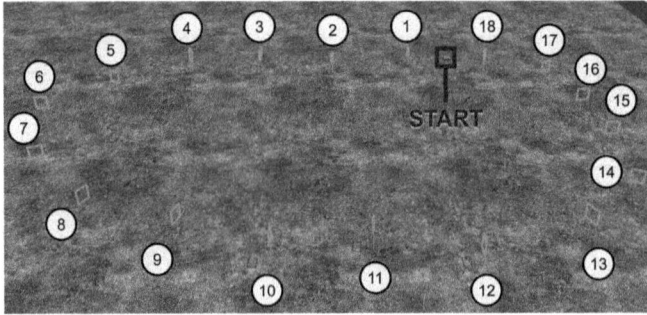

Figure 8.8 Racetrack, composed of 18 gates at different heights and orientations in the RotorS simulator. The racetrack extends over $68m \times 49m$ and has gate spacing of $6m$ to $15m$. Note that we highlighted the gates to indicate the order in which the gates have to be crossed.

$30m \times 20m$. The second racetrack is an elliptical shape included in the official Deep-Pilot repository [3]. This track has a reduced crossing space, increasing the difficulty of racing, with spacing between gates ranging from 6 to 15 metres. The racetrack features 18 gates positioned at different heights and angles, as shown in Figure 8.8, within an area of $68m \times 49m$. In addition, Gates 1, 2, 3, 5, 9, 12 and 14 have a height of 2m, Gates 4, 7, 8, 11, 13, 15, 16 and 18 have a height of 2.5m, and Gates 6, 10 and 17 are 3m high. Note that this track involves significant turns for the drone, making it challenging to maintain a constant speed, as in real racetracks.

8.4.4 EXPERIMENTS CONSTRAINTS

We have defined three constraints for our experiments. The first constraint, No prior information about the racetrack; this means that the area, size, shape, and number

---

[3]https://github.com/QuetzalCpp/DeepPilot/tree/Noetic

of the gates are unknown to DeepPilot. The second constraint is that DeepPilot has no knowledge of the dimensions, positions, and orientations of the gates. Finally, the third constraint is that DeepPilot navigates without external feedback.

8.4.5   RESULTS

In this section, we present the results obtained. We compare the performance of MPC, PI, and the two new DeepPilot models. First, we evaluate the trajectories generated by DeepPilot against the controller from which it learned flight commands. Then, we compare the flight commands executed by the controllers and DeepPilot. Next, we show a comparison of the trajectories of the controllers and the DeepPilot models against the performance of a human pilot. Finally, we report the time and speed it took to complete each of the evaluation tracks.

After obtaining the DeepPilot models using the PI and MPC datasets, we conducted three evaluations on each test track. We obtained a set of waypoints using the method described in Section 8.3.2 for each test racetrack to use only as a reference by PI and MPC controllers. First, we the performance of the model trained with the PI dataset by comparing DeepPilot (PI) (light green line) against the PI controller (green line) on the first evaluation racetrack, which was a curve composed of five gates placed at a height of 2m. Figure 8.9 shows the trajectory graphs, with the first image displaying a top view and the second image showing a side view. The graphs indicate that DeepPilot (PI) exhibits repetitive behaviour and slight roll variations while also showing some similarities in trajectory shape compared to the PI controller. For example, the movements between gates one and gate three.

Figure 8.10 shows a comparison of the flight commands (roll, pitch, yaw, and altitude) executed by the PI controller and DeepPilot (PI). In particular, DeepPilot (PI) not only learned how the PI controller crosses the gates, but also maintains

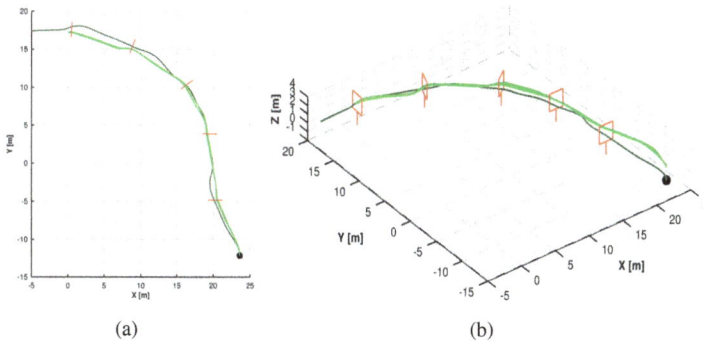

(a)   (b)

Figure 8.9   Comparison between ten trajectories executed by DeepPilot trained with PI flight commands (light green) and PI controller (green) using the waypoints discovered using the strategy described in Section 8.3.2. The black point indicates the start point. Figure (a) shows a top view of the comparison, and Figure (b) shows a side view.

(a) PI Controller

(b) DeepPilot Trained using PI Flight Commands

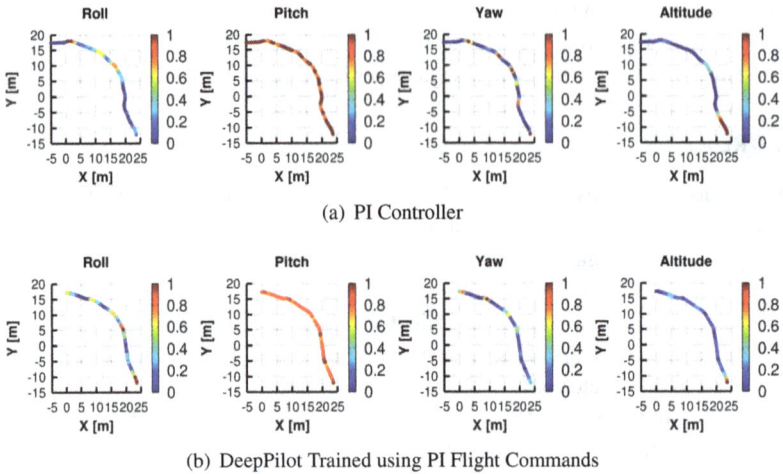

Figure 8.10 Comparison of the ranges of flight commands executed on the curve racetrack, where the maximum value is 1. Figure (a) shows the flight command value executed by the PI controller, and Figure (b) shows the range values executed by DeepPilot trained using PI flight commands.

its pitch speed at a constant value, close to the maximum, and reduces the pitch value when entering the blind spot zone, similar to the PI controller. Additionally, DeepPilot (PI) presents constant values close to 0 in the yaw flight command because the value only increases to 0.4 when it needs to change its direction, such as when crossing the gate. The height value remains at 1 at the beginning of the runway, and both navigators maintain its value at 0 since the gates do not require a height change. Finally, the roll value is the control command that presents more variations, particularly when approaching the gate.

The second evaluation involved testing the DeepPilot model trained with the MPC dataset, where we compared its performance against the MPC on the same curved track. In Figure 8.11, the left image shows ten trajectories executed by DeepPilot (MPC) in cyan blue and ten trajectories performed by the MPC. Since there were no external environmental disturbances, the controller generated ten equal trajectories. The right image shows a side view of the trajectories, where DeepPilot (MPC) exhibits variations in roll and altitude. This is more evident in Figure 8.12, where it can be seen that DeepPilot (MPC) frequently adjusts its height value, while the MPC maintains a constant altitude. The yaw command also presents variations along the track for DeepPilot (MPC), whereas the MPC's yaw value only changes orientation in the last two gates. Roll exhibits slight oscillations, and in one instance, DeepPilot (MPC) failed to cross gate 3, as it oscillated between gates. Finally, as with the MPC, the pitch command keeps its velocity constant, with a value close to 1.

In Figure 8.13, we show the comparison of the performance of five pilots on the curve racetrack: a human pilot (magenta), PI controller (green), DeepPilot (PI)

(a)   (b)

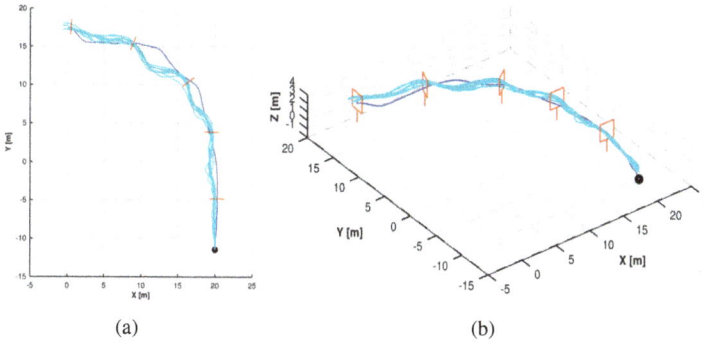

Figure 8.11 Comparison between ten trajectories executed by DeepPilot trained with PI flight commands (cyan) and MPC (blue) using the waypoints discovered using the strategy described in Section 8.3.2. The black point indicates the start point. Figure (a) shows a top view of the comparison, and Figure (b) shows a side view.

(light green), MPC (blue), and DeepPilot (MPC) (cyan blue). We provide data on the trajectory executed by each pilot. The human pilot has extensive experience flying real quadrotors indoors and outdoors and is well-versed in flying drones using the RotorS simulator. They had the same number of opportunities (ten) as the artificial pilots to fly on the racetrack used in these experiments. In addition, we show the average number of waypoints discovered in ten runs during the exploration process

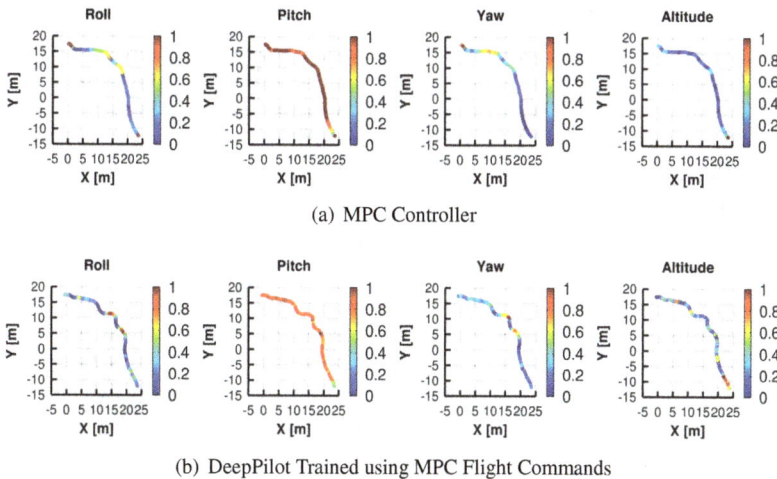

(a) MPC Controller

(b) DeepPilot Trained using MPC Flight Commands

Figure 8.12 Comparison of the ranges of flight commands executed on the curve racetrack, where the maximum value is 1. Figure (a) shows the flight command value executed by the MPC, and Figure (b) shows the range values executed by DeepPilot trained using MPC flight commands.

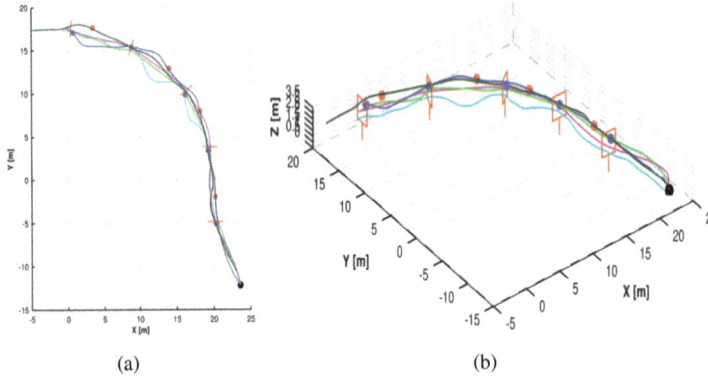

(a)   (b)

Figure 8.13 Comparison of trajectories performed by the human pilot (magenta), PI controller (green), DeepPilot (PI) (light green), MPC (blue), and DeepPilot (MPC) (cyan). The left figure shows a top view, and the right figure shows a side view.

using the strategy presented in [525]. It is important to note that the performance of both DeepPilot (PI) and DeepPilot (MPC) shows a significant improvement over the original DeepPilot and smoother movements when cornering. It even does not require the drone to be perpendicular to the gate.

To evaluate the performance of the DeepPilot models, we chose a more challenging track with gates at varying orientations and heights. We first evaluated the DeepPilot (PI) model, which demonstrated high repeatability and efficiency in completing the track, despite variations in altitude, see Figure 8.14. We compared the flight commands executed by the PI controller with those estimated by DeepPilot (PI). While pitch maintained a constant value close to the maximum, DeepPilot (PI) exhibited constant variations in roll, yaw, and altitude, see Figure 8.15.

DeepPilot (MPC) maintained a constant altitude but showed oscillations between gates. In Figure 8.16, it can be observed that both MPC and DeepPilot (MPC) kept values below 0.6, only passing this threshold when a gate had a height of 3m. Yaw remained stable and only increased in the curved zone. Roll had the most significant variations, reaching its maximum value when the gate was at the edge of the camera view. However, the maximum speed of 0.7 affected pitch, which was only able to use 40% of its original value to complete the track, given the drone's high speed and inertia.

Figure 8.18 compares the trajectories performed by a human pilot (magenta), a PI controller (green), DeepPilot (PI) (light green), MPC (blue), and DeepPilot (MPC) (cyan blue) on the second racetrack. Additionally, the entry (blue) and exit (red) points discovered in the blind spot zone of each gate are shown. These points were discovered with one lap. The figure shows that the human pilot performs a more stable trajectory than the other pilots. However, the PI controller and MPC exhibit stable movements similar to those of the human pilot. On the other hand, DeepPilot (MPC) and DeepPilot (PI) show some oscillations between gates.

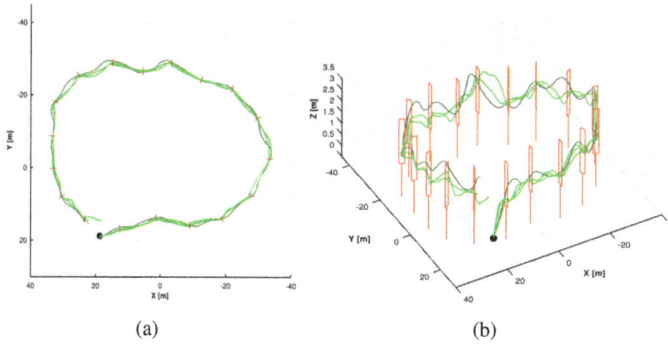

(a)   (b)

Figure 8.14   Comparison between ten trajectories executed by DeepPilot trained with PI flight commands (light green) and PI controller (green) using the waypoints discovered using the strategy described in Section 8.3.2. The black point indicates the start point. Figure (a) shows a top view of the comparison, and Figure (b) shows a side view.

In order to compare the performance of the different pilots, including the Human Pilot, PI controller, DeepPilot (PI), MPC, and DeepPilot (MPC), we summarised their results for each evaluation racetrack in Table 8.5. The Human Pilot achieved the fastest completion time, while the second fastest was the PI controller, completing the curve racetrack in 19.14 seconds and the ellipse racetrack in 72.04 seconds. Table 8.6 also shows the speed of each pilot, with the PI controller reaching a top speed of 2.81 m/s and the Human Pilot achieving 2.65 m/s. However, the Human

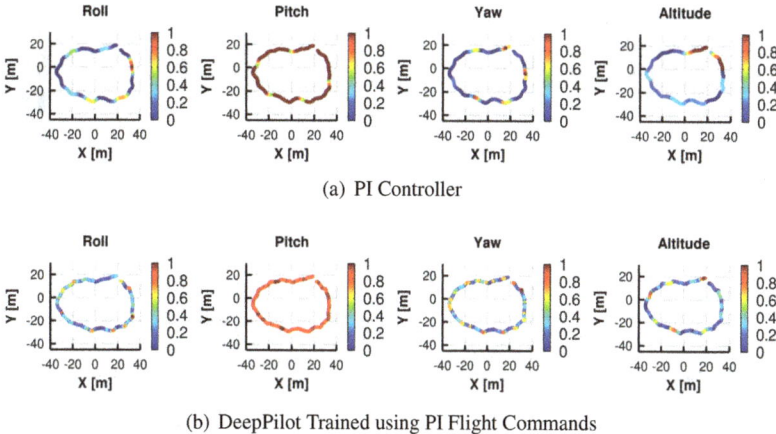

(a) PI Controller

(b) DeepPilot Trained using PI Flight Commands

Figure 8.15   Comparison of the ranges of flight commands executed on the elliptical racetrack, where the maximum value is 1. Figure (a) shows the flight command value executed by the PI controller, and Figure (b) shows the range values executed by DeepPilot trained using PI flight commands.

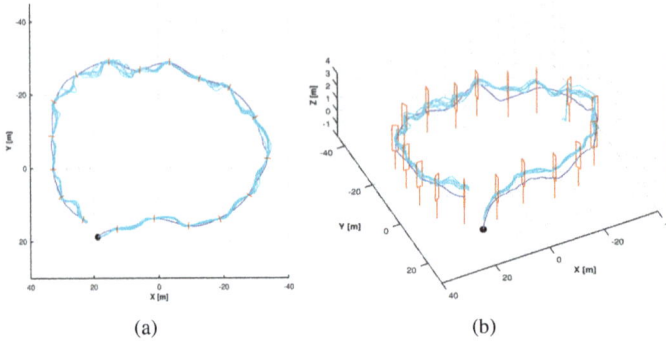

(a)   (b)

Figure 8.16 Comparison between ten trajectories executed by DeepPilot trained with MPC flight commands (cyan) and MPC (blue) using the waypoints discovered using the strategy described in Section 8.3.2. The black point indicates the start point. Figure (a) shows a top view of the comparison, and Figure (b) shows a side view.

Pilot reached a higher top speed of 3.0 m/s in the ellipse racetrack. Additionally, the new DeepPilot models, including DeepPilot (PI) and DeepPilot (MPC), showed significant improvements, with top speeds of 0.90 m/s and 0.89 m/s, respectively, in the ellipse racetrack, compared to the original DeepPilot which only achieved 0.41 m/s.

The results show that DeepPilot is able to learn flight commands irrespective of the source, be it a human pilot, a line follower controller, or a trajectory follower. This success is due to the set of images that provide temporal information (mosaic image),

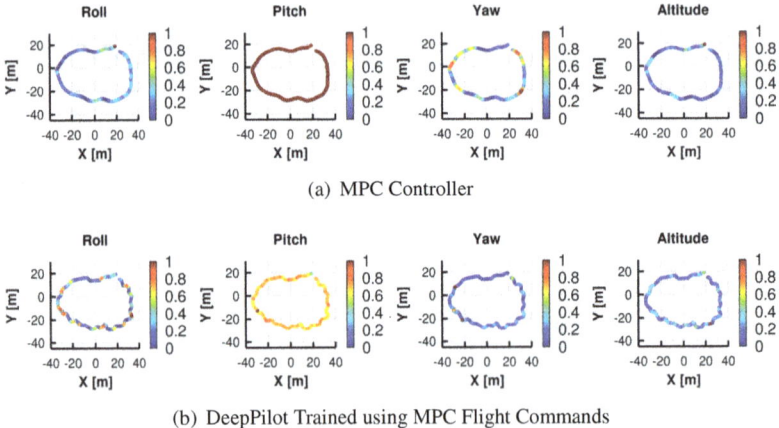

(a) MPC Controller

(b) DeepPilot Trained using MPC Flight Commands

Figure 8.17 Comparison of the ranges of flight commands executed on the elliptical racetrack, where the maximum value is 1. Figure (a) shows the flight command value executed by the MPC, and Figure (b) shows the range values executed by DeepPilot trained using MPC flight commands.

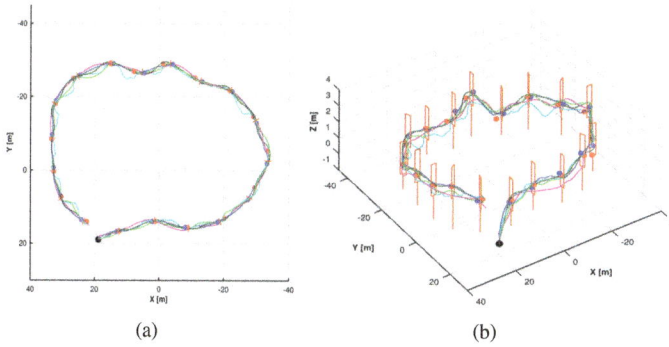

(a)   (b)

Figure 8.18 Comparison of trajectories performed by the human pilot (magenta), PI controller (green), DeepPilot (PI) (light green), MPC (blue), and DeepPilot (MPC) (cyan blue). The left figure shows a top view, and the right figure shows a side view.

which considers the current frame and previous frames. This image sequence enables DeepPilot to estimate flight commands in real time. However, DeepPilot learning from flight control systems provides the following five advantages: 1) it simplifies the collection of training data; 2) it cuts down the required training data by up to 57%; 3) it reduces the training time (20 times less); 4) it only requires one model to provide the four flight commands, and 5) it increases the speed of completing a racetrack by a factor of seven compared to the original model.

8.5   CONCLUSIONS

We have presented a learning-based methodology for the problem of Autonomous Drone Racing (ADR). In this approach, visual data is combined with control techniques to combine past, current, and future elements of the racetrack that is being traversed with the goal of learning control inputs to fly the drone autonomously without human intervention. This methodology follows three stages: exploration, navigation, and retraining. None of the stages provides information about the gates (i.e. their position, orientation, height, or number), only the drone's position on the racetrack.

Table 8.5

Time comparison between Human Pilot, PI controller, DeepPilot (PI), MPC, and DeepPilot (MPC), the values are in seconds.

Racetrack	Number of Gates	PI Ctrl	DP (PI)	MPC Ctrl	DP (MPC)	Human Pilot
Curve	5	19.14	69.27	21.08	49.43	16.35
Ellipse	18	72.04	206.90	76.49	223.13	62.64

Table 8.6

Speed comparison between Human Pilot, PI controller, DeepPilot (PI), MPC, and DeepPilot (MPC), the values are in m/s.

Racetrack	Number of Gates	PI Ctrl	DP (PI)	MPC Ctrl	DP (MPC)	Human Pilot
Curve	5	2.81	0.61	2.06	0.90	2.65
Ellipse	18	2.6	0.90	2.39	0.89	3.0

In the exploration stage, we use a single-shot detector for the drone to visually detect the gates on the racetrack and, in combination with the neural pilot, automatically discover the entry and exit positions of what we call the 'blind spot zone', where the gate is no longer visible during the crossing. Once these positions are discovered, we use the navigation stage where the waypoints found in the exploration stage serve as a reference for a flight controller to perform much faster flight. For this part, we carried out experiments testing with a proportional controller that follows line-segments formed by the waypoints. We also used a MPC to follow a trajectory built with splines calculated from the waypoints. For both controllers, we generated a new dataset to associate the temporal images captured in a mosaic image with flight commands. DeepPilot was trained with each dataset to assess the effect of each controller in combination with the visual data.

Our results show that DeepPilot substantially improves its performance, exhibiting a faster flight than when learning basic movements for one gate only. The results demonstrate that the Improved DeepPilot navigates naturally and smoothly, resembling human pilots' behavior. Furthermore, the new version does not require the drone to be perpendicular to the gate to cross it, as demonstrated by the provided examples. Additionally, the control signals prioritise yaw and pitch signals, eliminating ambiguity in roll and yaw. Therefore, unlike the original DeepPilot, which required three specialised models, the Improved DeepPilot only requires a single model to navigate the track. Also, we reduce training data by up to 57% and increase the speed of completing a racetrack by a factor of seven compared to the original model.

9 Optimization of UAV Flight Controllers for Trajectory Tracking by Metaheuristics

Jonathan Daniel Díaz-Muñoz
Department of Electronics, INAOE

Oscar Martínez-Fuentes
Department of Electronics, INAOE, Luis Enrique Erro No. 1, Santa María Tonantzintla, San Andrés Cholula, 72840, Puebla, México

Israel Cruz-Vega
Department of Electronics, INAOE

CONTENTS

9.1 INTRODUCTION

Unmanned aerial vehicles (UAVs) are considered aerial robots able to travel long distances at relatively high speeds. UAVs are preferred over other robotics structures due to their agility, reaching difficult access places, carrying some loads, capturing information at different heights, and autonomous exploration. Some of the applications of

DOI: 10.1201/9781003385615-9

UAVs are, for example, in the entertainment industry, to record aerial images, or in the transportation industry, where UAVs deliver packages [22, 348]. Another field of application is land surveillance, where the UAVs help explore and cover terrain [2] by using different communication systems [141, 204, 527].

However, UAVs are complex underactuated dynamic systems, presenting uncertainties in the mathematical model and suffering from external disturbances from weather conditions or other factors affecting the performance of the flight controller. Additionally, UAVs vary their physical characteristics, such as size, shape, weight, and the number of motors affecting their dynamic behavior. All this leads the research efforts to obtain accurate mathematical models and robust control techniques [97]. The Multi-rotor UAVs (Bicopter, Tricopter, Quadcopter,..., Octocopter), particularly Quadcopters, are the most commercial due to their easy manufacturing and vertical takeoff capacity. Due to this advantage, they are used in various applications.

The UAV flight control technique is a relevant part where the performance is mainly affected by the UAV mathematical model, controller parameter setting, external disturbances, and the complexity of the desired trajectory. There are several control techniques, where the choice depends on the system and the control specifications. Generally, the UAV flight control aims to reach a reference or to track trajectories. Common control techniques are PID control [283], fuzzy control [95], Model Predictive Control (MPC) [528], and Sliding Mode Control (SMC) [204] and their improvements [325, 326, 423, 447].

Some examples of research where intelligent control has excellent results appear in [429], where the authors employ reinforcement learning (RL) for a robot to learn a target location. In [584], authors use genetic algorithms (GA) to adjust the scan parameters based on the pheromone dispersion mobility model. In contrast, the particle swarm optimization (PSO) algorithm estimates the model parameters of an autonomous underwater vehicle (AUV) in [432]. On the other hand, the sliding mode control problem is combined with the dynamic differential evolutionary (DDE) algorithm to adjust some control parameters and gains [598], and the classical PID control problem is solved in [160] by using a reformative artificial Bee colony (RABC) algorithm to optimize the control parameters.

Inspired by our past efforts, in this work, we propose optimizing UAV flight control's performance by applying metaheuristics, comparing two of the most relevant AI techniques, differential evolution (DE) and accelerated particle swarm optimization (APSO) algorithms, also applying two control techniques, sliding mode Control (SMC) and proportional-Integral-derivative (PID) control.

The rest of this chapter is organized as follows: the UAV model development is presented in Section 9.2. In Sections 9.3 and 9.4, the control techniques and the optimization methodology by metaheuristics to adjust the controller parameters are developed, respectively. In the Section 9.5, we present the results and discussion of different tests for adjusting the controller parameters by metaheuristics and test trajectory tracking. Finally, the main conclusions are discussed in the Section 9.6.

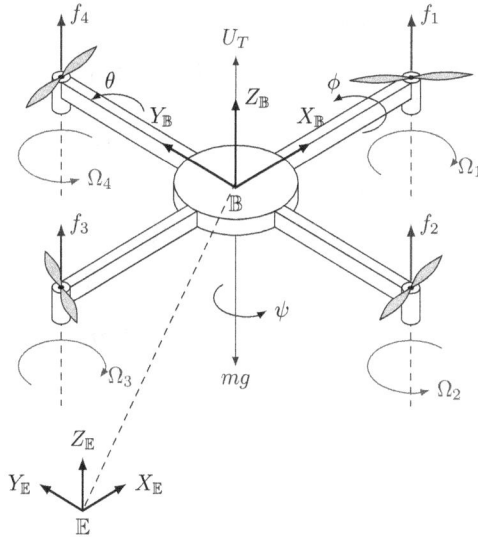

Figure 9.1   Quadrotor (UAV) reference scheme.

9.2   MATHEMATICAL MODEL

The task of the control engineer can be carried out jointly and interdisciplinary. The control objectives must be established, and then the variables to be controlled must be identified. In this section, a dynamic model associated with the UAV in Figure 9.1 is presented, describing its translational and rotational behavior. Although it is well known that there are various ways to perform mathematical models, in this type of system, the use of Euler-Lagrange formalism and the Newton-Euler method is widely used [6, 96, 714].

There are different configurations of UAVs. Ideally, a UAV of n rotors consists of n propulsors and their propellers, which are positioned symmetrically at the vertices of an n-sided polygon. In particular, for a quadrotor, the motors are positioned at the vertices of a square, where the propellers generate thrust in a direction perpendicular to the plane defined by the rotors. Consider the coordinate scheme associated with the quadrotor of Figure 9.1, with the inertial frame of reference $\{\mathbb{E}\}$, and $\{\mathbb{B}\}$, the frame of reference associated with the quadrotor. As we can see, $\{\mathbb{B}\}$ is fixed to the quadrotor center of mass, where $X_{\mathbb{E}}$, $X_{\mathbb{B}}$, $Y_{\mathbb{E}}$, $Y_{\mathbb{B}}$ and $Z_{\mathbb{E}}$, $Z_{\mathbb{B}}$ are the coordinates associated with the x, y, and z-axis, respectively, to each frame.

Let $\mathbf{q}$ be the vector of generalized coordinates of the quadrotor [96]:

$$\mathbf{q} = \left[\eta, \xi\right]^{\mathsf{T}} = \left[\phi, \theta, \psi, x, y, z\right]^{\mathsf{T}} \in \mathbb{R}^6, \tag{9.1}$$

where $\xi = \left[x, y, z\right]^{\mathsf{T}} \in \mathbb{R}^3$ denotes the position of the center of mass relative to the inertial frame, and $\eta = \left[\phi, \theta, \psi\right]^{\mathsf{T}} \in \mathbb{R}^3$ denoting the three Euler angles: ϕ (roll) around the x-axis, θ (pitch) around the y-axis, and ψ (yaw) around the z-axis, respectively. By using the Euler-Lagrange formalism, one has that

$$m\ddot{\xi} = \begin{bmatrix} c_\phi c_\psi s_\theta + s_\phi s_\psi \\ c_\phi s_\psi s_\theta - s_\phi c_\psi \\ c_\phi c_\theta \end{bmatrix} U_T - \begin{bmatrix} 0 \\ 0 \\ mg \end{bmatrix}, \tag{9.2}$$

with $c_\gamma = \cos(\gamma)$ and $s_\gamma = \sin(\gamma)$, and U_T represents the sum of the thrust forces generated by the rotors. Let $D_\xi = -\frac{1}{m}K_\xi\dot{\xi} + d_\xi$ be the external disturbances, then the translational movement of the quadrotor is given by [325]:

$$\ddot{x} = -\frac{1}{m}K_x\dot{x} + \frac{1}{m}\left(c_\phi c_\psi s_\theta + s_\phi s_\psi\right)U_T + d_x$$
$$\ddot{y} = -\frac{1}{m}K_y\dot{y} + \frac{1}{m}\left(c_\phi s_\psi s_\theta - s_\phi c_\psi\right)U_T + d_y \tag{9.3}$$
$$\ddot{z} = -\frac{1}{m}K_z\dot{z} + \frac{1}{m}\left(c_\phi c_\theta\right)U_T - g + d_z$$

On the other hand, the rotational dynamics associated with $\eta = [\phi, \theta, \psi]^\mathsf{T}$ is described using the Newton-Euler equation:

$$\begin{bmatrix} I_x\ddot{\phi} \\ I_y\ddot{\theta} \\ I_z\ddot{\psi} \end{bmatrix} = \begin{bmatrix} \dot{\theta}\dot{\psi}(I_y - I_z) \\ \dot{\phi}\dot{\psi}(I_z - I_x) \\ \dot{\phi}\dot{\theta}(I_x - I_y) \end{bmatrix} - \begin{bmatrix} K_\phi\dot{\phi}^2 \\ K_\theta\dot{\theta}^2 \\ K_\psi\dot{\psi}^2 \end{bmatrix} - \begin{bmatrix} J_r\Omega_r\dot{\theta} \\ -J_r\Omega_r\dot{\phi} \\ 0 \end{bmatrix} + \begin{bmatrix} \tau_\phi \\ \tau_\theta \\ \tau_\psi \end{bmatrix}, \tag{9.4}$$

where I_x, I_y, and I_z represent the moments of inertia about the axes x, y, and z, respectively. In addition, the friction coefficients are represented by K_ϕ, K_θ, and K_ψ, J_r is the rotor inertia and $\Omega_r = \Omega_1 - \Omega_2 + \Omega_3 - \Omega_4$. Considering the external disturbances $D_\eta = [d_\phi, d_\theta, d_\psi]^\mathsf{T}$, a state-variable description is easily obtained by introducing variables $\mathbf{X} = [x_1, x_2, x_3, x_4, x_5, x_6, x_7, x_8, x_9, x_{10}, x_{11}, x_{12}]^\mathsf{T} = [\phi, \dot{\phi}, \theta, \dot{\theta}, \psi, \dot{\psi}, x, \dot{x}, y, \dot{y}, z, \dot{z}]^\mathsf{T}$. Then, from (9.3) and (9.4):

$$\left. \begin{aligned} \dot{x}_1 &= x_2 \\ \dot{x}_2 &= k_1 x_4 x_6 + k_2 x_2^2 + k_3 x_4 + \rho_1\tau_\phi + d_\phi \\ \dot{x}_3 &= x_4 \\ \dot{x}_4 &= k_4 x_2 x_6 + k_5 x_4^2 + k_6 x_2 + \rho_2\tau_\theta + d_\theta \\ \dot{x}_5 &= x_6 \\ \dot{x}_6 &= k_7 x_2 x_4 + k_8 x_6^2 + \rho_3\tau_\psi + d_\psi \\ \dot{x}_7 &= x_8 \\ \dot{x}_8 &= k_9 x_8 + \frac{1}{m}\left(c_{x_1}s_{x_3}c_{x_5} + s_{x_1}s_{x_5}\right)U_T + d_x \\ \dot{x}_9 &= x_{10} \\ \dot{x}_{10} &= k_{10}x_{10} + \frac{1}{m}\left(c_{x_1}s_{x_3}s_{x_5} - s_{x_1}c_{x_5}\right)U_T + d_y \\ \dot{x}_{11} &= x_{12} \\ \dot{x}_{12} &= k_{11}x_{12} - g + \frac{1}{m}\left(c_{x_1}c_{x_3}\right)U_T + d_z \end{aligned} \right\} \tag{9.5}$$

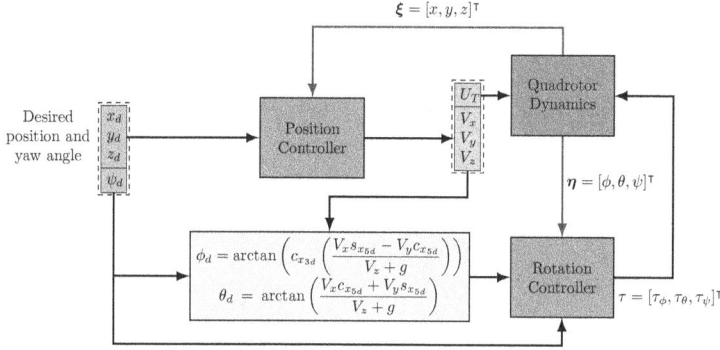

Figure 9.2 General control scheme for UAVs.

where $k_1 = \dfrac{I_y - I_z}{I_x}$, $k_2 = -\dfrac{K_\phi}{I_x}$, $k_3 = -\dfrac{J_r \Omega_r}{I_x}$, $k_4 = \dfrac{I_z - I_x}{I_y}$, $k_5 = -\dfrac{K_\theta}{I_y}$, $k_6 = \dfrac{J_r \Omega_r}{I_y}$, $k_7 = \dfrac{I_x - I_y}{I_z}$, $k_8 = -\dfrac{K_\psi}{I_z}$, $k_9 = -\dfrac{K_x}{m}$, $k_{10} = -\dfrac{K_y}{m}$, $k_{11} = -\dfrac{K_z}{m}$, $\rho_1 = \dfrac{1}{I_x}$, $\rho_2 = \dfrac{1}{I_y}$, $\rho_3 = \dfrac{1}{I_z}$. The numerical values associated to the parameters employed during this analysis are $g = 9.81 \,\text{m/s}^2$, $m = 0.74 \,\text{kg}$, $l = 0.2 \,\text{m}$, $I_x = I_y = 0.004 \,\text{kg m}^2$, $I_z = 0.0084 \,\text{kg m}^2$, $J_r = 2.8385 \times 10^{-5} \,\text{s N/m}$, $K_x = K_y = K_z = 5.5670 \times 10^{-4} \,\text{s N/m}$, $K_\phi = K_\theta = K_\psi = 5.5670 \times 10^{-4} \,\text{s N/rad}$, $b_d = 2.9842 \times 10^{-3} \,\text{N s}^2$ and $c_d = 3.2320 \times 10^{-2} \,\text{N m s}^2$.

9.3   CONTROL ALGORITHMS

Based on the desired trajectory design established in the previous section, we propose some control solutions in this section. Figure. 9.2 shows the general control scheme for UAVs. There we see that this system needs two control loops, one for position or translation and another for angles or rotation.

According to model (9.5), the control inputs are defined by τ_ϕ, τ_θ, and τ_ψ. These inputs directly influence the rotational dynamics and U_T for the translational dynamics. Furthermore, this dynamic is affected by the rotation angles, i.e., according to the value of the angles $x_1 = \phi$, $x_3 = \theta$. The system will have a translational displacement ($x_7 = x$, $x_9 = y$, $x_{11} = z$). For this reason, the terms with U_T are called the position's virtual control, given by

$$\begin{bmatrix} V_x \\ V_y \\ V_z \end{bmatrix} = \begin{bmatrix} \dfrac{1}{m}\left(c_{x_1} s_{x_3} c_{x_5} + s_{x_1} s_{x_5}\right) U_T \\ \dfrac{1}{m}\left(c_{x_1} s_{x_3} s_{x_5} - s_{x_1} c_{x_5}\right) U_T \\ \dfrac{1}{m}\left(c_{x_1} c_{x_3}\right) U_T \end{bmatrix}. \tag{9.6}$$

To control the position of UAVs, we can apply the control law to the virtual position control terms, i.e., obtain the values of V_x, V_y, and V_z by the control law, and with these values, calculate U_T and the angles ϕ, θ necessary to have the desired translational displacement. From expressions (9.6), it follows that

$$
\begin{aligned}
U_T &= m\sqrt{V_x^2 + V_y^2 + (V_z + g)^2} \\
x_{1d} &= \phi_d = \arctan\left(c_{x_{3d}}\left(\frac{V_x s_{x_{5d}} - V_y c_{x_{5d}}}{V_z + g}\right)\right) \\
x_{3d} &= \theta_d = \arctan\left(\frac{V_x c_{x_{5d}} + V_y s_{x_{5d}}}{V_z + g}\right)
\end{aligned}
\tag{9.7}
$$

Finally, the rotation controller receives these angles as desired states ($x_{1d} = \phi_d$, $x_{3d} = \theta_d$, $x_{5d} = \psi_d$) to compute the τ values with the control law. To establish the tracking problem, let $e_i = x_i - x_{id}$, for all $1 \leq i \leq 12$, be the tracking errors.

9.3.1 PID CONTROL

A Proportional-Integral-Derivative controller (PID controller) consists in a control action that uses the output feedback to calculate the difference between a desired trajectory (setpoint) and the measured variable (output). Then a correction based on terms associated with a proportional constant, the integral, and the derivate of the error is applied. Although there are many combinations to consider, generally, the PID control has the form

$$
u(t) = K_p e(t) + K_i \int_0^t e(\tau)\,d\tau + K_d \frac{de(t)}{dt},
\tag{9.8}
$$

with the gains K_p, K_i and K_d designed properly, assuring stability of the equilibrium point. By using the PID structure, then the position and rotation controller for the UAV, respectively, are

$$
\begin{aligned}
V_x &= K_{p_1} e_7 + K_{i_1}\int_0^t e_7\,d\tau + K_{d_1}\dot{e}_7
& \tau_\phi &= K_{p_2} e_1 + K_{i_2}\int_0^t e_1\,d\tau + K_{d_2}\dot{e}_1 \\
V_y &= K_{p_1} e_9 + K_{i_1}\int_0^t e_9\,d\tau + K_{d_1}\dot{e}_9
& \tau_\theta &= K_{p_2} e_3 + K_{i_2}\int_0^t e_3\,d\tau + K_{d_2}\dot{e}_3 \\
V_z &= K_{p_1} e_{11} + K_{i_1}\int_0^t e_{11}\,d\tau + K_{d_1}\dot{e}_{11}
& \tau_\psi &= K_{p_2} e_5 + K_{i_2}\int_0^t e_5\,d\tau + K_{d_2}\dot{e}_5
\end{aligned}
\tag{9.9}
$$

9.3.2 SLIDING MODE CONTROL (SMC)

In [325], the authors propose a sliding mode control methodology called the Adaptive Nonsingular Fast Terminal Sliding-Mode Control (ANFTSMC). This solution guarantees fast convergence (finite time), robust to external disturbances, and solve some chattering problems. For comparison purposes, using the deductive methodology exposed in [325], the controllers associated with the sliding modes technique are

$$
\left.\begin{aligned}
V_x &= -k_9 x_8 + \ddot{x}_d - \frac{1}{b_8 \beta_7} |e_8|^{2-\beta_7} \left[1 + b_7 \alpha_7 |e_7|^{\alpha_7 - 1}\right] \tanh(e_8) - c_7 \sigma_7 \\
&\quad - (a_{07} + a_{17}|e_7| + a_{27}|e_8| + h_7) \tanh(\sigma_7) \\
V_y &= -k_{10} x_{10} + \ddot{y}_d - \frac{1}{b_{10}\beta_9} |e_{10}|^{2-\beta_9} \left[1 + b_9 \alpha_9 |e_9|^{\alpha_9 - 1}\right] \tanh(e_{10}) - c_9 \sigma_9 \\
&\quad - (a_{09} + a_{19}|e_9| + a_{29}|e_{10}| + h_9) \tanh(\sigma_9) \\
V_z &= -k_{11} x_{12} + \ddot{z}_d - \frac{1}{b_{12}\beta_{11}} |e_{12}|^{2-\beta_{11}} \left[1 + b_{11}\alpha_{11}|e_{11}|^{\alpha_{11}-1}\right] \tanh(e_{12}) \\
&\quad - c_{11}\sigma_{11} - (a_{011} + a_{111}|e_{11}| + a_{211}|e_{12}| + h_{11}) \tanh(\sigma_{11})
\end{aligned}\right\}
\tag{9.10}
$$

$$
\left.\begin{aligned}
\tau_\phi &= \frac{1}{\rho_1}\left[-k_1 x_4 x_6 - k_2 x_2^2 - k_3 x_4 + \ddot{\phi}_d - \frac{1}{b_2 \beta_1}|e_2|^{2-\beta_1}\left(1 + b_1 \alpha_1 |e_1|^{\alpha_1 - 1}\right)\tanh(e_2)\right. \\
&\quad \left. -c_1 \sigma_1 - (a_{01} + a_{11}|e_1| + a_{21}|e_2| + h_1)\tanh(\sigma_1)\right] \\
\tau_\theta &= \frac{1}{\rho_2}\left[-k_4 x_2 x_6 - k_5 x_4^2 - k_6 x_2 + \ddot{\theta}_d - \frac{1}{b_4 \beta_3}|e_4|^{2-\beta_3}\left(1 + b_3 \alpha_3 |e_3|^{\alpha_3 - 1}\right)\tanh(e_4)\right. \\
&\quad \left. -c_3 \sigma_3 - (a_{03} + a_{13}|e_3| + a_{23}|e_4| + h_3)\tanh(\sigma_3)\right] \\
\tau_\psi &= \frac{1}{\rho_3}\left[-k_7 x_2 x_4 - k_8 x_6^2 + \ddot{\psi}_d - \frac{1}{b_6 \beta_5}|e_6|^{2-\beta_5}\left(1 + b_5 \alpha_5 |e_5|^{\alpha_5 - 1}\right)\tanh(e_6)\right. \\
&\quad \left. -c_5 \sigma_5 - (a_{05} + a_{15}|e_5| + a_{25}|e_6| + h_5)\tanh(\sigma_5)\right]
\end{aligned}\right\}
\tag{9.11}
$$

where $\sigma_n = e_n + b_n |e_n|^{\alpha_n} sign(e_n) + b_{n+1}|e_{n+1}|^{\beta_n} sign(e_{n+1})$, for $n \in \{1,3,5,7,9,11\}$. In addition, the parameters $1 < \beta_n < 2$ and $\alpha_n > \beta_n$. In order to reduce the chattering effects, the $sign(x)$ function is replaced by $\tanh(x)$. The numerical parameters employed in this analysis are $a_{01} = a_{03} = 0.5 \times 10^{-4}$, $a_{05} = 7.5 \times 10^{-3}$, $a_{07} = a_{09} = 2 \times 10^{-3}$, $a_{011} = 1 \times 10^{-3}$, $a_{11} = a_{13} = 0.25 \times 10^{-4}$, $a_{15} = 0.25 \times 10^{-3}$, $a_{17} = a_{19} = 1 \times 10^{-3}$, $a_{111} = 0.75 \times 10^{-3}$, $a_{21} = a_{23} = 5 \times 10^{-4}$, $a_{25} = 7.5 \times 10^{-3}$, and $a_{27} = a_{29} = a_{211} = 0.5 \times 10^{-3}$.

9.4 GAIN OPTIMIZATION BY AI TECHNIQUES

In order to show the comparatives of both controllers, in this section, we present an optimization methodology by using metaheuristics, employing the differential evolution and accelerated particle swarm optimization methods. In both cases, we define the aptitude function and some constraints that guarantee the control of the system.

Fitness Function: In control theory, the settling time t_s of the system is defined as the value of time t from which the system response remains within a threshold value established by criteria such as 2% or 5% of its final value.

The value of t_s is a measure that quantifies how fast the system reaches a desired value. In the UAV model, we have 6 outputs (x_n with $n = 1,3,5,\ldots 11$). In this case,

t_s is the maximum time establishment between these outputs, such that the fitness function is given by (9.12), with the criterion of 2% of x_d. This fitness function must be minimized by the optimization algorithms.

$$f(t_s) = \max \left\{ t_{s1}, \quad t_{s3}, \quad t_{s5}, \quad t_{s7}, \quad t_{s9}, \quad t_{s11} \right\} \tag{9.12}$$

Constraints: To help the optimization algorithms, we add some constraints to the individuals of the population so that they provide an adequate control response. The first restriction added is related to the overshoot M_p of the response given by (9.13), which is defined as the percentage value of the maximum peak that exceeds the desired final value of the system response:

$$M_p = \frac{\max\{x_n\} - x_{dn}}{x_{dn}} \times 100\%, \tag{9.13}$$

where M_p must satisfy the constraint given by

$$M_p \leq 10\%. \tag{9.14}$$

The error can be quantified with the value of the root mean square error (RMSE), which measures the dispersion of the system response to the desired value [102]. The RMSE value is given by

$$RMSE = \sqrt{\frac{1}{N} \sum (x_n(t) - x_{dn}(t))^2}, \quad \forall \ t \geq t_{sn}, \tag{9.15}$$

where $n = 5, 7, 8, 11$. In addition, the *RMSE* value must satisfy the constraint given by

$$RMSE \leq 0.01. \tag{9.16}$$

The Sliding-Mode controller of equations (9.10) and (9.11), has an extra constraint mentioned in Section 9.3.2. This restriction is:

$$\alpha_n > \beta_n, \quad \forall \ n = 1, 3, 5, \ldots, 11. \tag{9.17}$$

The constraints and the fitness function are evaluated for each individual of the meta-heuristic in function *Evaluate()*, whose pseudo-code is shown below:

```
 1. def Evaluate(_n, vpar):
 2.     f_ts = 100.0
 3.     # Constraint 1: alpha value
 4.     const1 = alpha_n > beta_n
 5.     if const1 == 'True':
 6.         # Calculate system response
 7.         X = Eval.EvalSimulation(vpar)
 8.         # Contraint 2: overshoot
 9.         M_p = Eval.EvalOvershoot(X,X_d)
10.         const2 = M_p <= 10
11.         if const2 == 'True':
12.             # Contraint 3: RMSE
13.             RMSE = Eval.EvalRMSE(X,X_d)
14.             const3 = RMSE <= 0.01
15.             if const3 == 'True':
16.                 # Calculate the fitness function value
17.                 f_ts = Eval.Evalts(X,X_d)
18. return f_ts
```

The first step is to check the constraint (9.17) or α_n value. We evaluate the simulation to obtain the system's response if this constraint is achieved. Subsequently, we evaluate the overshoot to check the constraint (9.14). If this constraint is achieved, we evaluate the RMSE value to check the constraint (9.16). Finally, if the RMSE value satisfies the condition, we evaluate the fitness function (9.12) and return the result.

9.4.1 DIFFERENTIAL EVOLUTION (DE)

Differential Evolution (DE) is part of the metaheuristics that perform a directed random search, use real numbers for representation, and offer a better approximation than *Genetic Algorithms*. Generally, the DE requires the following input parameters: population size, number of generations, recombination constant (CR), difference constant (F), and threshold value s [613].

Algorithm 7 shows the DE pseudocode to minimize the settling time t_s of the UAV flight control system. In the algorithm, the generations are updated by creating new vectors by performing mutation (9.18) and crossover (9.19) operations. Parameters a, b and c are different numbers randomly chosen; g denotes the current generation; $F \in [0,2]$ is the differential constant; $j = \{1,2,...,D\}$; $randb(j) \in [0,1]$ is the $j-th$ evaluation of a generated random number; $CR \in [0,1]$ is a crossover coefficient chosen by the user; and $rnbr(i) \in [0,D-1]$ is an index randomly generated [613].

$$v_i^{g+1} \leftarrow x_c^g + F(x_a^g - x_b^g) \tag{9.18}$$

$$u_{ij}^{g+1} \leftarrow \begin{cases} v_{ij}^{g+1} & if\,[randb(j) \leq CR]\,or\,rnbr(i) = j \\ x_{ij}^{g+1} & if\,[randb(j) > CR]\,and\,rnbr(i) \neq j \end{cases} \tag{9.19}$$

9.4.2 ACCELERATED PARTICLE SWARM OPTIMIZATION (APSO)

Similar to the DE, the APSO algorithm is part of the metaheuristics that perform a directed random search. Generally, APSO requires the following input parameters: population size, number of generations, β, and α [613]. Algorithm 8 shows the APSO pseudocode to minimize the settling time t_s of the UAV flight control system. In APSO, the particle behavior is defined by velocity (9.20) and position (9.21), where i is the index of the particle; j its dimension; p_i the best position found in i; p_g the best position found during the optimization; $\alpha \in \mathbb{R}$ is the inertial weight; $\beta \in \mathbb{R}$ the acceleration constant; and $U(\bullet)$ a random number generator with uniform distribution [613].

$$v_{ij}^{t+1} \leftarrow \alpha v_{ij}^t + U(0,\beta)(p_{ij} - x_{ij}^t) + U(0,\beta)(g_j - x_{ij}^t) \tag{9.20}$$

$$x_{ij}^{t+1} \leftarrow x_{ij}^t + v_{ij}^{t+1} \tag{9.21}$$

Algorithm 7 Differential Evolution Algorithm.

1: Initialize the population randomly ($\mathbf{x}$)
2: Evaluate $f(t_s)$ (9.12): $f_{ts} \leftarrow Evaluate(\mathbf{x})$
3: Save the results of evaluating f_{ts} in $score$
4: **for** $(counter = 1; counter \leq G; counter++)$ **do**
5:     **for** $(i = 1; i \leq N_p; i++)$ **do**
6:         Select three different indexes randomly (a, b and c in (9.18))
7:         **for** $(j = 1; j \leq D; j++)$ **do**
8:             **if** $U(0,1) < CR || j = D$ **then**
9:                 $trial_j \leftarrow x_{aj} + F(x_{bj} - x_{cj})$
10:             **else**
11:                 $trial_j \leftarrow x_{ij}$
12:             **end if**
13:         **end for**
14:         $f_{ts} \leftarrow Evaluate(trial)$
15:         **if** f_{ts} is better than $score_i$ **then**
16:             $score_i \leftarrow f_{ts}$
17:             $x_i \leftarrow trial$
18:         **end if**
19:     **end for**
20: **end for**
21: **return** x and $score$

9.5   NUMERICAL RESULTS AND DISCUSSION

We use Python programming language for simulations of the UAV model, controllers and metaheuristics. In the case of the UAV model (9.5), to obtain the response of the system, we use the Runge-Kutta numerical integration method of order 4 with a step size $h = 0.001$, with a total simulation time of $t = 10\,$s with initial conditions $x_n(0) = 0$.

In the PID control, we calculate the integral action with the Adams-Bashfoth integrator of order 2 given by $x_{i+1} = x_i + h \left[\frac{3}{2} f(x_i, t_i) - \frac{1}{2} f(x_{i-1}, t_{i-1}) \right]$. The derivative action is calculated by using the definition of Euler, i.e., $\frac{df}{dt} \simeq \frac{f(x_i, t_i) - f(x_{i-1}, t_{i-1})}{h}$. On the other hand, the SMC employs the first and second derivatives of the reference. These values are also obtained with the definition of Euler, where $\frac{d^2 f}{dt^2} \simeq \frac{f(x_i, t_i) - 2f(x_{i-1}, t_{i-1}) + f(x_{i-2}, t_{i-2})}{h^2}$.

In the process of optimization by metaheuristics, the desired states or reference signals of the controllers are unit steps, such that $x_{d5} = 1$, $x_{d7} = 1$, $x_{d9} = 1$, and $x_{d11} = 1$. It is important to remember that the values of x_{d1} and x_{d3} are given by expressions (9.7). In addition, we added two individuals in the initial population of the metaheuristics with which the controllers do not perform well. Tables 9.1 and 9.2

present these individuals, as well as the search range of the controller parameters. Initially, these ranges have larger values, but later they are reduced to improve the convergence of the algorithms since many of the values outside the ranges of the tables do not control the system. In addition, Table 9.3 shows the input parameters mentioned in Sections 9.4.1 and 9.4.2 of the DE and APSO algorithms, respectively.

9.5.1 SOLUTIONS FOR PID CONTROL PROVIDED BY METAHEURISTICS

By using the individuals added to the initial population of Table 9.1, and the input parameters of the DE given in Table 9.3, we execute the Algorithm 7 to minimize the settling time t_s of the PID control system for the UAV flight. Table 9.4 shows the four best solutions for PID control provided by DE. In the solutions, we can see that the constant of integration for the position controller has small values and even zero in some solutions. On the other hand, the two best solutions given by DE are 1 and 2, which have lower t_s and M_p. We can consider solution 1 the best because it has better values of M_p and RMSE. Figure 9.3 shows the UAV outputs (rotation and position) controlled with PID with parameters of 'Solution 1' from Table 9.4. As we can see,

---

Algorithm 8 Accelerated Particle Swarm Optimization Algorithm.

1: Initialize the particle's position randomly ($\mathbf{x}$)
2: Initialize the velocity of the particles v
3: Evaluate $f(t_s)$ (9.12): $f_{ts} \leftarrow Evaluate(\mathbf{x})$
4: Save the results of evaluating f_{ts} in $score$ and $p \leftarrow x$
5: Find the best value from p and save it in g
6: **for** ($counter = 1$; $counter \leq G$; $counter + +$) **do**
7:   **for** ($i = 1$; $i \leq N_p$; $i + +$) **do**
8:     **for** ($j = 1$; $j \leq D$; $j + +$) **do**
9:       $v_{ij} \leftarrow \alpha v_{ij} + U(0,\beta)(p_{ij} - x_{ij}) + U(0,\beta)(g_j - x_{ij})$ ▷ This evaluates the new velocity using (9.20)
10:       $x_{ij} \leftarrow x_{ij} + v_{ij}$ ▷ This evaluates the new position using (9.21)
11:     **end for**
12:     $f_{ts} \leftarrow Evaluate(x_i)$
13:     **if** f_{ts} is better than $score_i$ **then**
14:       $score_i \leftarrow f_{ts}$
15:       $p_i \leftarrow x_i$
16:       **if** p_i is better than g **then**
17:         $g \leftarrow p_i$
18:       **end if**
19:     **end if**
20:   **end for**
21: **end for**
22: **return** x, p, g and $score$

Table 9.1

Individuals added to the initial population of the PID control and search range

Parameter	Individual 1	Individual 2	Range
K_{p1}	2.75	2.73	$[0.0-50.0]$
K_{i1}	0.0	0.0	$[0.0-2.0]$
K_{d1}	2.25	2.23	$[0.0-15.0]$
K_{p2}	1.5	1.4	$[0.0-50.0]$
K_{i2}	5×10^{-6}	5×10^{-6}	$[0.0-50.0]$
K_{d2}	2.25	2.23	$[0.0-10.0]$
$f(t_s)$	**5.8590**	**6.2220**	
$\max(M_p)$	**12.349**	**13.097**	
$\max(RMSE)$	**0.008481**	**0.009129**	

the control response is fast, reaching the reference and satisfying the restrictions of M_p and RMSE.

In the same way, for the case of the APSO with the individuals added to the initial population and the input parameters of the APSO (Table 9.3), the Algorithm 8 is

Table 9.2

Individuals added to the initial population of the SMC and search range

Parameter	Individual 1	Individual 2	Range
b_n	0.03	0.03	$[0.0-50.0]$
b_{n+1}	0.2	0.2	$[0.0-30.0]$
α_n	2.0	2.0	$[1.0-2.0]$
β_n	1.6667	1.6667	$[1.0-2.0]$
c_n	1.21	1.21	$[0.0-10.0]$
h_n	0.5	0.5	$[0.0-20.0]$
b_j	8.5	8.7406	$[0.0-50.0]$
b_{j+1}	0.4	0.4838	$[0.0-30.0]$
α_j	1.2	1.2	$[1.0-2.0]$
β_j	1.0885	1.0885	$[1.0-2.0]$
c_j	15	15	$[0.0-50.0]$
h_j	0.5	0.5	$[0.0-50.0]$
$f(t_s)$	**8.8140**	**9.5310**	
$\max(M_p)$	**4.2359**	**4.2147**	
$\max(RMSE)$	**0.008809**	**0.009112**	

Table 9.3
Metaheuristics input parameters for control optimization

DE input parameters		APSO input parameters	
Parameter	Value	Parameter	Value
Population size	30	Population size	100
Number of generations	100	Number of generations	30
Difference constant	0.3	β	0.3
Recombination constant	0.4	α	0.7

done, in order to minimize the settling time t_s. Table 9.5 shows the four best solutions for PID control provided by APSO. Again, the solutions have a small integration constant for the position controller. The two best solutions given by APSO are 1 and 2, both have a similar value of t_s. However, solution 2 has better values of M_p and RMSE.

Figure 9.4 depicts the results associated to the controlled UAV with PID with the 'Solution 2' parameters from Table 9.5. Although the control response is fast, it is clear that the steady-state error decreases. The solutions provided by both metaheuristics are very similar, especially in the values of t_s, and in both cases, the constant of integration for the position controller has small values. Unlike DE, APSO did not return any values equal to those added to the initial population because the characteristics of this metaheuristic cause small changes between generations. On the other hand, the results provided by APSO have higher M_p and RMSE than those offered by DE.

Table 9.4
Four best solutions for PID control provided by DE

Parameter	Solution 1	Solution 2	Solution 3	Solution 4
K_{p1}	44.91190	22.72824	49.73190	23.63797
K_{i1}	0.0	0.35911	0.29583	0.0
K_{d1}	12.54893	7.10535	11.38001	8.88362
K_{p2}	37.98538	19.65010	27.94196	37.77004
K_{i2}	4.68955	16.02527	2.44093	42.65924
K_{d2}	2.25	0.85815	4.30680	0.76276
$f(t_s)$	1.07299	1.33499	0.83600	1.32399
$\max(M_p)$	0.68750	3.83354	2.05340	6.84326
$\max(RMSE)$	0.00163	0.00569	0.00901	0.00437

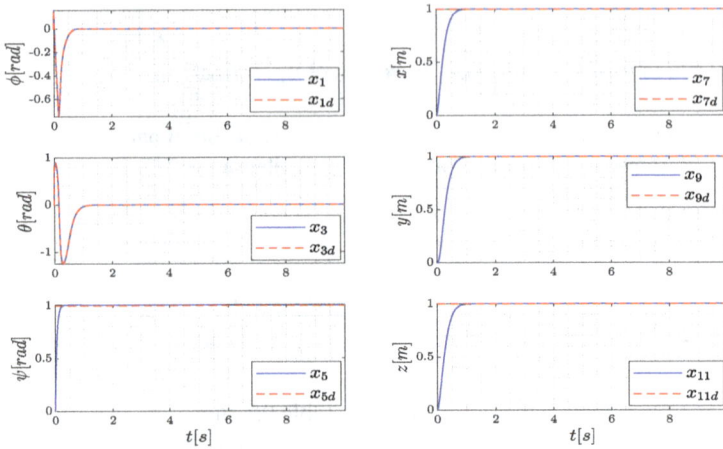

Figure 9.3 Results of the PID control with *'Solution 1'* parameters provided by DE. (red line: reference signals, blue line: UAV outputs).

9.5.2 SOLUTIONS FOR SMC PROVIDED BY METAHEURISTICS

In the case of the SMC, Table 9.2 shows the individuals added to the initial population, and Table 9.3 shows the input parameters of the DE. Again, we execute the algorithm described in Section 9.4.1 to minimize the settling time t_s. Table 9.6 shows the four best solutions for SMC provided by DE. We can see that the solutions have great variation between their values because there are more parameters and the possible values increase. The two best solutions given by DE are 1 and 3 because all 4

Table 9.5
Four best solutions for PID control provided by APSO

Parameter	Solution 1	Solution 2	Solution 3	Solution 4
K_{p1}	40.34137	36.53113	45.20234	30.78077
K_{i1}	1.15870	0.82433	1.47721	0.28652
K_{d1}	10.17815	9.60870	13.49124	8.60004
K_{p2}	35.54109	42.11706	19.17181	26.85436
K_{i2}	26.52664	22.53225	33.09432	14.82749
K_{d2}	1.76952	2.37472	1.45289	1.34536
$f(t_s)$	0.95400	0.98800	1.17499	1.04299
$\max(M_p)$	3.10287	2.56505	8.79023	2.73055
$\max(RMSE)$	0.00776	0.00677	0.00946	0.00616

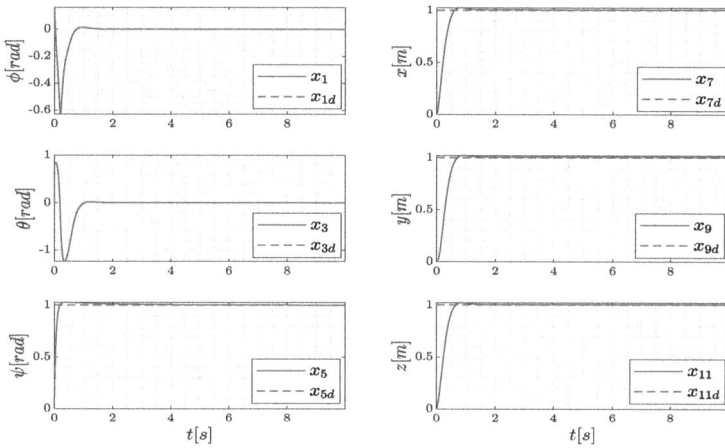

Figure 9.4 Results of the PID control with *'Solution 2'* parameters provided by APSO. (red line: reference signals, blue line: UAV outputs).

solutions have similar t_s values, but 1 and 2 have lower M_p and RMSE values. Here we can consider solution 1 as the best because its results are better.

Figure 9.5 shows the results associated to the controlled UAV by using the SMC with the 'Solution 1' parameters from Table 9.6. Compared to PID control, the con-

Table 9.6
Four best solutions for SMC provided by DE

Parameter	Solution 1	Solution 2	Solution 3	Solution 4
b_n	46.39280	44.91259	27.05800	26.04811
b_{n+1}	3.91446	10.37728	3.09141	6.96358
α_n	1.30016	1.11985	1.04837	1.48643
β_n	1.05557	1.04896	1.03251	1.06600
c_n	0.13862	0.32405	0.03088	0.16067
h_n	12.49016	3.73925	9.91126	6.18091
b_j	33.91494	39.69149	40.46043	43.38813
b_{j+1}	7.28023	2.62951	9.34793	3.67149
α_j	1.71886	1.62170	1.22081	1.54240
β_j	1.24673	1.26161	1.18297	1.25271
c_j	42.18140	20.89751	26.46219	43.92880
h_j	29.35933	34.72730	23.02794	33.00217
$f(t_s)$	**1.85599**	**1.87499**	**1.87399**	**2.23899**
$max(M_p)$	**1.11642**	**1.51765**	**1.43122**	**1.45858**
$max(RMSE)$	**0.00390**	**0.00576**	**0.00480**	**0.00559**

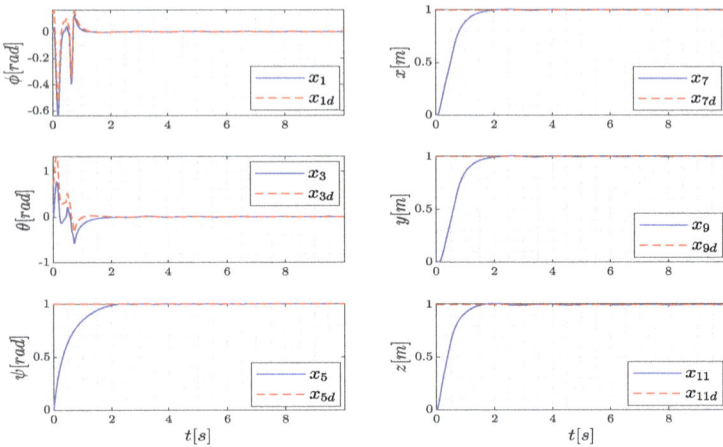

Figure 9.5 Results of the SMC with *'Solution 1'* parameters provided by DE. (red line: reference signals, blue line: UAV outputs).

trol response is slower due to the rotation controller. However, the values of M_p and RMSE are smaller, which implies less the steady-state error.

Finally, the APSO algorithm described in Section 9.4.2 is realized to minimize the settling time t_s of the SMC controller for the UAV flight. The results are presented in Table 9.7, where the four best solutions for SMC provided by APSO are showed. Although the solutions are very similar, solutions 3 and 4 have better results. However, both solutions do not equal the results of the solutions provided by DE. In conclusion, solutions 1 and 3 given by DE are better than APSO results. Figure 9.6 shows the results with SMC provide by the parameters of 'Solution 4' from Table 9.7. In this case, the control response is slower than the PID, but it keeps small values of M_p and RMSE. As we mentioned before, for the SMC, the metaheuristic that provided the best solutions was DE, this result may be caused by the number of controller parameters, which complicates the convergence of the APSO. On the other hand, the response of the PID controller is faster than that of the SMC. The proportional action is essential because it provides the initial impulse necessary to achieve the desired value. However, the overshoot values (M_p) and the steady-state error quantized with the RMSE value are lower with the SMC.

9.5.3 TRAJECTORY TRACKING

One of the classical problems in UAVs control is to design an overall system such that the output $y(t)$ approaches $\mathbf{r}_d(t)$ as t approaches infinity (asymptotic tracking) or t approaches a finite-time $t > 0$ (finite-time tracking). Sometimes, the control algorithms employed to solve these tracking problems need extra information, such as the first or second derivative of the desired trajectory (velocity and acceleration). However, if

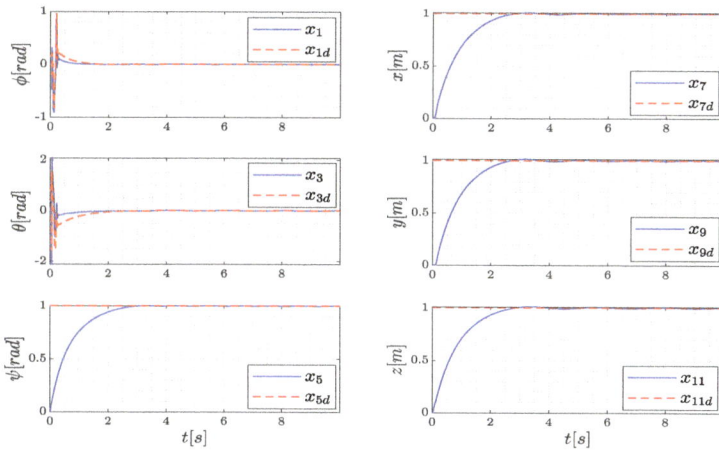

Figure 9.6 Results of the SMC with '*Solution 4*' parameters provided by APSO. (red line: reference signals, blue line: UAV outputs).

the reference signal is non-differentiable, then calculating velocity and acceleration may be difficult, with possible effects influencing negatively sudden changes in the position or orientation of the quadrotor. In order to establish desired trajectories such that their derivatives can be calculated (smooth trajectories), in [421], a methodology

Table 9.7
Four best solutions for SMC provided by APSO

Parameter	Solution 1	Solution 2	Solution 3	Solution 4
b_n	9.42768	12.09916	17.11371	21.58818
b_{n+1}	7.45883	9.09044	13.09237	14.55005
α_n	1.52583	1.50922	1.57044	1.41533
β_n	1.17389	1.34626	1.45521	1.31589
c_n	2.37183	0.79765	1.00637	0.75360
h_n	3.03640	5.82699	6.72177	6.85932
b_j	33.41827	26.07647	20.01328	29.47084
b_{j+1}	3.81886	14.64296	8.11599	13.54946
α_j	1.45741	1.61213	1.63789	1.49986
β_j	1.19254	1.40232	1.20347	1.19675
c_j	43.80254	13.87508	18.53280	12.79587
h_j	37.55390	32.53011	31.97970	21.40161
$\mathbf{f(t_s)}$	**2.47699**	**2.74899**	**2.46299**	**2.49699**
$\max(\mathbf{M_p})$	**1.97555**	**1.95485**	**1.68242**	**1.36751**
$\max(\mathbf{RMSE})$	**0.00821**	**0.00821**	**0.00696**	**0.00532**

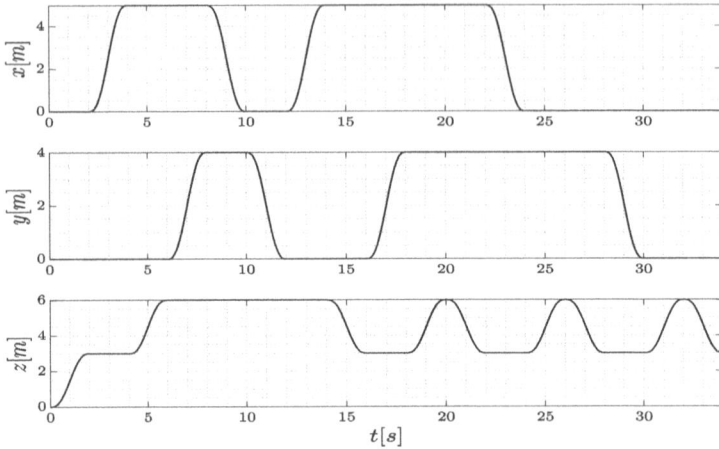

Figure 9.7   Desired trajectories of the UAV associated to x, y and z.

for the design of trajectories based on the calculation of variations is proposed so that a cost function with restrictions is minimized.

Consider the desired trajectory depicted in Figure 9.7, with the initial position $(0,0,0)$ and $x = x_{7d}$, $y = x_{9d}$, and $z = x_{11d}$. For simulation purposes, the desired angle ψ remains fixed at $x_{5d} = 1$ rad.

From the results shown in Tables 9.4 to 9.7, we select the two best solutions for each controller. These solutions are for the PID control $set1 \leftarrow$ Solution 1 provided by DE (Table 9.4), and $set2 \leftarrow$ Solution 2 provided by APSO (Table 9.5). For the SMC $set1 \leftarrow$ Solution 1, and $set2 \leftarrow$ Solution 3 provided by DE (Table 9.6). With these sets of solutions, we simulate in Matlab the trajectory tracking control.

Table 9.8 shows the results of the simulations. In this case, we quantify the trajectory tracking results with the RMSE, MAE (Mean Absolute Error), and R2 (Coefficient of Determination) values. In the RMSE values, we can see that the PID control of the rotation has smaller errors than the SMC, i.e., the response with the SMC has high error peaks, since this metric is very sensitive to large error values. However, the position controller with SMC has better results, i.e., the PID control for the position has high error peaks. The MAE value confirms these results, here the values are not so different between both controllers because this metric is not very sensitive to occasional high errors, but again, the rotation controller is better with PID control than with SMC, and the position controller is better with SMC than with PID control. Finally, with the R2 value, we quantify how well the UAV adjusts or follows the desired trajectory. This value with both controllers is higher than 99.8%, being this value the lowest obtained with PID, and 99.99% the highest obtained with SMC.

The simulation results associated with the trajectory tracking are shown in Figures 9.8 and 9.9, respectively, for the PID control and the Sliding-Mode-Control.

Table 9.8

Results of trajectory tracking errors

Error type	State	PID control		SMC	
		set1	set2	set1	set2
RMSE	x_1	0.00038	0.00034	0.00187	0.00211
	x_3	0.00038	0.00034	0.03046	0.02817
	x_5	0.03034	0.02961	0.08479	0.08112
	x_7	0.04902	0.06333	0.01296	0.01376
	x_9	0.03922	0.05066	0.01674	0.01770
	x_{11}	0.04447	0.05710	0.00675	0.00756
MAE	x_1	0.00022	0.00020	0.00119	0.00135
	x_3	0.00022	0.00020	0.02050	0.01915
	x_5	0.00335	0.00306	0.02884	0.02685
	x_7	0.02301	0.02946	0.00934	0.01001
	x_9	0.01842	0.02362	0.01187	0.01259
	x_{11}	0.03059	0.03910	0.00525	0.00591
R2	x_1	0.99999	0.99999	0.99993	0.99991
	x_3	0.99999	0.99999	0.98561	0.98886
	x_5	–	–	–	–
	x_7	0.99963	0.99938	0.99997	0.99997
	x_9	0.99963	0.99938	0.99992	0.99991
	x_{11}	0.99934	0.99891	0.99998	0.99998

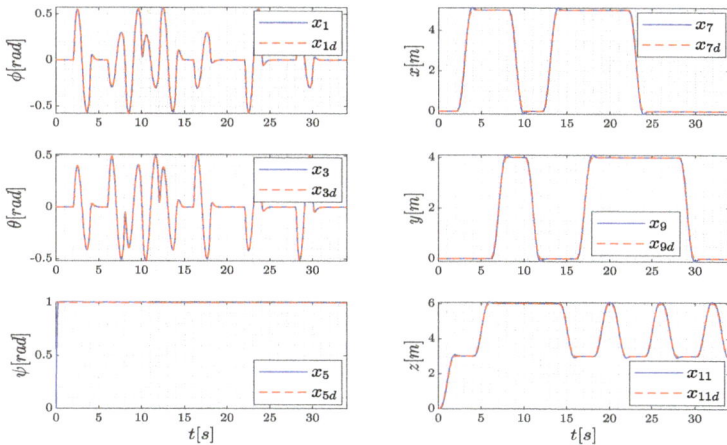

Figure 9.8 Result of trajectory tracking with PID control and *'set1'* parameters in time. (red line: desired trajectory, blue line: UAV outputs).

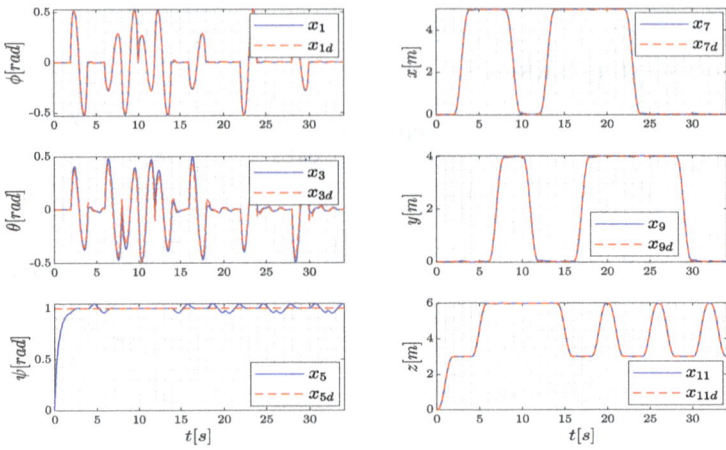

Figure 9.9 Result of trajectory tracking with SMC and *'set1'* parameters in time. (red line: desired trajectory, blue line: UAV outputs).

According to the results, the figures show a better performance of the PID control for the rotation and a better performance of the SMC for the position. These results are mainly caused by the fact that the PID control is fast due to its proportional action but causes high overshoot values on direction changes. Furthermore, the SMC is a non-linear controller, so the values provided by the metaheuristics perform well at points far from the tuning ($x_d = 1$). On the other hand, the metaheuristics had to adjust more parameters of the SMC than of the PID control. Hence, the convergence in the parameters of the rotation controller is more complicated. However, the SMC has better trajectory tracking performance, and this controller adjusts better to the desired trajectory. On the other hand, for better visualization, Figure 9.10 shows the

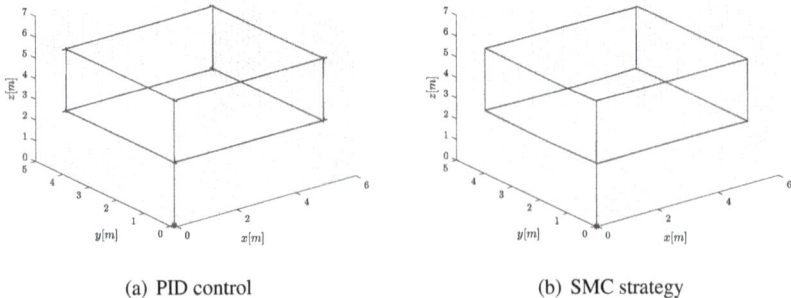

(a) PID control   (b) SMC strategy

Figure 9.10 Result of trajectory tracking with *'set1'* parameters.

trajectory tracking in 3D, and according to the results, the UAV follows the proposed cube as a desired trajectory.

9.6   CONCLUDING REMARKS

A successful comparison between the PID control and Sliding Mode Controller (SMC) was presented in this chapter. Based on the metrics used, such as settling time, overshoot, and errors, we implemented the intelligent control approach to improve the performance of the controllers by using techniques such as Differential Evolution (DE) and Accelerated Particle Swarm Optimization (APSO). The numerical results showed the advantages and reliability of the AI techniques in calculating gains of the classical control techniques, which suggests designing hybrid control algorithms to optimize and solve UAVs' flight control problems.

10 Development of a synthetic Dataset Using Aerial Navigation to Validate a Texture Classification Model

J. M. Fortuna-Cervantes
Tecnológico Nacional de México, Instituto Tecnológico de San Luis Potos, Av. Tecnológico s/n, 78437 San Luis Potosí, México; juan.fc@slp.tecnm.mx

M. T. Ramírez-Torres
Coordinación Académica Región Altiplano Oeste, Universidad Autónoma de San Luis Potos, Carretera Salinas-Santo Domingo 200 Salinas, 78600 San Luis Potosí, Mexico; *Corresponding Author: tulio.torres@uaslp.mx

M. Mejía-Carlos
Instituto de Investigación en Comunicación Óptica, Universidad Autónoma de San Luis Potosí, álvaro Obregón 64, 78000 San Luis Potosí, Mexico; marcela.mejia@uaslp.mx

Jose Martinez-Carranza
Department of Computational Science, Instítuto Nacional de Astrofísica, Óptica y Electrónica (INAOE), 72840 Puebla, México— University of Bristol, Bristol, UK; carranza@inaoep.mx

CONTENTS

DOI: 10.1201/9781003385615-10

10.1   INTRODUCTION

Texture analysis is a traditional problem in computer vision because it involves obtaining information that describes the image content. In robotics, object detection is a problem for robots that perform tasks in real scenarios and in real-time, given the lighting conditions, indeterminate orientations, object identity, shape, color, and texture. Furthermore, the information may differ in outdoor and indoor environments, which varies the target information [83]. Providing the resources to the robot by integrating sensors can improve object detection.

Microaerial vehicles (MAVs) have been used in different environments due to their easy control and implementation of algorithms through computer vision in tasks of the classification, detection, and localization of the target. There are different vision methods in the area, such as optical flow, segmentation, edge detector, morphological operations, and feature extractor for other tasks. These methods have been combined to improve detection performance while the MAV executes its aerial navigation (recognition) or autonomous flight. However, these methods can be computationally expensive to perform real-time detection, affecting the overall system performance.

In image processing, texture can be defined from neighboring pixels, and intensity distribution over the image [639]. Besides, some classification methods for texture analysis include statistical, geometric, model, and spectral. If we focus on spectral methods, these methods describe the texture in the frequency domain. They are based on the decomposition of a signal in terms of basis functions. Furthermore, they use the expansion coefficients as elements of the feature vector.

Deep learning has become a helpful tool for image classification, object detection, and segmentation. Especially if we talk about convolutional neural networks (CNNs), these achieve learning multiple features to recognize targets without reference to their position, indeterminate orientations, scale, and target rotation. VGG16, VGG19, AlexNet, SSD, and YOLO are architectures that perform well in image classification and object detection tasks [317, 372, 510, 572].

Therefore, we decide to merge these methodologies (deep learning and wavelet features) as a solution for texture classification. The objective is that the MAV performs the aerial navigation (inside the virtual environment) for the classification system to recognize the object, see Figure 10.1. This work focuses on preview information (in data collected by MAV) and structural recognition of the object (with a

Figure 10.1 We designed a system for texture classification in aerial navigation based on knowledge inference over the DTD database. See at https://youtu.be/d41kgBw7Y_c.

particular texture) within a region of interest in the image plane.

The implementation of our system is developed with the fusion of two approaches. The first is in the spatial domain, using transfer learning. We take as a baseline the VGG16 architecture with the features of the ImageNet database. The second approach focuses on the spectral domain, applying the Haar wavelet transform in two dimensions to obtain features at different scales [689]. The VGG16 network has been selected for its fast performance and implementation with transfer learning and adaptability with wavelet analysis. Internally, the system is divided into two stages: the first corresponds to feature extraction and the second to the classification stage. We used the Describable Texture Database (DTD) to train our model, which contains 47 texture classes with 120 images per class. On average, we have tested with some textures for the classification task in the virtual environment, and the prediction can be performed correctly, with an average processing speed of 2 fps. The test gives rise to a new Synthetic Aerial Dataset of Textured Objects (SADTO).

The rest of the chapter is organized in Section 10.2 related work; Section 10.3 introduces the methodology to approach the texture classification problem. Section 10.4 shows the results with the DTD dataset and the experimental part to test our model. Finally, Section 10.5 presents the conclusions.

10.2 RELATED WORK

In recent years, MAVs applications for object detection tasks have been studied and developed [48] [486]. Several approaches using deep learning give excellent performance in applications. For example, in some tasks for autonomous navigation, we find in the literature a methodology for obstacle detection and avoidance using an architecture called AlexNet that allows classifying the images captured by the camera onboard the drone [154]. The learning of this architecture is transferred from the ImageNet database to improve the classification performance [317]. Moreover, to detect objects and autonomous landing, in [83], a detection system is presented to solve one of the missions in the IMAV2019 indoor competition. They involve the implementation of the SSD7 onboard the MAV. This SSD7 network is chosen for its fast performance on low-budget microcomputers with no GPU. The method proposed in [510] is an architecture called YOLO, which presents an actual performance in real-time image detection and processing at 45 frames per second. Besides, in object detection tasks, in [278], the authors propose a deep learning approach to robustly estimate the object's center. Also, generating a line of sight as a guide proves to be a solution to avoid collision with other objects due to complications such as varying illumination conditions, object geometry, and overlapping in the image plane.

On the other hand, many projects employ deep learning and wavelet analysis in visual processing. For example, on image classification, the method proposed in [669] converts images from the CIFAR-10 and KDEF database to the wavelet domain, thus obtaining temporal and frequency features. The different representations are added to multiple CNN architectures. This combination of information in the wavelet domain achieves higher detection efficiency and faster execution times than the spatial domain procedure. In this sense, the authors in [670] mention that although CNN is a universal extractor, in practice, it needs to be clarified whether CNN can learn to perform spectral analysis. In [32], the authors propose an architecture called CNN Texture to have this approach within the CNN. Their idea focuses on the fact that the information extracted by convolutional layers is of minor importance in texture analysis. They use a statistical energy metric in the feature extraction stage. This information is concatenated with the classification stage, the fully connected layer. Specifically, the architecture shows an improvement in performance and a reduction in computational cost.

In terms of texture classification in image processing applications [182], the authors propose an architecture called wavelet CNN to generalize spectral information lost in conventional CNNs. This information is beneficial for texture classification as it usually contains details about the object's shape. Furthermore, the model allows us to have fewer parameters than traditional CNNs, so it is possible to train with less memory. In general, through a state-of-the-art review, we have observed that computational intelligence algorithms improve detection strategies in Micro Aerial Vehicle applications.

10.3   MATERIALS AND METHODS

This chapter proposes an approach based on transfer learning and wavelet features. This system allows to prediction or classifies the texture in images transmitted by the on-board camera of the drone, whose objects are in an outdoor scenario (in Gazebo), a virtual simulation environment. We are only interested in texture recognition, mainly to know one of the characteristics of the object. So, we

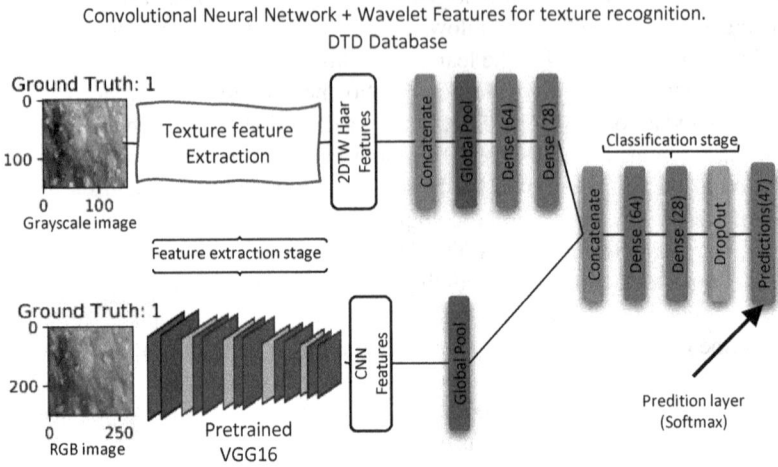

Figure 10.2 Texture classification system.

limit the image plane (640×360) to a region of interest (300×300 pixels). As a result, the system will have the image in RGB and a grayscale version. These two images are the inputs for our proposed classification system; see Figure 10.2.

10.3.1 TRANSFER LEARNING

A practical approach in deep learning with small datasets is using pre-trained networks. In this case, we consider a CNN previously trained with the ImageNet database, which has 1,000 classes [317]. A CNN architecture comprises two main parts, feature extraction and classification stages for image classification tasks. The first stage is called the convolutional base. In this case, feature extraction involves using previously extracted representations during training.The purpose is to adapt to the new features of the new dataset. Therefore, these features are used in a new classifier trained from scratch, as shown in Figure 10.3.

The level of representations extracted by the convolutional layers depends on the depth of the layer in the model. Earlier layers in the model extract local and generic feature maps (such as visual edges, colors, and textures), while higher layers extract more abstract concepts. Therefore, if a new dataset differs significantly from the dataset on which the original model was trained, it may be better to use only the earlier layers of the model for feature extraction rather than the full convolutional basis.

The VGG16 architecture, among others, comes pre-packaged in Keras Framework. These models can be used for prediction, feature extraction, and fine-tuning. The list of image classification models (all pre-trained on the ImageNet dataset) can be imported from the Keras.applications module.

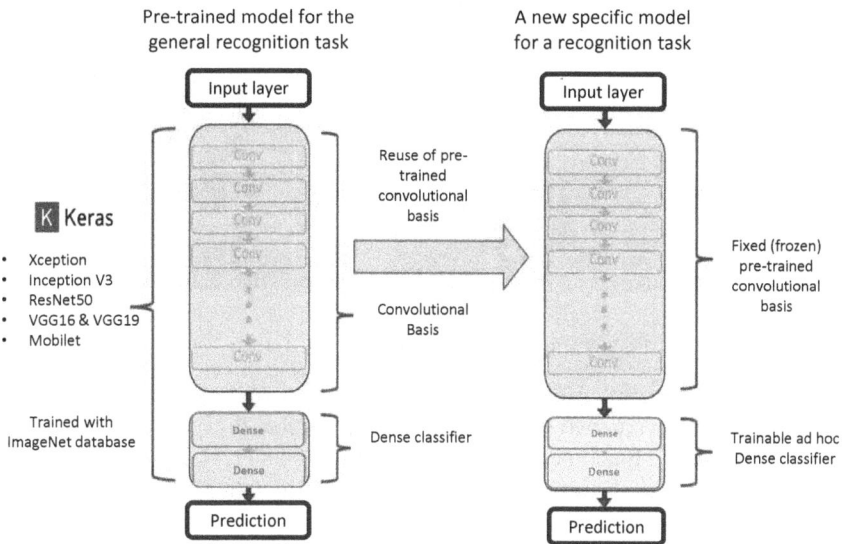

Figure 10.3 Classifiers using the same convolutional basis: A pre-trained model for a general recognition task is taken as a basis. We reuse the feature extraction stage from the base architecture and cancel the dense classifier stage. Therefore, a new model is obtained for a specific recognition task [513].

10.3.2 DESCRIBABLE TEXTURES DATASET

Describable textures dataset (DTD) was selected to be used. It contains 47 classes of 120 images in the wild, meaning that the images have been acquired in uncontrolled conditions [120]. This dataset includes ten divisions with 40 training images, 40 validation images, and 40 test images for each class. Our experiment will create a new dataset, with a distribution of 70% for training, 15% for validation, and the remaining 15% for testing. Figure 10.4 shows some images from this set. One limitation of the dataset is the number of images per class, so it is decided to use the transfer learning method to improve the classification performance of our model. The synaptic weights are based on the ImageNet database, which will feed the feature extraction stage of the base architecture VGG16.

10.3.3 WAVELET ANALYSIS

Wavelets represent functions as simpler, fixed building blocks at different scales and positions [669]. The one-dimensional wavelet transform can be easily extended to a Two-Dimensional Wavelet Transform (2D-WT), which is widely applied to two-dimensional signals such as images [213, 652]. It has greatly impacted image processing tasks such as edge detection, image recognition, and image compression [639].

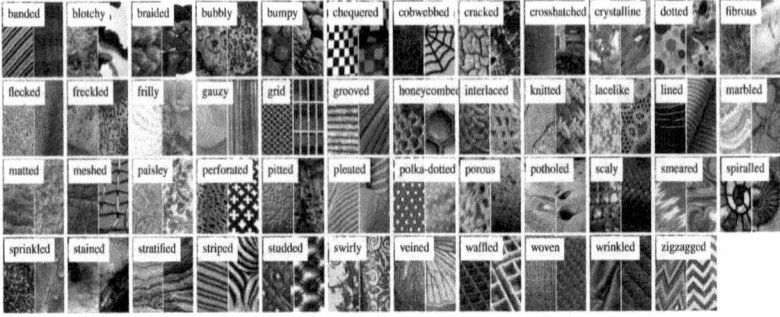

Figure 10.4 DTD dataset example images [120].

2D-WT considers a two-dimensional scale function $\Phi(x,y)$, and three two-dimensional wavelet functions $\Psi^H(x,y)$, $\Psi^V(x,y)$ and $\Psi^D(x,y)$, resulting in a lower resolution image than original, as well as detailed information on the horizontal (H), vertical (V) and diagonal (D) perspectives. Each function corresponds to the product of a function of scale φ and its corresponding wavelet ψ; in this way we have:

$$\Phi(x,y) = \varphi(x)\,\varphi(y) \tag{10.1}$$

$$\Psi^H(x,y) = \psi(x)\,\varphi(y) \tag{10.2}$$

$$\Psi^V(x,y) = \varphi(x)\,\psi(y) \tag{10.3}$$

$$\Psi^D(x,y) = \psi(x)\,\psi(y) \tag{10.4}$$

Scale and translation base functions are defined by:

$$\Phi_{j;m,n}(x,y) = 2^{\frac{j}{2}}\Phi\left(2^j x - m, 2^j y - n\right) \tag{10.5}$$

$$\Psi_{j;m,n}(x,y) = 2^{\frac{j}{2}}\Psi^d\left(2^j x - m, 2^j y - n\right) \tag{10.6}$$

where $j,m,n \in \mathbb{Z}$, and the superindex d assume the values H, V, and D to identify the directional wavelets given in equations 10.2–10.4. Considering that the equations 10.5–10.6 constitute an orthonormal basis for $\mathbf{L}^2\left(\mathbb{R}^2\right)$, the expansion of a finite energy function $f(x,y)$ is then defined as:

$$f(x,y) = \frac{1}{\sqrt{MN}}\sum_m \sum_n a_{j0;m,n}\Phi_{j0;m,n}(x,y)$$

$$+ \frac{1}{\sqrt{MN}}\sum_{d=H,V,D}\sum_{j=j0}\sum_m \sum_n d^d_{j;m,n}\Psi^d_{j;,m,n}(x,y) \tag{10.7}$$

where the scale coefficients $a_{j;m,n}$ and wavelet $d^d_{j;m,n}$ are defined by:

$$a_{j;m,n} = \iint f(x,y), \Phi_{j;m,n}(x,y)\,dxdy \tag{10.8}$$

$$d^d_{j;m.n} = \iint f(x), \Psi^d_{j;m,n}(x,y)\,dxdy \tag{10.9}$$

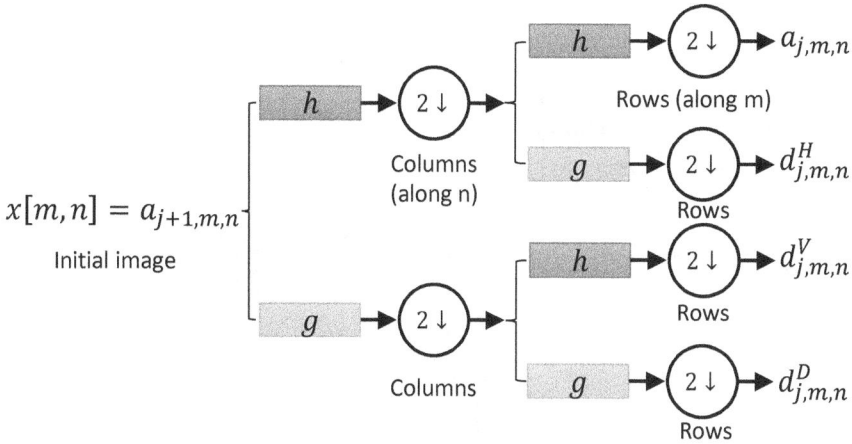

Figure 10.5 The first level of decomposition applied to an image using the filter bank [638].

Expressions 10.7 and 10.8–10.9 represent the equations of synthesis and analysis of the original image, and together they constitute the Two-Dimensional Discrete Wavelet Transform (2D-DWT) [638].

10.3.4 MULTIRESOLUTION ANALYSIS

The Mallat-Multiresolution Analysis (M-MA) algorithm makes it possible to calculate the coefficients numerically $a_{j;m,n}$ and $d_{j;m,n}$ of the two-dimensional functions or image (denoted by $x[m,n]$), with a theoretical basis of the Fast Two-Dimensional Wavelet Transform (2D-FWT) [397]. Also, the algorithm provides a connection between the wavelets and the filter banks [448]. The multiresolution decomposition of an image is represented by a series of approximations and details in sub-images. In the first level of decomposition, two filters are applied, respectively, one low-pass (h) and one high-pass (g), each followed by a subsampling operation by a factor of 2, as illustrated in Figure 10.5.

The result of applying three levels of wavelet decomposition to an image ($x[m,n]$), sized $M \times N$, is illustrated in Figure 10.6. After the two-dimensional signal passes through the filter bank structure shown in Figure 10.5, four sub-images with $M/2$ rows and $N/2$ columns are obtained; i.e., each one of the four sub-images has a quarter of the pixels of the input image. The approximation sub-image is achieved by the approximation calculations along the rows of the original image, followed by the approximation calculations across the columns. This sub-image is an average version of the image ($x[m,n]$), with a quarter resolution and statistical properties similar to the original signal [652]. The rest of the sub-images shows specific characteristics of the original image in a particular direction, providing the detail coefficients: horizontal, vertical, and diagonal. The same wavelet transform is applied only to the approximation sub-image to determine the next level of decomposition. Therefore,

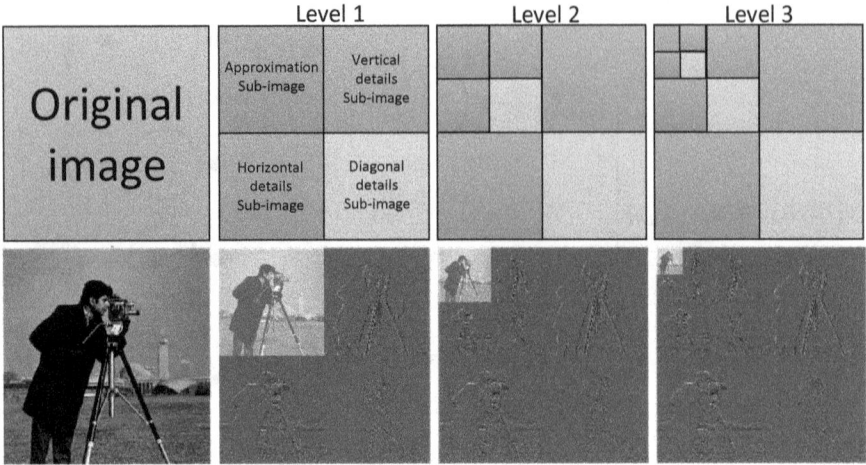

Figure 10.6 Decomposition process applying three levels of the filter bank, which results are some approximation and detail sub-images.

we get four sub-images but now with dimensions of $M/2^2$ rows and $N/2^2$ columns. This iteration is repeated until the desired resolution level or until the level allowed by the image's dimensions [638].

In general, multiresolution decomposition for a two-dimensional signal reveals differences in resolution levels. It shows details in different orientations and properties that indicate that the 2D-DWT is well-suited for detecting important information from the original two-dimensional signal or image.

The Haar wavelet is the most straightforward and is one of the primary examples to illustrate the wavelet theory better. It is widely used when there are signals with abrupt changes. It has dramatically impacted image processing tasks such as edge detection, image recognition, and image compression [639].

The concept of the wavelet transform is easy to generalize to two-dimensional functions such as images. 2D-DWT can be done in several ways. Only two are mentioned here: the standard decomposition and the non-standard or pyramidal decomposition. The process with the bank filter describes the non-standard decomposition. It is the basis of this research work. Because most of the existing techniques for image management are usually complex, challenging to implement, and have a high computational cost, recently, wavelet methods have been applied to different stages of image processing because they are very *efficient* computationally and *easy* to implement numerically [637].

Before training, the Haar wavelet transforms in two dimensions and is applied to one level; see Figure 10.2. The factor of one represents the level of image decomposition. This new spectral information is essential for classification. Therefore, four sets (in the wavelet domain) are automatically generated to determine the characteristic attributes of each texture. This information can be combined with the spatial information of the VGG16 architecture. Also, it is essential to mention that this

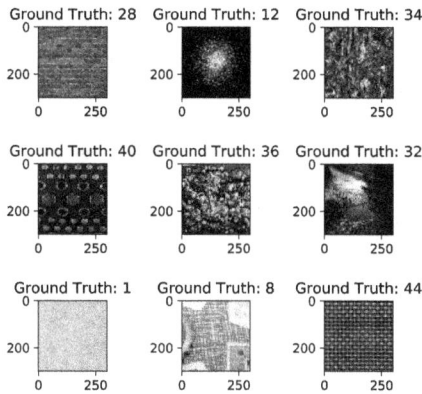

Figure 10.7   Images textures that have been decoded (Class) to train the classification model.

process is only applied to the image previously converted to grayscale, performing the decomposition for a single channel; see Figures 10.7 and 10.8.

10.3.5   SYNTHETIC AERIAL DATASET OF TEXTURED OBJECTS (SADTO)

We use the Gazebo simulator to generate synthetic images (SADTO) from virtual worlds and models designed in the simulator. From this, we propose a dataset of images with an environment where a textured object is located in order to validate our classification system.

As an initial setup, we represent the virtual world with a real-world scene. We use a MAV available in the simulator to fly through the areas where the textured objects are located. The MAV is equipped with an RGB camera which obtains color images. These images were taken through the ROS — a system that allows us to connect cross-platform between different devices employing topics and messages on a data bus. The camera onboard the MAV is positioned in front of its structure, which allows it to have a frontal view. The MAV front camera's image resolution is 720 pixels at a sample rate of 30 fps, with a field of view of 92°. As for the model in the simulator, it maintains the original configuration, but for our case study, a region of interest is implemented. The new configuration results in a set of quadratic images. In other words, images with a resolution of 300×300 pixels. The quadratic images will serve as input to the classification system.

The images are generally generated from designed models and synthetic photo-realistic scenes. Figure 10.10 shows the overall distribution of images by class — the images from the validation set for our learning model. Different positions of the synthetic scene and 3D models with textures were used. The images were acquired by the MAV flying at a constant speed, at an average altitude of 1 m, and captured at an average of 2 fps. Thus, it produces an overlapping of images. Due to the use of

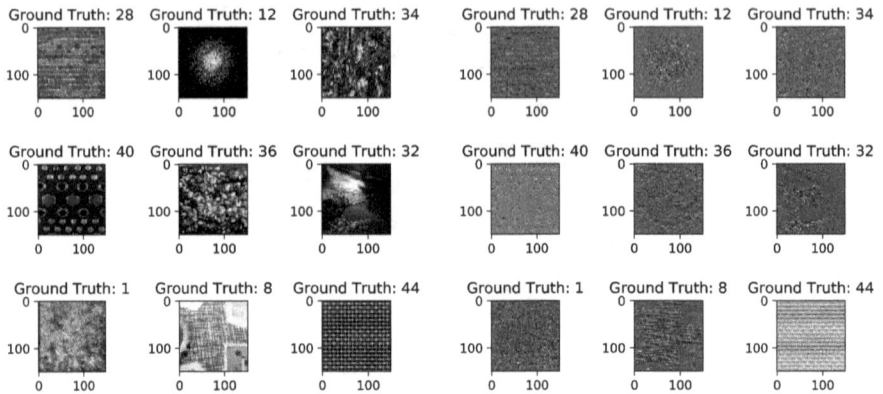

(a) Approximation.

(b) Horizontal details.

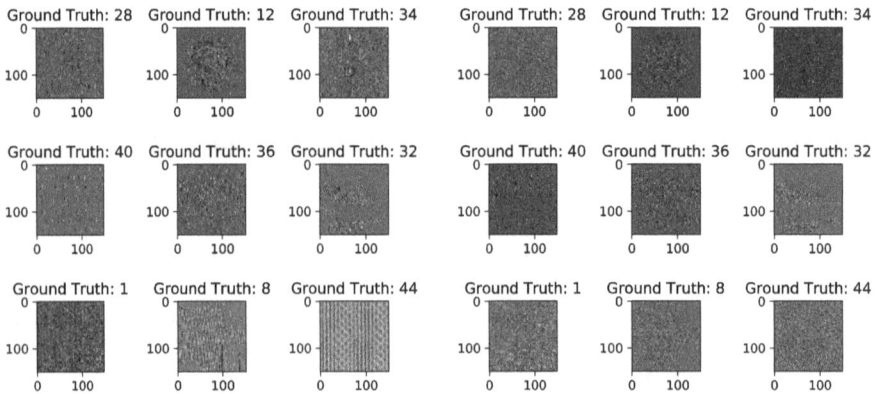

(c) Vertical details.

(d) Diagonal details.

Figure 10.8 Approximation and details set of wavelet features.

different structures within the virtual world background, our dataset offers diverse images, highlighting the textured object. The dataset contains ten classes with 1000 RGB images. The initial proposal of 10 classes is due to training our model with the DTD database. Therefore, they are the classes with the best classification performance.

We used ten 3D cube-shaped models obtained from the Gazebo simulator model base. The Gazebo developer community proposes the design and implementation of the cube. The 3D models initially had no defined texture, so each was redesigned with a specific texture. Figures 10.11–10.12 show some photogrammetric and synthetic scenes used to generate the validation dataset. The test set is available to the public at https://github.com/JanManuell/Synthetic-data-using-aerial-navigation. For the test stage in MAV, a scenario is designed in the gazebo simulator. The virtual scenario is created with ten cubes with certain textures (Figure 10.1). These textures are selected due to the performance achieved in the model testing stage. Therefore, the chosen classes have a performance above 70% accuracy (Table 10.3).

10.4   EXPERIMENTS AND RESULTS

The experiment to train our learning model was carried out with the Keras API with Tensorflow as Backend [114]. Besides, the OpenCV libraries are used for image processing due to their ease of use and adaptability in programming. Also, we use the Pywt library [336] from which the Haar wavelet transform was chosen as the feature extractor method. An aerial navigation experiment was performed using the ROS framework and Gazebo simulation environment to validate the classification system and its learning generalization. This section describes the results obtained in each experiment.

10.4.1   MODEL TRAINING

In the first instance, the VGG16 network was trained from scratch. As a result, it is not able to generalize its learning. Therefore, it is possible to use the transfer learning methodology. Table 10.1 shows the achieved performance of the pre-trained network and our proposal with the wavelet feature fusion. It shows the accuracy performance of the three sets to validate the model (training, validation, and test). In the case of the pre-trained network, slight overfitting is observed. The model will be adjusted to learn specific instances and will be unable to recognize new textures. One way to improve the performance of the model is to integrate the wavelet features. In this case, we eliminate overfitting and homogeneity between the three sets. Besides, the value of the test set is highlighted because these are images the model has never seen.

In summary, the classification system has 14,778,735 synaptic learning weights. Sixty-four thousand forty-seven are trainable parameters, of which 16,832 correspond to wavelet features. Table 10.2 summarizes the achieved performance of our classification system, as well as a comparison to AlexNet (trained from scratch), T-CNN, and Wavelet CNN [32, 182].

Table 10.1

Classification results for the pre-trained VGG16 network and our model indicated as accuracy (%).

	Training	Validation	Test
Pre-trained model	68.15	50.41	54.49
Our model	57.67	51.22	**53.19**

Table 10.2

Classification results and comparison with other state-of-the-art pre-trained architectures with ImageNet, in terms of accuracy (%).

	AlexNet	T-CNN	Wavelet CNN	Our model
DTD	22.7	55.8	59.8	**53.19**

10.4.2 TEXTURE CLASSIFICATION DTD

Other metrics evaluate the performance of the DTD dataset classes. The metrics such as precision, recall, and f1-score are given when applying the classification_report method, where it is necessary to involve the true and prediction labels of the model. Table 10.3 shows the classes that performed above 70% classification. Also, Table 10.4 shows three random categories that perform above 50% classification. This class selection analysis provides the basis for the design of the textured cubes of the Gazebo environment. On the other hand, we can observe the similarity and correlation between classes (about the test set) by performing the prediction. Figure 10.9 shows the true and prediction labels at the top of each texture.

10.4.3 TEXTURE CLASSIFICATION IN AERIAL NAVIGATION

Navigation and aerial recognition tested the classification model. Using the ROS framework, we created a virtual environment with the Gazebo simulator, controlling and sending information from the on-board camera of the drone. In the world presented in Figure 10.1, we positioned the ten cubes with the selected textures in a row. Therefore, the position of the cubes allows the evaluation of the prediction

Table 10.3

Classes (test set) that results with precision above 70%.

Class	precision	recall	f1-score	support
bubb	0.73	0.61	0.67	18
cheq	1.00	0.78	0.88	18
fibr	0.73	0.61	0.67	18
fril	0.72	0.72	0.72	18
stri	0.77	0.94	0.85	18
stud	0.70	0.78	0.74	18
zigz	0.75	0.67	0.71	18

Table 10.4

Classes (test set) that results with precision above 50%. They are chosen from the easy human visual perception of the texture.

Class	precision	recall	f1-score	support
hone	0.58	0.61	0.59	18
line	0.50	0.28	0.36	18
polk	0.65	0.61	0.63	18

model during aerial exploration. The idea of the model is that it generalizes its learning to textured objects. In total, 1000 image captures (SADTO) were performed in a navigation recognition for each class. The proposed texture sets (bubbly, chequered, honey, striped, studded) obtain a high correlation with their original label above 60% accuracy; see Figure 10.10.

Some images (Figures 10.11 and 10.12) of the recognition set are shown with their original label and their prediction label. However, we can observe that the five test images incorrectly predict the frilly, lined, polka-dotted, and zigzagged set. These five images relate to the whole recognition set, except for fibrous, polka-dotted, and zigzagged, which achieve at least 3% accuracy. This result allows us to understand the generalization of learning between the model and textured objects.

The experimental development is positive because of the ten classes; the worst performing was frilly, with a high correlation with the bubbly, cobwebbed,

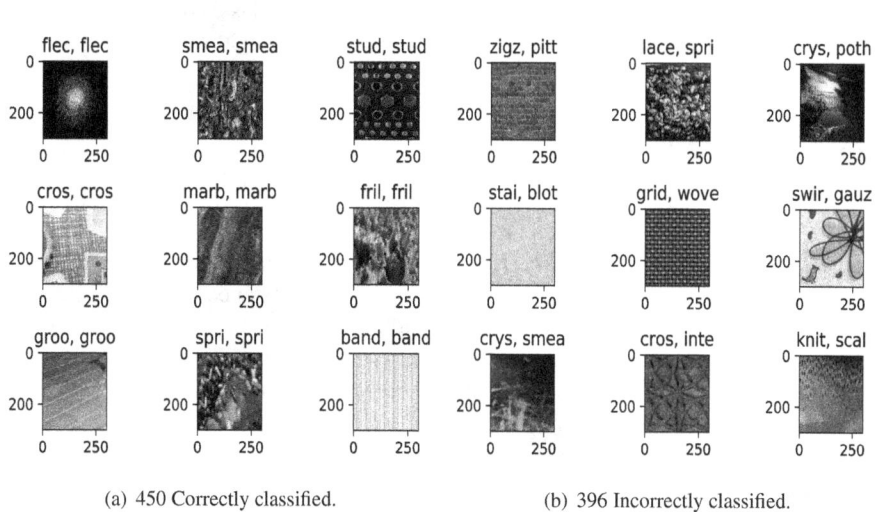

(a) 450 Correctly classified.   (b) 396 Incorrectly classified.

Figure 10.9   Classification of textures randomly (from a total of 846 images) using the DTD prediction model.

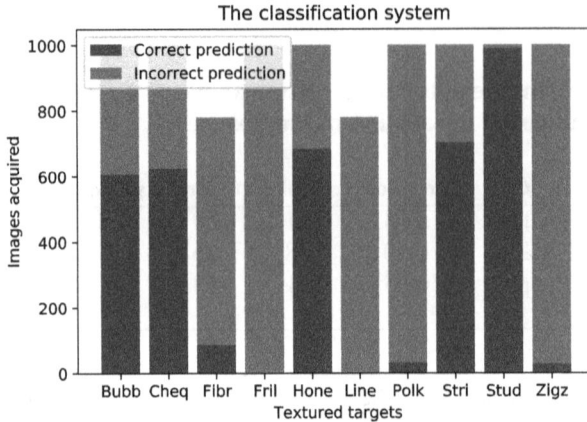

Figure 10.10 The number of images with textures obtained with the onboard camera while flying recognition.

freckled, and studded classes. In contrast, the fibrous class is highly correlated with the cobwebbed class. On the other hand, lined has a high correlation with the zigzagged, braided, striped, and swirly classes. In the case of the polka-dotted class, it performed poorly. However, it is observed that most of the predicted labels were of the dotted class.

Therefore, the two classes were observed to have almost the same characteristic patterns. In addition, the studded class obtained nearly 100% *Accuracy*. Bubbly, chequered, honey, and striped at least achieved 60% *Accuracy*. On the other hand, the striped class shows a low correlation with the cobwebbed class. Also, the zigzagged class has a high correlation with the meshed class and a slight correlation with the grid class.

10.5 CONCLUSION

The localization and object detection tasks using visual information are challenging, particularly when objects exhibit repetitive texture. However, these tasks open the opportunity for various applications using MAVs equipped with on-board cameras to be used for object detection and recognition, for instance, for parcel pick-up, place recognition, landing zone detection and many more. Seeking to improve the detection and recognition stage, in this chapter, we have investigated the use of spectral analysis in combination with deep neural networks. In particular, in this proposal, we merged the (additionally created) spectral feature maps to CNN learning. Also, it is shown that the model used achieves to eliminate overfitting and better accuracy in the classification of textures with a significant increase in the number of parameters to be trained. The tests performed in the simulation show some interesting results. The prediction model shows generalized learning of the texture attached to the objects in the proposed synthetic dataset (SADTO). Furthermore, despite having a low

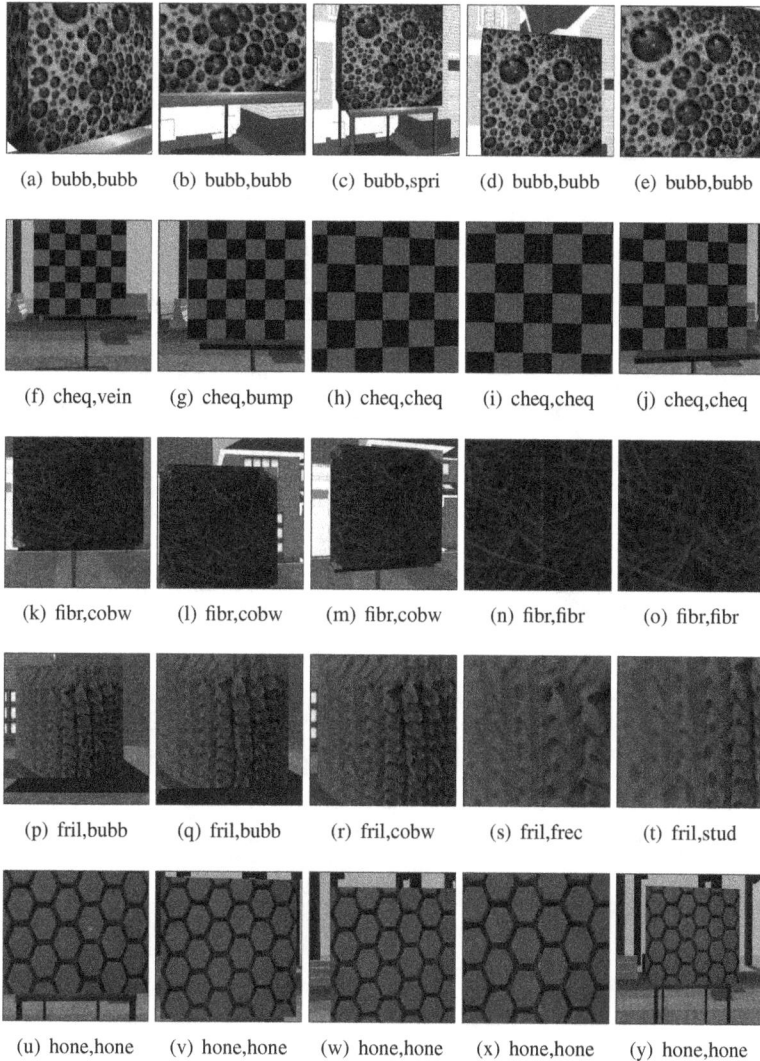

(a) bubb,bubb   (b) bubb,bubb   (c) bubb,spri   (d) bubb,bubb   (e) bubb,bubb

(f) cheq,vein   (g) cheq,bump   (h) cheq,cheq   (i) cheq,cheq   (j) cheq,cheq

(k) fibr,cobw   (l) fibr,cobw   (m) fibr,cobw   (n) fibr,fibr   (o) fibr,fibr

(p) fril,bubb   (q) fril,bubb   (r) fril,cobw   (s) fril,frec   (t) fril,stud

(u) hone,hone   (v) hone,hone   (w) hone,hone   (x) hone,hone   (y) hone,hone

Figure 10.11 Image sequence acquired by the camera onboard the drone. The classification system has a good inference on the texture in the first, second, and fifth rows.

classification rate, it is shown that the model correctly classifies most of the test classes.

As future work, this will test with other texture features, also seeking to conduct tests in real-world scenarios.

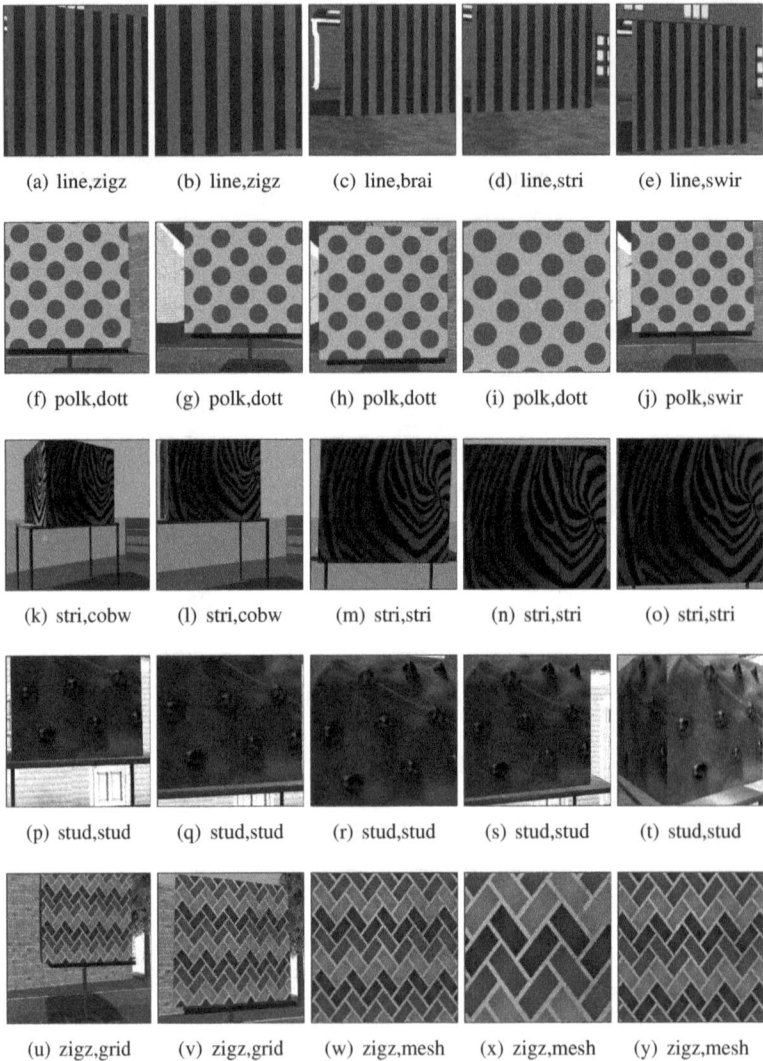

(a) line,zigz   (b) line,zigz   (c) line,brai   (d) line,stri   (e) line,swir

(f) polk,dott   (g) polk,dott   (h) polk,dott   (i) polk,dott   (j) polk,swir

(k) stri,cobw   (l) stri,cobw   (m) stri,stri   (n) stri,stri   (o) stri,stri

(p) stud,stud   (q) stud,stud   (r) stud,stud   (s) stud,stud   (t) stud,stud

(u) zigz,grid   (v) zigz,grid   (w) zigz,mesh   (x) zigz,mesh   (y) zigz,mesh

Figure 10.12 Image sequence acquired by the camera onboard the drone. In the second, third, and fourth rows, the classification system gets good classification performance.

11 Coverage Analysis in Air-Ground Communications under Random Disturbances in an Unmanned Aerial Vehicle

Esteban Tlelo-Coyotecatl
Centro de Investigación y Estudios Avanzados del IPN
(CINVESTAV), Míexico

Giselle Monserrat Galván-Tejada
Centro de Investigación y Estudios Avanzados del IPN
(CINVESTAV), Míexico

Manuel Mauricio Lara-Barrón
Centro de Investigación y Estudios Avanzados del IPN
(CINVESTAV), Míexico

CONTENTS

DOI: 10.1201/9781003385615-11

11.1 INTRODUCTION

In recent years, the use of unmanned aerial vehicles (UAVs) has increased in various sectors, such as in industrial and recreational use. In the first case, which is the focus of interest of this work, it was decided to incorporate the use of UAVs because they can perform tasks that it would be difficult for humans to perform or improve existing services. On the one hand, one of the technologies that has begun to incorporate the use of this type of vehicle is fifth generation mobile communication (5G), so that the coverage is not only provided through terrestrial base stations (BS), but this service can be incorporated in UAVs for areas where good coverage is not available or is non-existent. On the other hand, the UAVs would perform the work of an air base station (*ABS*), but the problem is that random disturbances affect the behavior of the coverage provided by the ABS. In this manner, one problem is oriented to analyze the behavior of the ABS to know what the coverage percentage is over a simulation time.

The coverage analysis in an ABS can be studied in different ways, not only in a geometric way, it can be studied by measuring the power levels received by each user in a given area, thus it is possible to know the percentage of effective coverage area provided by an ABS. Furthermore, in this chapter, the coverage analysis in air-ground communications is performed by simulating a quadrotor-type UAV, which is subjected to random disturbances in order to observe the variation in power levels received by ground users, which are reflected in the coverage area.

From the point of view of the authors in [556], UAVs are expected to be an important component for the fifth generation (5G) of cellular architectures, which can facilitate wide communication *(broadcasting)* without connection or point-to-multipoint transmissions, improving uplinks and downlinks. The main benefits for which the use of drones in 5G is being promoted are highlighted in [556]:

They can operate in dangerous or disastrous environments.

They can easily change location depending on the use that is being given.

They can offer Line-of-Sight (LoS) propagation conditions with terrestrial users.

They can adjust their height to improve the Quality of Service (QoS) depending on the user's requirements, that is, they can improve the signal strength, the desired data rate and avoid interference.

Not all drones have the same characteristics, therefore not all can fulfill the same tasks, which is why they are distinguished by their limitations, determined by their size, weight and power by their acronym SWAP (*Size, Weight, and Power*). These limitations impact the maximum altitude at which they operate, communication, coverage capabilities, calculations, and resistance of a drone [556].

Nowadays, UAVs play a very important role in 5G, showing growth in various fields due to their high mobility, relatively low cost, and easy development. Because of this, UAV vehicles aided by Air-to-Ground communication (A2G) hold promise for the next generations of mobile communications. For instance, some authors such as [42], [466], [531], [607], and [680] stress the importance of this term, which generally refers to communication between an air terminal and one on land. Within the framework of the integration of UAVs in 5G, drones serving as the aerial segment of A2G communications can operate as an ABS or as user earth (UE). In the first case, A2G communications will be between an ABS and terrestrial UEs, while in the second case, A2G communications will be between air UEs and terrestrial BSs.

According to [297], there are two popular frequency bands, 2.4 GHz and 5.8 GHz, used by companies for flight control operations. However, the frequency bands can be used for additional applications, for example, to transfer videos from the UAV to the BS a frequency of 3.4 GHz is used. For most scenarios, the 5.8 GHz band is a better choice than 2.4 GHz , because it has less interference. However, due to the lower losses that occur in a signal transmitted with a frequency of 2.4 GHz and for coverage study purposes, this frequency will be used in the simulations of this chapter.

11.2 CHARACTERIZATION OF UAV DISTURBANCES

11.2.1 DEFINITION OF COVERAGE

The term coverage refers to the geographical area within which a service is available, for example, speaking of a cellular mobile network, it would be the equivalent of the area in which the signal strength is above a threshold given by the network specifications, having, in principle, better coverage in the center of the area than at the edges. This coverage is provided by a BS, as illustrated in Figure. 11.1. The author of [3] talks about expanding coverage. In Figure. 11.1, the delimitations or maximum ranges of some base stations are shown, and an expansion system is proposed with which there would be better coverage so that the devices can be linked together. This expansion system consists of placing several BSs to obtain a larger coverage area.

According to [621], the service area is an area in which a telecommunications operator provides network communication, that is, there may be a coverage area in which the operator can provide services, but if it is not necessary to use all the area, then the service area would also be delimited according to the needs of the user. Figure 11.2 exemplifies this concept.

11.2.2 EFFECTS OF RANDOM DISTURBANCES FROM UAVS ON COVERAGE

As mentioned above, UAVs are used in some radio communication systems due to the flexibility provided by their mobility and their relatively low cost of development. What is sought to be obtained using UAVs in 5G communications are ABS, repeaters or that the UAVs themselves operate as users. In the case of ABS or repeaters, the objective is to have an alternative to expand coverage, with a good level of reliability and improving capacity. For instance, in A2G communication links, the use of ABSs is proposed considering the place where they will be positioned, with the aim of expanding the coverage provided by a BS, as sketched in Figure. 11.3. Besides, in flight dynamics, an UAV experiences random disturbances due to airflow, vibrations

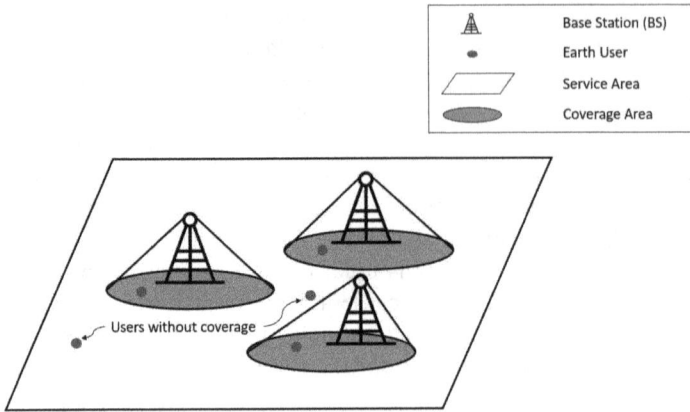

Figure 11.1 Example of the coverage concept.

in its own body, among other factors. This causes twists and distortions of the coverage area in the ABSs and difficulties in the A2G wireless communication links, as described in [730], where the simulations are based on a geometric analysis of the coverage area.

Figure 11.4 shows the model of the system, as presented in [730], where the angle α denotes the inclination of the UAV with respect to the axis z. When there is an angle of inclination, the coverage area is affected, so that terrestrial users who are on the edge of the coverage area may enter to an area without service. It is important to mention that the energy radiated by the ABS can be represented graphically by a

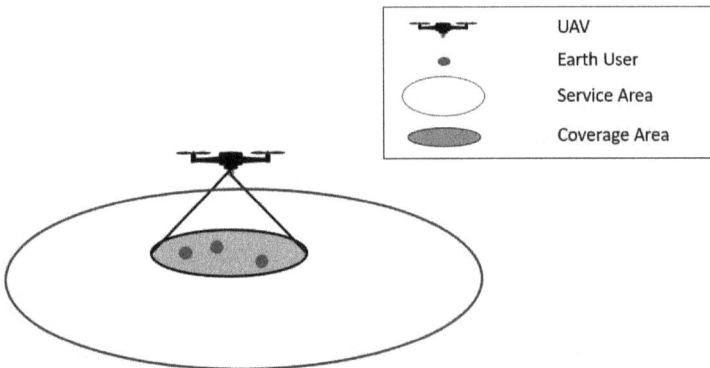

Figure 11.2 UAV coverage area.

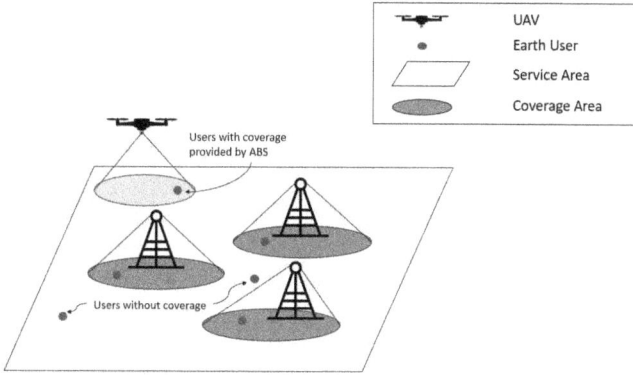

Figure 11.3 Complement of a certain coverage operator with UAV.

conical beam with a width ω_m, as shown in Figure 11.4, where the base of the cone corresponds to the coverage area. It is important to mention that the coverage area is modified due to the angle of inclination of the ABS.

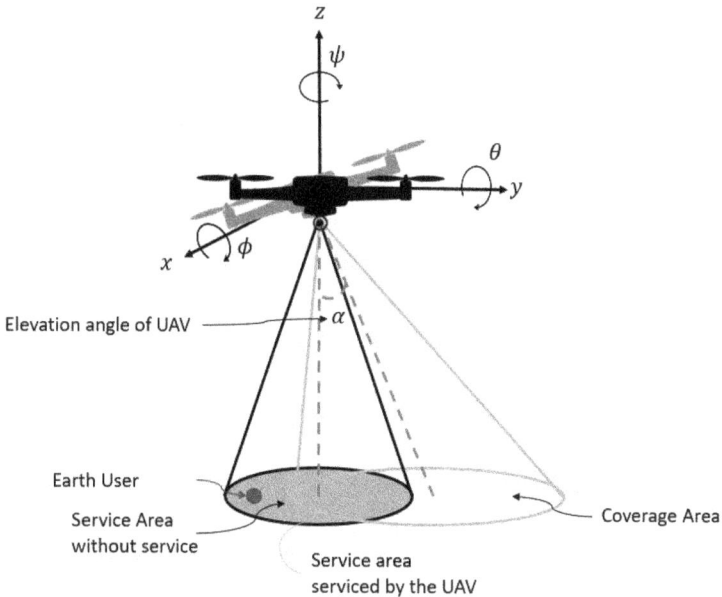

Figure 11.4 System model based on [730].

11.2.3   COVERAGE AREA REFLECTED THROUGH RECEIVED POWER LEVELS

The way in which the use of ABS becomes possible is by installing a 5G transceiver in the UAV. However, the signal intensity received by the user on the ground will be affected by the propagation conditions, the operating frequency, and the type of antenna used both in the ABS and in the user equipment. There are different types of antennas, which can have different radiation patterns as mentioned in [260]. There are theoretical and practical radiation standards (such as those that can be found in manufacturers' specification sheets). In order to illustrate the effect of antenna characteristics on the coverage concept, two cases are shown below, simulating the received power levels in a given area using an isotropic antenna with unity gain at the transmitter and a practical antenna operating at 2.4 GHz. These simulations were performed under free space propagation conditions. [171], i.e., no obstructions in the transmitter-receiver path. The free space propagation model is given by (11.1), where P_R is the received power in W, P_T is the transmit power in W, G_T is the transmitting antenna gain, G_R is the receiver antenna gain, λ is the wavelength in m, given by the quotient c/f, where f is the frequency in Hz, c is the speed of light constant in m/s and d is the transmitter-to-receiver separation distance in m.

$$P_R = P_T G_T G_R \left(\frac{\lambda}{4\pi d} \right)^2 \tag{11.1}$$

For these simulations, the transmitter was located in the center of an area testing two cases of transmit power: 1 W equivalent to 30 dBm, which is the transmit power of an ABS, see [535] and 0.316 W equivalent to 25 dBm transmit power used for ABSs in the 5G network [400]. The receiving points over this area (test points) were considered to have isotropic antennas with unity gain. To have a fair comparison with the second case of the directive antenna, the operating frequency was left fixed at 2.4 GHz. The results of these simulations are shown in Figures 11.5 to 11.8, where a range of colors can be seen starting in the center with warm colors (yellow scale) and as they move away from the center the colors become colder and colder (blue scale). This color scale represents received power levels and therefore relates to coverage as discussed at the beginning of this section. Thus, warm colors have a better coverage than cool colors, this is clearly seen in the color scale shown on the right of each figure. For the second case, it was considered that the transmitter radiated with a practical antenna taken from [243], with a directional radiation pattern thus identifying what is known as main lobe and secondary lobes. In this case, the orientation of this antenna (and antennas with similar radiation patterns) plays an important role in coverage.

Figure 11.5 shows the simulation results of the received power levels within a perimeter of 1000 m × 1000 m, placing an isotropic antenna in the center and having ideal conditions, that is, in free space conditions, where there is no factor that alters the signal propagation, other than the distance itself. The parameters used to obtain the results of Figure 11.5 were a frequency $f = 250$ MHz, $P_T = 1$ W, and for the case shown in Figure 11.7, the values were those given by the manufacturer: $f = 185$ MHz and the G_T is variable according to the angle formed between the transmitter and

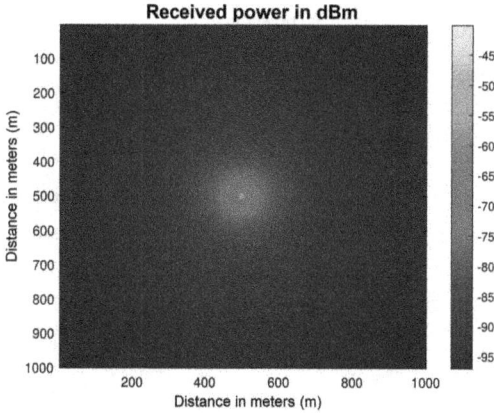

Figure 11.5 Received power levels under free space propagation conditions. Transmitter radiating with an isotropic antenna.

receiver, for simplicity $P_T = 1$ W as transmission power value. Figures 11.6 and 11.8 exemplify what is a coverage area where, for illustrative purposes, an area delimited between -80 dBm and -90 dBm is shown to have acceptable coverage [110].

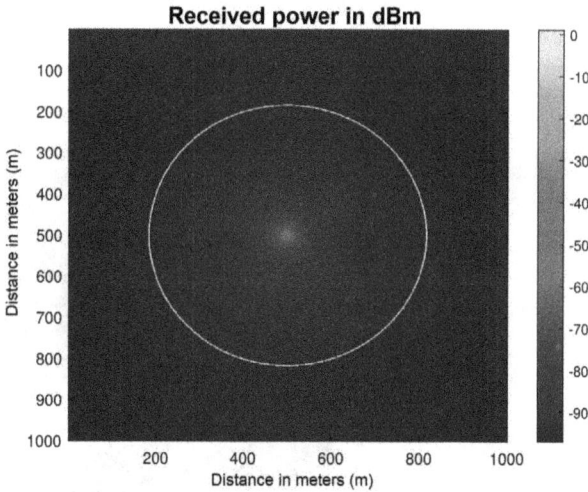

Figure 11.6 Received power levels under free space propagation conditions. Transmitter radiating with an isotropic antenna showing the coverage area.

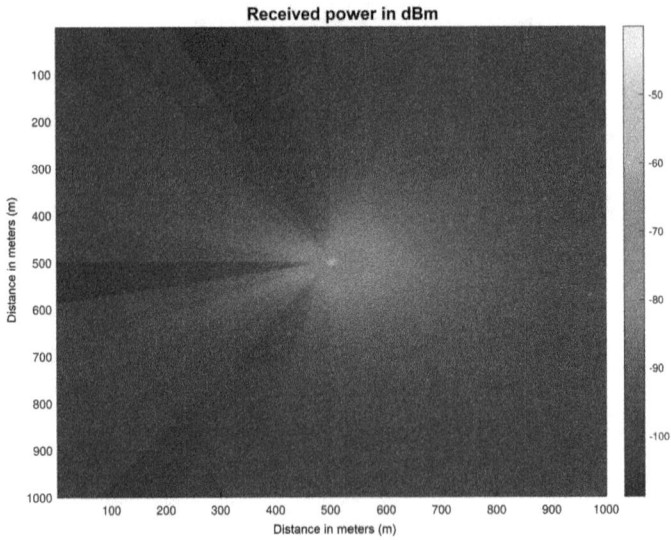

Figure 11.7   Received power levels using a commercially available radiation pattern antenna at the transmitter.

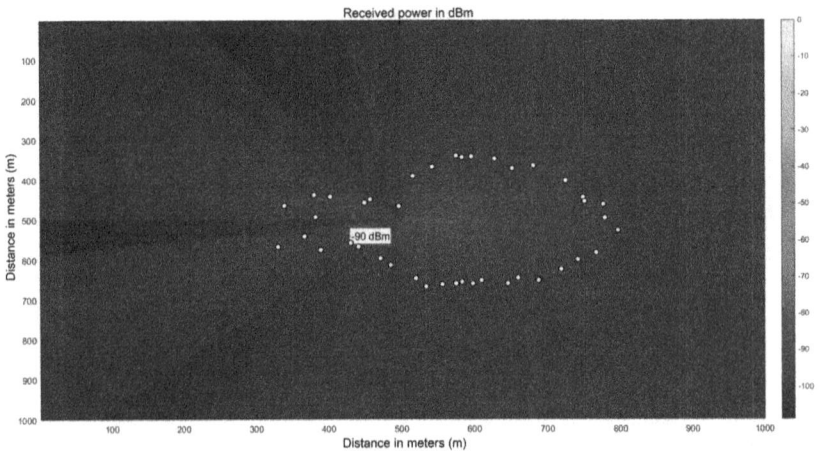

Figure 11.8   Received power levels simulating a commercial antenna on the transmitter.

11.2.4 DRYDEN WIND MODEL

To represent the random disturbances introduced in the UAV, the Dryden wind model (DWM) is used herein and described in Simulink, since it generates random turbulence and presents versatility at the time of implementation [25]. In the DWM model, the linear and angular velocity components of continuous bursts are treated as spatially varying stochastic processes that specify the power spectral density of each component. A noise component is applied, which is modeled with known spectral properties such as velocity perturbations and angle variation to the vehicle body axes. The effect of turbulence is captured during the discrete-time simulations.

The noise spectrum for each of the perturbations is described by a turbulence scale length (L) in m, the air velocity (V) in m/s and the turbulence intensity $(sigma)$. The turbulence scale length L is the length of the turbulence field expressed on the longitudinal, lateral and vertical axis $(L_u, L_v$ and $L_w)$. Turbulence intensity $sigma$ is the magnitude of the turbulence expressed on the longitudinal, lateral and vertical axis $(L_u, L_v$ and $L_w)$. In DWM, a white noise signal (band-limited white Gaussian noise) is used to excite the waves of the power spectrum, which passes through filters, called *Dryden continuous filters*. White noise is a random signal that has the same intensity at different frequencies, which gives it a constant power spectral density.

According to [225], Dryden continuous filters for velocity spectra and Dryden continuous filters for angular variations are expressed by the transfer functions derived from the square roots of the Dryden power spectrum equations. Then, the output of the filter is made up of 3 linear velocity components (u_g, v_g, w_g) and 3 angle variation components (p_g, q_g, r_g). These outputs will go to the UAV model as a perturbation signal that enters as a vector, to later be added to the translational and rotational dynamic model as an external perturbation and thus to vary the UAV position.

Figure 11.9 shows a block diagram that illustrates how the velocity components given by the DWM are introduced to the UAV model. Basically, these linear velocity components obtained from the DWM will be introduced into the UAV, added to the mathematical model of the UAV shown in the next section and with that the simulations of Section 11.5.

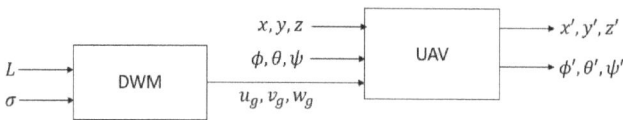

Figure 11.9 Diagram of UAV with disturbance

11.3 MATHEMATICAL MODELING OF A QUADCOPTER UAV

The following are the dynamic equations of a quadrotor derived from the Newton-Euler and Euler-Lagrange equations [209]. As a control technique, a proportional-derivative (PD) controller is used to control the flight path of the quadrotor as well as to reduce the effect of fluctuations caused by random external forces. For the mathematical modeling of the quadrotor, one can designate the vector ξ for the translational coordinates and the vector η for the rotational coordinates representing the Euler angles: *roll* ϕ, *pitch* θ and *yaw* ψ.

$$\xi = \begin{bmatrix} x \\ y \\ z \end{bmatrix}, \quad \eta = \begin{bmatrix} \phi \\ \theta \\ \psi \end{bmatrix} \tag{11.2}$$

In order to obtain the rotational control of the UAV, it is necessary to obtain the Coriolis matrix, whose effect is very important when modeling a UAV, since an object moving on any rotating system undergoes an additional acceleration (in this case the reference frame is the Earth). [482]). For the rotational and translational control, the Coriolis matrix obtained in [209] that will be an important part of the dynamics, is:

$$\mathbf{C} = \begin{bmatrix} C_{11} & C_{12} & C_{13} \\ C_{21} & C_{22} & C_{23} \\ C_{31} & C_{32} & C_{33} \end{bmatrix} \tag{11.3}$$

Where the components are

$C_{11} = 0$

$C_{12} = (Iyy - Izz)(\dot{\theta}\cos(\phi)\sin(\phi) + \dot{\psi}\sin^2(\phi)\cos(\theta)) + (Izz - Iyy)\dot{\psi}\cos^2(\phi)\cos(\theta) - Ixx\dot{\psi}\cos(\theta)$

$C_{13} = (Izz - Iyy)\dot{\psi}\cos(\phi)\sin(\phi)\cos^2(\theta)$

$C_{21} = (Izz - Iyy)(\dot{\theta}\cos(\phi)\sin(\phi) + \dot{\psi}\sin^2(\phi)\cos(\theta)) + (Iyy - Izz)\dot{\psi}\cos^2(\phi)\cos(\theta) + Ixx\dot{\psi}\cos(\theta)$

$C_{22} = (Izz - Iyy)\dot{\phi}\cos(\phi)\sin(\phi)$

$C_{23} = -Ixx\dot{\psi}\cos(\theta)\sin(\theta) + Iyy\dot{\psi}\sin^2(\theta)\cos(\theta)\sin(\theta) + Izz\dot{\psi}\cos^2(\theta)\cos(\theta)\sin(\theta)$

$C_{31} = (Iyy - Izz)\dot{\psi}\cos^2(\theta)\cos(\theta)\sin(\theta) - Ixx\dot{\theta}\cos(\theta)$

$C_{32} = (Izz - Iyy)(\dot{\theta}\cos(\phi)\sin(\phi)\sin(\theta) + \dot{\phi}\sin^2(\phi)\cos(\theta)) + (Iyy - Izz)\dot{\theta}\cos^2(\phi)\cos(\theta) + Ixx\dot{\psi}\cos(\theta)\sin(\theta) + Iyy\dot{\psi}\sin^2(\theta)\cos(\theta)\sin(\theta) - Izz\dot{\psi}\cos^2(\theta)\cos(\theta)\sin(\theta)$

$C_{33} = (Iyy - Izz)\dot{\phi}\cos(\phi)\sin(\phi)\cos^2(\theta) - Iyy\dot{\psi}\sin^2(\theta)\cos(\theta)\sin(\theta) - Izz\dot{\psi}\cos^2(\theta)\cos(\theta)\sin(\theta) + Ixx\dot{\theta}\cos(\theta)\sin(\theta)$

The Lagrangian is defined as [209]:

$$(\mathbf{q},\dot{\mathbf{q}}) = \frac{1}{2}m\dot{\xi}^T\dot{\xi} - mgz + \frac{1}{2}\dot{\eta}^T\mathbf{J}\dot{\eta} \tag{11.4}$$

Where m is the mass of the quadcopter, g is the gravitational acceleration and z is the height of the vehicle, m is the mass of the quadcopter, g is the gravitational acceleration, and z is the height of the vehicle, being $\mathbf{q}$ equal to:

$$\mathbf{q} = \begin{bmatrix} \xi \\ \eta \end{bmatrix} \tag{11.5}$$

The Lagrangian with respect to the rotational coordinates is defined as:

$$(\eta,\dot{\eta}) = \frac{1}{2}Ixx(\dot{\phi} - \dot{\psi}\sin(\theta))^2 + \frac{1}{2}Iyy(\dot{\theta}\cos(\phi) + \dot{\psi}\sin(\phi)\cos(\theta))^2 + \frac{1}{2}Izz(\dot{\theta}\sin(\phi)$$
$$+ \dot{\psi}\cos(\phi)\cos(\theta))^2 \tag{11.6}$$

Note that in (11.6) now the Lagrangian is only a function of η and its derivative, which is due to the fact that neither the translational kinetic energy nor the potential energy is considered. This is because the position of the UAV is not required to obtain the rotational control, but only its orientation is used. Then, by calculating the partial derivatives of the Lagrangian as a function of η in (11.6), the results are

$$\frac{\partial(\eta,\dot{\eta})}{\partial\phi} = (\dot{\psi}^2\cos(\phi)\cos^2(\theta)\sin(\phi) + \dot{\psi}\dot{\theta}\cos^2(\phi)\cos(\theta) - \dot{\psi}\dot{\theta}\cos(\theta)\sin^2(\phi)$$
$$- \dot{\theta}\cos(\phi)\sin(\phi))Iyy + (\dot{\psi}\dot{\theta}\cos(\theta)\sin^2(\phi) - \dot{\psi}\dot{\theta}\cos^2(\phi)\cos(\theta)$$
$$- \dot{\psi}^2\cos(\phi)\cos^2(\theta)\sin(\phi) + \dot{\theta}^2\cos(\phi)\sin(\phi))Izz \tag{11.7}$$

$$\frac{\partial(\eta,\dot{\eta})}{\partial\theta} = (\cos(\theta)\sin(\theta)\dot{\psi}^2 - \dot{\phi}\cos(\theta)\dot{\psi})Ixx +$$
$$(-\cos(\theta)\sin(\theta)\dot{\psi}^2\sin^2(\phi) - \dot{\theta}\cos(\phi)\sin(\theta)\dot{\psi}\sin(\phi))Iyy \tag{11.8}$$
$$+ (\dot{\psi}\dot{\theta}\cos(\phi)\sin(\phi)\sin(\theta) - \dot{\psi}^2\cos^2(\phi)\cos(\theta)\sin(\theta))Izz$$

$$\frac{\partial(\eta,\dot{\eta})}{\partial\psi} = 0 \tag{11.9}$$

Likewise, the partial derivatives of the Lagrangian were calculated as a function of $\dot{\eta}$, from (11.6):

$$\frac{\partial(\eta,\dot{\eta})}{\partial\dot{\phi}} = (\dot{\phi} - \dot{\psi}\sin(\theta))Ixx \tag{11.10}$$

$$\frac{\partial(\eta,\dot{\eta})}{\partial\dot{\theta}} = (\dot{\theta}\cos^2(\phi) + \dot{\psi}\cos(\theta)\sin(\phi)\cos(\phi))Iyy + \\ (\dot{\theta}\sin^2(\phi) - \dot{\psi}\cos(\phi)\cos(\theta)\sin(\phi))Izz \tag{11.11}$$

$$\frac{\partial(\eta,\dot{\eta})}{\partial\dot{\psi}} = (\dot{\psi}\sin^2(\theta) - \dot{\phi}\sin(\theta))Ixx + (\dot{\psi}\cos^2(\theta)\sin^2(\phi) + \\ \dot{\theta}\cos(\phi)\cos(\theta)\sin(\phi))Iyy + (\dot{\psi}\cos^2(\phi)\cos^2(\theta) - \\ \dot{\theta}\cos(\phi)\cos(\theta)\sin(\phi))Izz \tag{11.12}$$

Finally, calculating the derivatives with respect to time of Eqs. (11.10), (11.11) and (11.12), one gets the following results:

$$\frac{d}{dt}\left(\frac{\partial(\eta,\dot{\eta})}{\partial\dot{\phi}}\right) = Ixx(\ddot{\phi} - \ddot{\psi}\sin(\theta) - \dot{\phi}\dot{\psi}\cos(\theta)) = \tau_\phi \tag{11.13}$$

$$\frac{d}{dt}\left(\frac{\partial(\eta,\dot{\eta})}{\partial\dot{\theta}}\right) = Iyy(\ddot{\theta}\cos^2(\phi) - 2\dot{\theta}\dot{\phi}\cos(\phi)\sin(\phi) + \ddot{\psi}\cos(\phi)\sin(\phi)\cos(\theta) - \\ \dot{\psi}\dot{\phi}\sin^2(\phi)\cos(\theta) + \dot{\psi}\dot{\phi}\cos^2(\phi)\cos(\theta) - \dot{\psi}\dot{\theta}\cos(\phi)\sin(\phi)\sin(\theta)) \\ + Izz(\ddot{\theta}\sin^2(\phi) - 2\dot{\theta}\dot{\phi}\cos(\phi)\sin(\phi) + \ddot{\psi}\cos(\phi)\sin(\phi)\cos(\theta) - \\ \dot{\psi}\dot{\phi}\sin^2(\phi)\cos(\theta) + \dot{\psi}\dot{\phi}\cos^2(\phi)\cos(\theta) - \dot{\psi}\dot{\theta}\cos(\phi)\sin(\phi)\sin(\theta)) \\ = \tau_\theta \tag{11.14}$$

$$\frac{d}{dt}\left(\frac{\partial(\eta,\dot{\eta})}{\partial\dot{\psi}}\right) = Ixx(-\ddot{\phi}\sin(\theta) - \dot{\phi}\dot{\theta}\cos(\theta) + \ddot{\psi}\sin^2(\theta) + 2\dot{\psi}\dot{\theta}\sin(\theta)\cos(\theta)) \\ + Iyy(\ddot{\theta}\cos(\phi)\sin(\phi)\cos(\theta) - \dot{\theta}\dot{\phi}\sin^2(\phi)\cos(\theta) + \\ \dot{\theta}\dot{\phi}\cos^2(\phi)\cos(\theta) - \dot{\theta}^2\cos(\phi)\sin(\phi)\sin(\theta) + \ddot{\psi}\sin^2(\phi)\cos^2(\theta) + \\ 2\dot{\psi}\dot{\phi}\cos(\phi)\sin(\phi)\cos^2(\theta) - 2\dot{\psi}\dot{\theta}\sin^2(\phi)\cos(\theta)\sin(\theta)) + \\ Izz(-\ddot{\theta}\cos(\phi)\sin(\phi)\cos(\theta) + \dot{\theta}\dot{\phi}\sin^2(\phi)\cos(\theta) \\ - \dot{\theta}\dot{\phi}\cos^2(\phi)\cos(\theta) + \dot{\theta}^2\cos(\phi)\sin(\phi)\sin(\theta) + \ddot{\psi}\sin^2(\phi)\cos^2(\theta) \\ - 2\dot{\psi}\dot{\phi}\cos(\phi)\sin(\phi)\cos^2(\theta) - 2\dot{\psi}\dot{\theta}\sin^2(\phi)\cos(\theta)\sin(\theta)) = \tau_\psi \tag{11.15}$$

Once the Coriolis matrix has been calculated, it will be used in the next section to perform the rotational control of the UAV, i.e., to steer the vehicle in the desired direction considering the Coriolis effects.

11.3.1   ROTATIONAL CONTROL (ORIENTATION)

The dynamic orientation model is briefly presented in [209]. In this case, the Euler-Lagrange equations for rotational motion are

$$\tau = \frac{d}{dt}\left(\frac{\partial(\eta,\dot{\eta})}{\partial\dot{\eta}}\right) - \frac{\partial(\eta,\dot{\eta})}{\partial\eta} \tag{11.16}$$

Deriving with respect to time the expression in parentheses one gets:

$$\tau = J\ddot{\eta} + \dot{J}\dot{\eta} - \frac{1}{2}\frac{\partial}{\partial\eta}\left(\dot{\eta}^{T}J\eta\right) \tag{11.17}$$

Writing the Euler-Lagrange equations in terms of rotational coordinates and using the Coriolis matrix, since $\mathbf{C}(\eta,\dot{\eta}) = \dot{\mathbf{J}} - \frac{\partial}{\partial\eta}\left(\frac{1}{2}\dot{\eta}^{T}\mathbf{J}\right)$, one obtains:

$$J\ddot{\eta} + C(\eta,\dot{\eta})\dot{\eta} = \tau \tag{11.18}$$

When the respective angular accelerations of (11.18) are considered [209], the rotation dynamics is defined as follows:

$$\ddot{\phi} = \frac{I_{xx} + I_{yy} - I_{zz}}{I_{xx}}\dot{\psi}\dot{\theta} + \frac{\tau_{\phi}}{I_{xx}} \tag{11.19}$$

$$\ddot{\theta} = \frac{-I_{xx} - I_{yy} + I_{zz}}{I_{yy}}\dot{\psi}\dot{\phi} + \frac{\tau_{\theta}}{I_{yy}} \tag{11.20}$$

$$\ddot{\psi} = \frac{I_{xx} - I_{yy} + I_{zz}}{I_{zz}}\dot{\theta}\dot{\phi} + \frac{\tau_{\psi}}{I_{zz}} \tag{11.21}$$

The following control signal is used to perform the orientation control:

$$\tau = \mathbf{C}(\eta,\dot{\eta})\dot{\eta} + \mathbf{J}\tau_{new} \tag{11.22}$$

where τ_{new} is an auxiliary signal given by:

$$\tau_{new} = k_{1}\dot{e}_{\eta} + k_{2}e_{\eta} + \ddot{\eta}_{d} \tag{11.23}$$

It is important to mention that η_{d} is the desired orientation vector. On the other hand, the orientation error is defined as [209]:

$$e_{\eta} = \eta_{d} - \eta \tag{11.24}$$

The dynamics of the error is desired to be:

$$\ddot{e}_{\eta} + k_{1}\dot{e}_{\eta} + k_{2}e_{\eta} = 0 \tag{11.25}$$

Since at the time of the simulation the error is updated and the Equation (11.25) can be reduced by using offsetting gains k_1 and k_2, then it tends to zero, and the control signal can be described as follows:

$$\tau = \mathbf{C}(\eta,\dot{\eta})\dot{\eta} + \mathbf{J}(k_1\dot{e}_\eta + k_2 e_\eta + \ddot{\eta}_d) \tag{11.26}$$

For each angle (ϕ, θ, ψ), a rotational orientation control signal is defined in Matlab based on Eq. (11.26), such that the different signals resulted as (in Matlab code):
Rotational control signal for ϕ:

```
1    u(i+1,1) = k1_phi*e_phi_p+k2_phi*e_phi+phi_d2p(i);
```

Rotational control signal for θ:

```
1    u(i+1,2) = k1_theta*e_theta_p+k2_theta*e_theta+theta_d2p(i);
```

Rotational control signal for ψ:

```
1    u(i+1,3) = k1_psi*e_psi_p+k2_psi*e_psi+psi_d2p(i);
```

11.3.2 TRANSLATIONAL CONTROL

The dynamic model of the quadrotor [209] is given by:

$$\frac{d}{dt}\left(\frac{\partial}{\partial \dot{q}}\right) - \frac{\partial}{\partial q} = \begin{bmatrix} F_\xi \\ \tau \end{bmatrix} \tag{11.27}$$

Where τ represents the moments *roll*, *pitch*, and *yaw* and F_ξ represents the resultant force of translational motion.

The Euler-Lagrange equations for the translational coordinates are as follows [209]:

$$F_\xi = \frac{d}{dt}\left(\frac{\partial(\xi,\dot{\xi})}{\partial \dot{\xi}}\right) - \left(\frac{\partial(\xi,\dot{\xi})}{\partial \xi}\right) \tag{11.28}$$

$$F_\xi = m\ddot{\xi} + mg \tag{11.29}$$

Rewriting Equation (11.29) as a function of the state vector ξ, one gets:

$$\ddot{\xi} = \frac{1}{m}R*F - \begin{bmatrix} 0 \\ 0 \\ F \end{bmatrix} \tag{11.30}$$

Where $F_\xi = R \times F$ and R is the direct cosine matrix and F the translational force vector.

To carry out the translational control part, the following position control equations can be used [209]:

$$\ddot{x} = \frac{f}{m}(\cos\phi\sin\theta\cos\psi + \sin\phi\sin\psi) \tag{11.31}$$

$$\ddot{y} = \frac{f}{m}(\cos\phi\sin\theta\cos\psi - \sin\phi\sin\psi) \qquad (11.32)$$

$$\ddot{z} = \frac{f}{m}(\cos\phi\cos\theta - g) \qquad (11.33)$$

Where m is the mass of the UAV, f the resultant force of the translational motion, and g is the force of gravity. In the same way, it is necessary to consider the height control, which is given by (11.34), where f_{new} is a vector of forces that are being updated as the simulation progresses.

$$f = \frac{m}{\cos\phi\cos\theta}(g + f_{new}) \qquad (11.34)$$

For the case of the position on the axis z of the UAV, it is used f_{new}, which is described by:

$$f_{new} = \frac{m}{\cos\phi\cos\theta}(g - k_{1z}\dot{e}_\psi - k_{2z}e_\psi + \ddot{\psi}_d) \qquad (11.35)$$

11.4   SIMULATION INTERFACE

To carry out the simulation interface developed herein, one can chose to use the model of a PVTOL quadcopter in positive configuration, since these UAVs can be used as ABSs because if a fixed-wing drone is used, it would not be possible to perform stationary flights. Therefore, it would not be possible to perform work that requires the drone to be flying fixed at a certain altitude, such as acting as ABS in a certain area. [179]. Rotational control of a quad-rotor vehicle using rotational dynamics was described in the past Section 11.3.

11.4.1   INTEGRATION OF DWM AND QUAD-ROTOR UAV MODELS

To carry out the simulation, the rotational and translational dynamics of the system must be placed, since these will be varying according to the displacement or orientation of the quadrotor vehicle in any direction. The dynamics of the system for translation and rotation are

For translation:

$$\ddot{x} = \frac{f}{m}(\cos\phi\sin\theta\cos\psi + \sin\phi\sin\psi) \qquad (11.36)$$

$$\ddot{y} = \frac{f}{m}(\cos\phi\sin\theta\cos\psi - \sin\phi\sin\psi) \qquad (11.37)$$

$$\ddot{z} = \frac{f}{m}(\cos\phi\cos\theta - g) \tag{11.38}$$

For rotation:

$$\ddot{\eta} = \mathbf{J}^{-1}(\tau_B - \mathbf{C}(\eta, \dot{\eta})\dot{\eta}) \tag{11.39}$$

The vectors η and ξ are the orientation and position, respectively, so that as the simulation progresses one have knowledge of the position and orientation on the UAV. Now, for the purposes of this work, the integration of a perturbation model with the UAV model was sought. In this sense, in Section 11.2, the operation of the DWM model has been described, which provides the linear velocity components as outputs. ug, vg and wg. These components are the elements of a disturbance vector. Per given as:

$$\text{Per} = \begin{bmatrix} u_g \\ v_g \\ w_g \end{bmatrix} \tag{11.40}$$

Thus, this vector is introduced to the mathematical model of the quadrotor UAV, as follows:

$$\dot{\xi} + \text{Per} = \begin{bmatrix} \dot{x} \\ \dot{y} \\ \dot{z} \end{bmatrix} + \begin{bmatrix} u_g \\ v_g \\ w_g \end{bmatrix} \tag{11.41}$$

11.4.2   DESCRIPTION OF THE SIMULATION PLATFORM

The simulation platform is organized as shown in the general diagram shown in Figure 11.10. It was chosen to perform the simulation of the UAV platform in Simulink, which is a simulation environment integrated to Matlab that offers a wide range of tools to simulate almost any system using block diagrams, this is very useful since in this environment the Dryden perturbation was introduced to the modeling of the UAV platform.

The output of the dynamic model of the UAV is a vector $\mathbf{x}$, which contains ϕ, θ, ψ, $\dot{\phi}$, $\dot{\theta}$, $\dot{\psi}$, x, y, z, $\dot{x}$, $\dot{y}$, and $\dot{z}$, the variables to be used are x_7, x_8, and x_9, corresponding to the displacement of the UAV in the axes x, y, and z, respectively. The following variables are also used x_1, x_2, and x_3, corresponding, respectively, to ϕ, θ, and ψ. These values are extracted to calculate the received power levels in an external program depicted in the block shown in Figure. 11.11 and implemented in MATLAB.

The calculation of the received power is carried out under free space propagation conditions, i.e., when the transmitter and receiver have a clear and unobstructed line

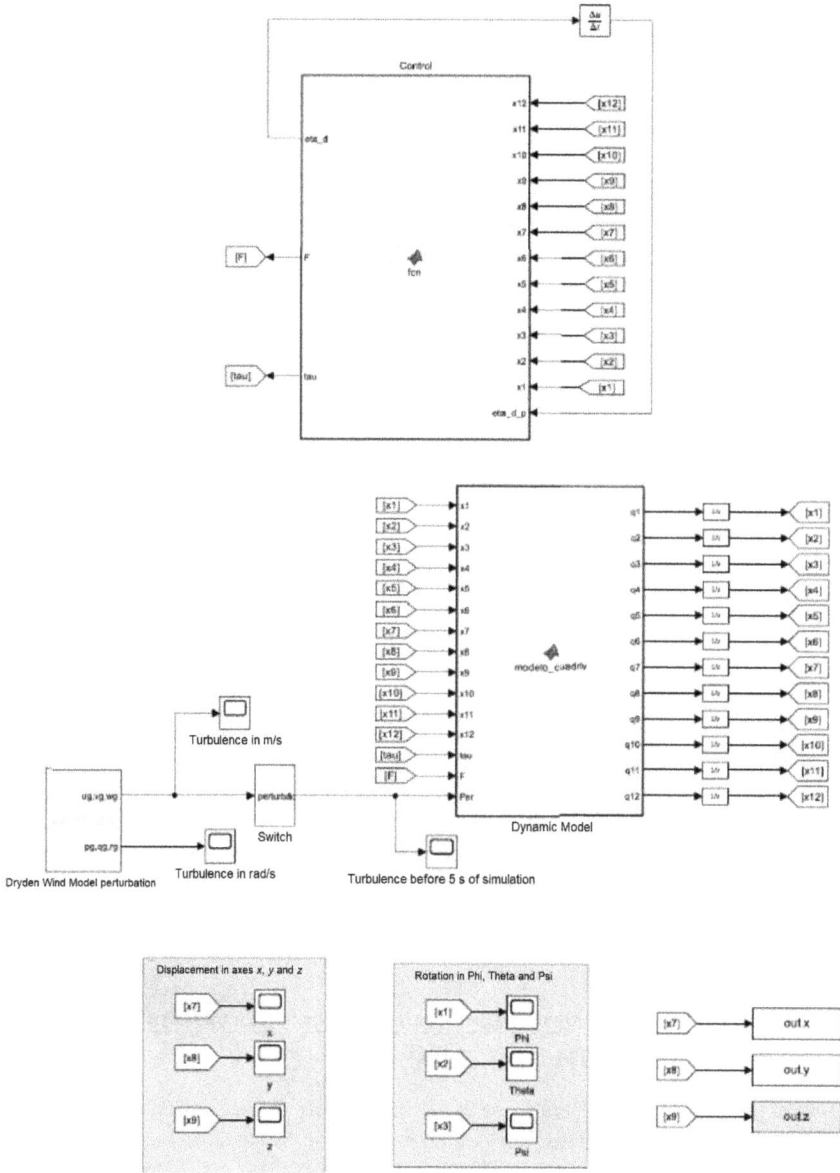

Figure 11.10 Block diagram of the simulation platform.

of sight between them, for which Eq. (11.1) can be used. In this equation, it is necessary to know the distance between the transmitter and receiver. In Figure. 11.11, one
can see this at z, which is considered as input to the block "Calculation of received
power". This will give the power matrix received on the ground, depending on the

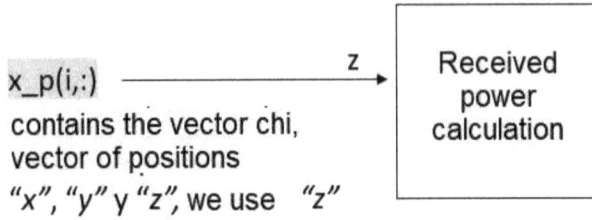

Figure 11.11   Input variables to the received power calculation block.

case, a matrix of n × n data, for the simulations shown in Section 11.5, where power level matrices were obtained from 1000 × 1000 data, and where each position in the matrix corresponds to 1 m on the ground.

In the general diagram of Figure 11.10, the control block is included. Note that in the dynamic model, a disturbance block is used following the DWM model explained in Section 11.2 and the result of which is explained in Section 11.5 in order to observe variations in the UAV positions. Once the positions on the UAV vary, the coverage will also change position because the received power levels will not be the same for the users on the ground due to the position and tilt angle α the UAV has. In Figure 11.10, a yellow label is being marked which represents the height z which will be used as input to the received power program shown in Figure. 11.11 (and labeled in yellow as well).

In addition, in the block diagram shown in Figure 11.10, 2 blocks are shown at the bottom: "Displacement in x and z axes" and "Rotation of the UAV Phi Theta and Psi" whose variables are extracted to be used in the free space program implemented in MATLAB.

11.5   SIMULATION AND ANALYSIS OF RESULTS OF THE UAV PLATFORM WITH DISTURBANCES

General simulation parameters were used, which were chosen based on the existing literature, labeled as: transmission power P_t of 0.316 W, which is equivalent to 25 dBm that is used for ABSs in 5G [400] networks. The ABS and the receiving points over this area (test points) were considered to have isotropic antennas with unity gain and the operating frequency was left fixed at 2.4 GHz.

As for the DWM simulations, the parameters used to generate the perturbations are: h (height) equal to the output corresponding to z of the simulation platform, moderate wind 6.9444 m/s and strong wind 13.8889 m/s in the x, y, and z axes.

The results of the UAV displacement once a random perturbation was introduced are shown below, using the block representing the DWM model explained in Section

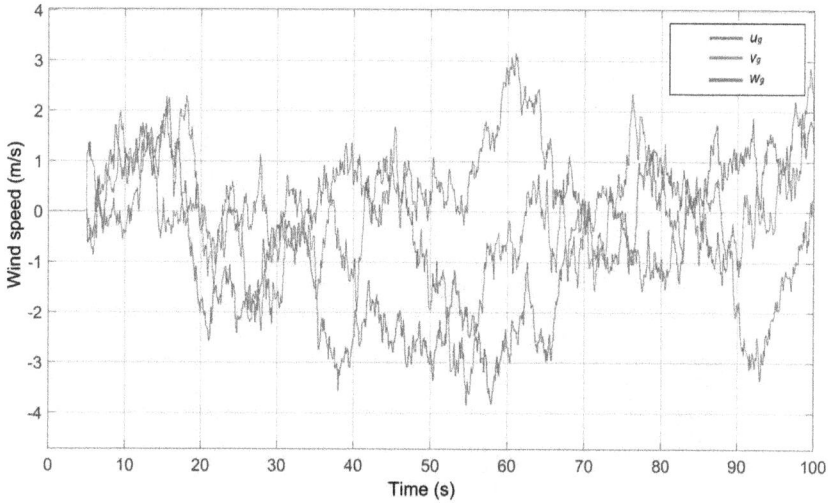

Figure 11.12 Simulation results of the perturbation vector based on the DWM initialized at 5 seconds.

11.2, which generates random turbulence. These perturbations were initiated once the UAV is in *hover* mode, which for this case was 5 s after starting the simulation. Figure 11.12 presents the result of this simulation where three graphs can be seen, each one of them corresponds to a linear velocity component as mentioned in Section 11.2. The green colored plot represents u_g, the blue colored plot corresponds to v_g, and finally the orange colored plot shows the behavior of w_g.

The displacements obtained in each axis of the ABS are used to perform the simulation of received power levels, distributing in an area of 1000 m to 10,000 ground users, which remain in fixed positions while the ABS changes its position due to the random perturbation.

These random perturbations modify the behavior of the UAV in terms of position, in an analogous way there is a modification in the angles *phi*, *theta*, and *psi*, which modify the parameters with which the received power levels are calculated.

EFFECTS OF DISTURBANCES ON ABS COVERAGE

The simulation results of the effect of disturbances in the coverage area that a UAV operating as an ABS can give are shown and explained below. The results are plotted using the command *imagesc* in MATLAB, which produces a color matrix, where zero is the maximum value that can be obtained; this value starts next to the position where the ABS is located, also at the coverage threshold that was defined. In the figures shown in this section [XY] represents the position in the *x, y* plane of the ABS, Index shows the received power value in dBm and [R,G,B] indicates the color scale of the plane at the coordinate where the ground user is located.

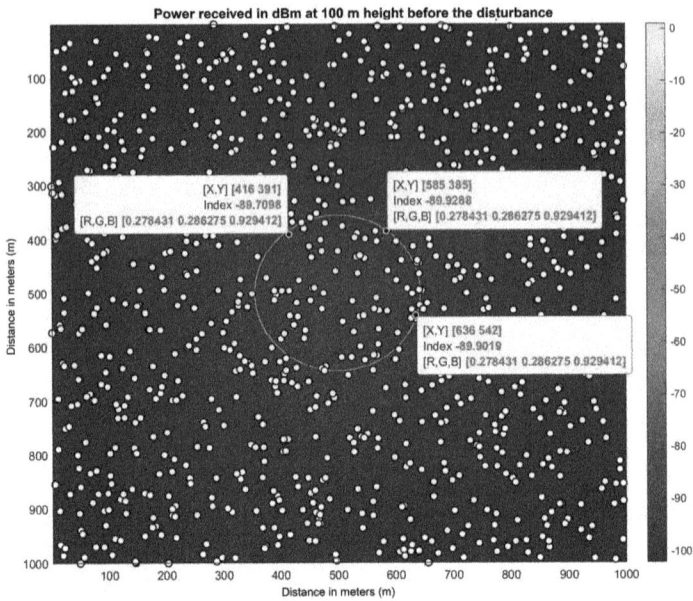

Figure 11.13   Users distributed within the coverage area prior to the disturbance.

A quantitative analysis was carried out to determine the percentage of the number of users who still have service coverage after the ABS has been disturbed. For this purpose, 1,000 users were distributed over an area of 1000 m × 1000 m. The results of this analysis are shown in Figures. 11.13 and 11.14 for the cases before and after the perturbation, respectively. As one can see, there are users that are at the limit of the coverage area, so that when the UAV moves, they may or may not be within the coverage area. For this analysis, only users within the coverage area before the disturbance will be considered. It can be seen that in Figure 11.14 only a user who was at the limits of the coverage area, which is at the coordinates [585 385] is no longer receiving a power level above -90 dBm, so he is no longer in the coverage area limited by this power threshold.

Under these conditions and in order to illustrate the above, the percentage of users affected by the change of coverage of a UAV due to disturbances was analyzed. Initially, there were 74 fixed users within the UAV coverage area without disturbances. Then, at the end of the disturbance, there were 73 users within the area, which implies a reduction to 98.64% of users. It is important to note that, in order to obtain this percentage, new users that could remain in the new coverage area generated by the UAV disturbance were not considered.

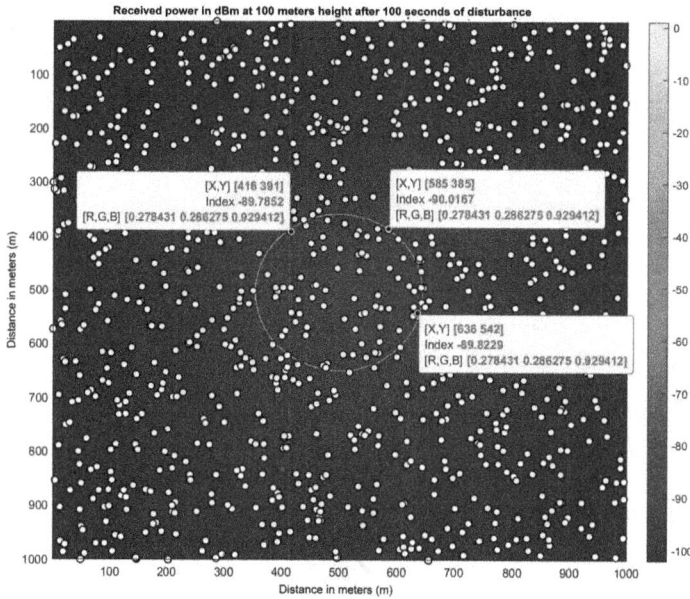

Figure 11.14   Users distributed within the coverage area after disturbance.

11.6 CONCLUSIONS

The quantitative study of the coverage provided by an ABS is of great importance due to its integration in 5G communication networks. In this work, simulations were performed to observe how the coverage area varies when introducing random pertur-bations to the ABS to affect the angle α of the ABS.

As for the height of the ABS, this has a greater impact on coverage, for example, for moderate wind, when the UAV is closer to the ground, there is no loss of cov-erage greater than 5%, while the higher the height of the UAV, the maximum loss of coverage is 6%. On the other hand, when the wind speed is higher, it generates more displacements in the UAV, even so, the percentage of users within the coverage remains above 90%.

By taking into account how the percentage of the coverage area is being affected, further work can consider other characteristics of the UAV in terms of size, weight, and power (SWAP), as these are the basic characteristics for use cases of a UAV operating as ABS.

12 A Review of Noise Production and Mitigation in UAVs

Caleb Rascon
Instituto de Investigaciones en Matematicas Avanzadas y en Sistemas, Universidad Nacional Autonoma de Mexico, Mexico City 04510, Mexico

Jose Martinez-Carranza
Computer Science Department, Instituto Nacional de Astrofísica, Óptica y Electrónica (INAOE), Puebla 72840, Mexico.

CONTENTS

12.1 INTRODUCTION

The growth of the drone market for leisure and civilian applications has raised concerns regarding the public acceptance of drone technology. These concerns include safety, health, animal well-being, and noise pollution issues. The lack of specific standards for drone noise underscores the need to establish means to measure its effects and develop guidelines for deployment in urban settings.

Public acceptance of drones is affected by factors such as annoyance caused by sound levels, societal safety, and environmental impacts. The continuous presence of noisy drones flying over homes and workplaces may lead to decreased public acceptance due to safety and health concerns, similar to living or working near airports.

In this context, this chapter aims to explore the aforementioned issues, touching on three main subjects: 1) the impact of UAV noise on society, 2) noise modeling, and 3) noise mitigation techniques.

DOI: 10.1201/9781003385615-12

Each section aims to provide the reader with a concise description of recent works delving into each subject. From this review, we argue that the topic of noise produced by UAVs or drones (as they are commonly known) is an area with potential research and development opportunities seeking to achieve the concept of urban air mobility (UAM), for which drones are the fundamental element. In this context, drones are envisaged as commercial products for different civilian applications, from catastrophe response, research, and civil protection, to parcel delivery and air taxis.

However, such applications are questioned by the public around the world due to different concerns, namely, criminal activities, misuse, and privacy breaching. Even applications such as parcel delivery and air taxi, with clear benefits to individuals, are perceived with skepticism. In fact, in the case of air taxis, a human pilot is demanded by the users before thinking of making use of such technology.

The above has to be faced on different fronts and one of the most obvious is that of public acceptance, which will increase if concerns around commercial applications are mitigated. This implies the development of guidelines, regulations, liability and insurance policies, and mitigation mechanisms to guarantee the safe use of drones in civilian spaces.

From our review, we noted that public acceptance tends to see as positive those applications that are non-profitable, such as emergency response, research, and civil protection. More advanced commercial applications are also well received, as long as flight operations are not performed near to their home or workplace. This is known as the "not in my backyard" response, which is strongly related to safety, liability, and privacy concerns.

We will also discuss that, even when drone noise does not seem to be a priority, it becomes relevant if the UAM concept becomes a reality with the deployment of hundred or thousand of drones whose collective noise may become worse than urban traffic or aircraft noise in airports. In this respect, drone noise may become a health risk to the public, and therefore, it is essential to develop studies to understand the impact of noise on the public. However, adequate metrics and experimental frameworks must be developed for the latter to enable a thorough evaluation.

Finally, we also discuss noise mitigation techniques, which could enable the development of friendly drones that could be deployed seamlessly in urban applications without calling the attention of civilians. For instance, drones without noise could be used in sports and social events and even for studies of wildlife, since it has also been demonstrated that animals note the presence of drones and may become stressed by them.

We conclude the chapter with a discussion and final remarks, reflecting on the fact that the study of drone noise will contribute to the acceptance of drone technology by the public.

12.2   UAV NOISE AND ITS IMPACT ON SOCIETY

The growth of the drone market for leisure and civilian applications, namely, aerial video recording and photography, precision agriculture, and outdoor monitoring, among many others, has raised concerns regarding population acceptance of this technology. Several issues arise when unmanned aerial vehicles, known informally as drones, from safety concerns to health and pollution issues from the noise produced during flight. Motivated by these issues, in this section, we will briefly revise relevant works on drone noise and its effect on public acceptance, including the impact of noise on wildlife, including some proposals to evaluate and measure more effectively the effect of noise on the public.

First, it is important to mention that one of the main indicators, mentioned in several of the works in the literature, to attempt to measure a drone acoustic is that of sound pressure level (SPL), see Section 12.3, Equation 12.1.

SPL measures the intensity of sound waves relative to the minimum threshold of human hearing, using decibels (dB) as a unit. SPL is a logarithmic scale, this is, every increase of 10 dB represents a ten-fold increase in sound pressure.

Sound pressure levels are often used to describe the loudness of sounds in various environments, such as concerts, workplaces, and residential areas. Exposure to high sound pressure levels can have detrimental effects on human hearing, so it is important to monitor and control noise levels in various settings to prevent hearing damage. Some SPL and their corresponding sources are indicated in Table 12.1 [517]:

Drones are a relatively new source of noise in the environment. In 2019, 1.32 million small UAVs (weighing less than 25 kg) were registered for leisure, while 385 thousand were registered for professional applications in the USA. It is forecasted that by 2024, registered drones will increase by 12% (to 1.48 million) for leisure and 115% (to 0.83 million) for professional applications. The ANSI Standardization Roadmap of 2020 [299] highlighted that no specific standards for drone noise currently exist.

This underscores the significance of establishing means to measure the effects of drones and developing guidelines for their deployment in urban settings.

Table 12.1
Noise Levels from Different Sources taken from [517]

Sound Source (Noise)	SPL dB	Level
Aircraft at take off	120	Extremely Loud
Car horn	110	Extremely Loud
Subway	100	Very Loud
Truck, motorcycle	90	Very Loud
Busy crossroads	80	Very Loud
Noise level near a motorway	70	Loud
Busy street through open windows	60	Moderate
Light traffic	50	Faint
Quiet room	[40 30 20]	Faint
Desert	10	Faint
Hearing threshold	0	Faint

Addressing this issue is the primary objective of the study presented in [550]. The article highlights that only a handful of research teams have conducted empirical testing on the short-term impacts of drones on humans, despite the possibility of their acoustic or visual features adversely affecting people's well-being. Also, it is stated that discomfort towards drone noise strongly depends on the SPL as it happens with other transportation noise sources. This work highlights several key points: 1) on average, drones are more annoying than road vehicles in terms of SPL; 2) hovering multi rotor drones cause more annoyance than starting jet aircraft; and 3) soundscapes can increase annoyance even when road traffic noise is high. The SPL of a drone depends on its type, size, weight, number of propellers, and distance from the observer, ranging from 60 to over 100 dB. Therefore, the authors in [550] call for more comprehensive evaluations across a wider range of urban scenarios to better understand the noise annoyance caused by drones.

Apart from annoyance caused by sound levels, there are other concerns that affect the public acceptance of drones. These include societal safety and environmental impacts, which have been studied in [101] through various surveys in different countries, and keeping in mind the new concept of UAM, which considers the use of drones for commercial purposes heavily oriented to civilian applications. While the surveys do not indicate any clear trend in the level of acceptance of drones and urban air mobility from 2015 to 2021, the latest survey shows an 83% acceptance rate, the highest yet, as opposed to less than 63% in previous years. However, this rate is not conclusive for commercial use. The surveys reveal that most Americans find the concept of drone delivery attractive, but a significant portion are undecided, and different groups have varying levels of interest in drone delivery. The most pressing issues include drone malfunctions, misuse, privacy concerns, potential damage, and nuisance. The results of surveys also indicate that the costs of drone services and the amount of time saved have a significant impact on respondents' attitudes toward drones. Drone operations related to health and welfare generally receive higher levels of acceptance than those related to leisure or business.

Public opinions on drones vary over time and across countries. However, there seems to be a growing trend of acceptance for urban air mobility, with around half to three-quarters of the public supporting this idea. It is worth noting that people tend to react positively to business models involving drones, provided that safety is ensured and the drones are deployed away from residential or workplace areas, which is often referred to as the "not in my backyard" response. From [101], we can summarize a list of societal concerns:

Environment: noise impact; emissions impact; impacts on animals and flora; recycling; impact of climate change; visual pollution; loss of privacy/intrusion; nuisance.
Safety: safety concerns; security concerns; cybersecurity.
Fairness: lack of transparency; cost of services; competency; liability.
Economy: jobs; economic viability; demand.

Public opinions on drones vary across time and countries, but there is a growing trend of acceptance for urban air mobility, with between half and three-quarters of the public supporting this idea. However, people generally react positively to business models involving drones only if safety is ensured and the drones are deployed away from residential or workplace areas, a response commonly known as "not in my backyard". According to [101], societal concerns related to drones include noise impact, impacts on animals and flora, visual pollution, and loss of privacy or intrusion. The same work provides a list of mitigations to address these concerns, such as avoiding or limiting hovering drone flights, establishing no-fly zones and times, registering electronic equipment, obtaining flight permits and insurance, and implementing similar mitigation strategies used for other urban and aerial transports.

Similarly, the works in [165, 245] evaluated public acceptance of civilian drones for daily-life applications in a study conducted in Germany in 2018. Notably, the "not in my backyard" response was also observed among German participants in the study. The initial insights from the study revealed that applications such as package delivery or aerial taxis are compelling as long as they do not operate above residential areas. However, there were strong concerns about the increase in drone operations.

The study results showed the following general attitudes toward civil drones in Germany: 49% positive, 43% negative, and 8% undecided. It is noteworthy that the positive attitude is only slightly in favor, which could easily change after negative news involving drones. The respondents in the study expressed concerns about civil applications of drones in the following proportions, as shown in Table 12.2. Note that the most pressing concern is the potential for misuse for criminal activities, followed by privacy breaches, liability, safety and damages, and lastly, noise. The low level of concern regarding noise may reflect a lack of awareness about the noise annoyance produced by drones, which is not surprising given that drones are currently not as ubiquitous as other technologies such as mobile phones, GPS, or the Internet.

Table 12.3 provides a breakdown of the participants' attitudes toward various drone applications. In summary, the most accepted applications are those related to catastrophe response, research work, and civil protection, which are activities that may be seen as non-profitable in commercial terms. However, as the applications move towards a business model, the acceptance level decreases. Surprisingly, parcel

Table 12.2

Concerns about civil applications of drones

Concern	Percentage
Noise	53%
Animal welfare	68%
Damages and injuries	72%
Transport safety	73%
Liability and insurance	75%
Violation of privacy	86%
Crime and misuse	91%

Table 12.3

Attitudes toward different drone applications (DNK: Do Not Know; DF: Disagree Fully; DS: Disagree Slightly; AS: Agree Slightly, AF: Agree Fully.). Numbers are in percentage (%).

Applications	DNK	DF	DS	AS	AF
Catastrophe response	2	3	3	27	65
Science and Research	0	3	6	37	54
Life-saving and civil protection	0	5	8	27	60
Traffic, surveillance, infrastructure monitoring	3	7	11	36	43
Medicine, transport of samples	2	8	15	26	49
Farming	3	12	18	32	35
Pictures and videos (news)	2	18	22	39	19
Leisure time activities	0	23	30	34	13
Parcel delivery	2	27	32	25	14
Pictures and videos (advertisement)	1	44	29	17	9

delivery is one of the least popular applications. This may be related to the "not in my backyard" response, where an individual may recognize the advantages of aerial parcel delivery but also perceive it as a vulnerable activity that could raise concerns similar to those listed in Table 12.2.

It is important to note that even when the drone applications are for personal purposes, the acceptance trends remain the same. The acceptance percentages for five specific applications that the participants would use personally are shown in Table 12.4. The applications related to non-profit or social benefit receive a higher approval, but still not exceeding 50% as shown in Table 12.3, and approval rates decrease as the applications move toward a business model. Air taxi applications receive the largest rejection, which is understandable as people prioritize safety when traveling. The "not in my backyard" response reflects an attitude towards maintaining safety and privacy in their personal space.

In summary, the work described in [165, 245] found that four out of ten respondents had a negative attitude toward civil drones, while five out of ten had a positive attitude, with the rest being undecided. Attitudes were influenced by gender, age, and level of information. Noise concerns, especially among women, were confirmed as important factors for acceptance. Advanced air mobility for passenger transport had little acceptance, with 85% of respondents indicating they would not use air taxis, but if used, a pilot should be present, which is very relevant for autonomous drone applications given that such an opinion could affect the funding of this type of research. Establishing regulations for noise management is a fundamental challenge for sustainable UAM. Further research should continue to propose relevant regulation.

Table 12.4

Attitudes toward different drone applications for own purposes (DNK: Do Not Know; DF: Disagree Fully; DS: Disagree Slightly; AS: Agree Slightly, AF: Agree Fully). Numbers are in percentage (%).

Applications	DNK	DF	DS	AS	AF
Protection by police, fire brigades	1	7	9	34	49
First aid	2	11	10	31	46
Parcel delivery	0	44	27	17	12
Leisure time activity	0	47	28	15	10
Air taxi	1	55	30	9	5

Although previous research indicated that noise was one of the least concerning issues related to drones, it is important to note that the deployment of noisy drones in urban areas may have negative implications. While a single drone may not be an issue, the continuous presence of many drones flying over homes and workplaces may lead to decreased public acceptance due to safety and health concerns, similar to the impact of living or working near airports. Therefore, it is important to consider the noise impact of drones when designing and deploying them in urban areas.

The World Health Organization (WHO) has found significant evidence where aircraft noise is linked to cardiovascular disease, sleep disturbance, annoyance, and cognitive impairment. They strongly recommend avoiding exposure to levels above 45 dB, as this level has been associated with adverse health effects. It is also recommended to avoid night-time exposure to more than 40 dB, as it has been linked to adverse effects on sleep [459].

The concern over drone noise has prompted research groups to develop studies that call for the creation of evaluation frameworks and metrics. These frameworks aim to understand the effects of drone noise on humans under various scenarios such as urban, rural, and in different soundscapes, and under different drone deployment numbers. Several studies have been conducted in this area [120, 615–617]. Some researchers have even developed studies in virtual environments to simulate the deployment of thousands of drones, which could provide insight into potential noise effects in large areas [58].

Lastly, we highlight the research conducted in [415], which aimed to investigate the impact of drone noise on the behavior of 18 different species of megafauna. The findings revealed that the different species have varying levels of sensitivity to noise at different frequencies generated by drones. This indicates that drone noise can potentially cause behavioral changes in animals, highlighting the need to reduce drone impact on target species and to inform future drone use guidelines. This is a significant aspect that matters to the public as urban and rural areas are home to various animal species, including pets and strays, whose well-being can also have an impact on the public. Additionally, animal welfare is widely promoted, making it essential to consider their protection and care in drone operations.

12.3 NOISE MODELING

As presented in Section 12.2, there is a clear impact of the noise emanating from UAVs toward humans from both the psychological and physiological standpoint, as well as other animal species. It is then of interest to get into how such noise is produced, so as to be able to mitigate it (the topic of Section 12.4).

Conventional UAV noise can be separated into several components [284]:

The vibration of the UAV inner components.
The flow of air through the UAV, pushed by the propellers.
The air turbulence created by the propellers.

The "vibration" component is usually considered considerably smaller compared to the other two components, to the point that is rarely even mentioned. In fact, of the cited references, only [627] mentions it, and concludes that it can just be ignored.

As for the other two components, it appears that the "flow of air" component tends to not be as "annoying" than the "air turbulence" component [550]. As such, the air turbulence caused by the UAV's propellers, at least from what we can gather from the literature, is the main culprit of its noise and, thus, the one with the most effort behind its modeling and mitigation. To this effect, in this work, we will refer to this component as "UAV noise" as a whole, with the understanding that the other two are present but not as imminent to be mitigated.

Several studies [32, 58, 120, 157, 182, 242, 312, 504, 550, 616, 675] have come to the conclusion that UAV noise has the following characteristics:

Tonal. A motor by itself, with its accompanying blades, produces noise that bears one tone: its fundamental frequency (ω_F). Although other tones (or frequencies) are present, the vast majority of which are harmonics located at frequencies $\omega_{h_i} = i * \omega_F$ (where $i > 1$ is an integer). The other frequencies may be attributed to the other aforementioned components, and not as prominent.

Broadband. As it can be trivially deduced, ω_F varies depending on the speed the motor is rotating. However, even though ω_F can range from low- to mid-frequency tones, its harmonics occupy a wide range of high-frequencies (up to a 2 orders of magnitude greater than ω_F), all of which bearing a considerable amount of power, decreasing to a level barely below the power at ω_F.

These characteristics in conjunction, result in a loud and "whiny" sound that, as mentioned before, is considered very "annoying" to humans, even when it's not as loud as other types of urban noises [550].

Furthermore, the power and tonality of the UAV noise appears to be affected by several variables. One of which, as previously mentioned, is the *speed of the motor* (measured in rotations per minute or RPM). It has been successfully used [675] to create a noise emission model based on multiple regression approach, where the power of the whole UAV noise is divided by the contribution of each motor based on its RPM at each moment in time. Another variable is the *number of propellers*, which minimizes the power of the noise the more they are [222]. However, the authors of [98] concluded that even though decreasing the number of blades does in fact increase the power of the noise (since the motor requires to turn faster to create the same amount of thrust), it shifts that power to lower frequencies, making it less "annoying". Yet another variable is the *shape of propellers*, the design of which have been inspired by owl feathers [453,667], optimized outright with different parameter constraints [222], specifically to modify the tonality and power the UAV noise, and even ones that shapes can be modified on-the-fly [182].

It is important to note that while UAV noise is cumbersome, because of the impact of its motor speed, number and shape of its propellers to its tonality, the combination of all these variables can also be considered unique to an specific UAV model. Meaning, the UAV noise can actually be used as a type of an acoustic fingerprint [503] just by extracting its frequency cepstral features. In fact, as it will be discussed later, even its own tonality can be used as part of a process to mitigate it [444].

There have several works that attempt to model how UAV noise is created:

[675]: as mentioned earlier, it is based on integrating the noise from several motors, each modeled based on its RPM throughout time.

[157]: it integrates the frequency-domain version of the Ffowcs-William and Hawkings equations (for drone noise prediction), with an aerodynamic model based on Blade Element Momentum Theory.

[300], based on Lattice Boltzmann method (LBM) with a simple mathematical ellipsoid model for steady loading noise; unsteady loading noises were not able to be accurately predicted. A similar model was also proposed in [470].

Although all these models were moderately successful in modeling the creation of UAV noise, it is also important to model how this noise propagates through the air. It is well established [284] that the magnitude of the SPL can be defined as $L = 10 log \frac{p}{p_r}$ in decibels (dB), where p is the sound pressure magnitude of the acoustic signal, and p_r is a reference pressure of $2 \times 10^{-5} \frac{N}{m^2}$. The SPL of a propagated signal $L(r)$ at a distance r from the source of such signal, can be modeled as:

$$L(r) = L_0 e^{-\frac{\alpha(\omega_F)r}{2}} \tag{12.1}$$

where L_0 is the SPL of the signal at its source. Thus, the farther we are from the source, the less SPL the signal has. Additionally, the air in which the signal propagates acts as a type of low pass filter, since higher frequencies tend to be absorbed by the vibration of its nitrogen and oxygen molecules. This is represented as the attenuation factor $(\alpha(\omega_F))$ that imposes further SPL diminishment as the fundamental frequency of the signal increases.

Of course, there are other elements at play, like reflections from objects in the UAVs surroundings (such as buildings or even the floor itself) or other noise sources, which may increase or decrease the captured SPL $L(r)$. All of this is incorporated in ENVironmental Acoustic Ray-tracing Code (EnvARC) that was used in [58] to simulate the noise of UAVs flying through a simulated urban area, and worked well.

In addition to the distance to the source, the measured SPL is also affected by one's vertical direction in relation to the UAV (aka. elevation angle). In [550], the authors integrate the findings of two other works, and concluded that: 1) the noise SPL increased toward the top and bottom of the UAV (meaning, as the absolute value of the elevation angle increased); and 2) the noise-SPL-to-elevation ratio increased as the noise ω_F increased.

The UAV's pitch angle is also to be considered, since when it is at 0°, both the pitch and the elevation angle share the same frame of reference. However, when the UAV is moving, caused by a difference in speed between its motors, the UAV's pitch changes and, in turn, the elevation angle as well. To this effect, in [488], the authors found a similar positive relation between noise SPL and pitch.

Interestingly, the authors of [24] found that, from the source, the noise emission is greater on elevation angles closer to the middle. This may seem as contrary to what was found in [488, 550]; however, it is important to remember that the SPL noise emitted from the source may not be indicative to the SPL that is measured from a distance. For example, a lower elevation angle is closer to the motors and, thus, it is expected to be louder. On the other hand, measuring the SPL from a distance toward the bottom of the UAV should be louder than from the sides since air is being pushed towards that direction, contributing to a higher SPL.

Finally, it is also important to mention that SPL is not the only manner in which one can measure the impact of UAV noise. As it was mentioned earlier, UAV noise is not actually louder than other types urban noise (such as traffic or airplanes), but it is considered more "annoying" [550]. To this effect, the authors of [617] found that measuring the noise in terms of Tonality, as well as a combination of Loudness and Sharpness, presented a closer match to a rating of a group of human listeners. This is consistent to the aforementioned tonal nature of UAV noise as being a prime characteristic of its unseemliness.

12.4 NOISE MITIGATION TECHNIQUES

As it can be deduced, mitigating UAV noise is of great interest, given that most of the backlash it causes to all the scenarios they could be employed, limiting its applicability. Having described the different characteristics of UAV noise and the efforts behind modeling it, we will now focus in presenting an overview of current approaches to mitigating it.

From what we could gather from the literature, there seem to be four types of mitigation techniques:

1. Trajectory optimization
2. UAV and motor blade design
3. Automatic noise canceling

Their advantages and disadvantages will be discussed specifically in this section, and generally in Section 12.5.

It is important to note that the World Health Organization, guided by the standard ISO 1996-1:2016 (section 3.6.4 of ISO 2016), has established a "day-evening-night-weighted sound pressure level" (L_{DEN}). This can be used as a type of threshold [293], or an objective to reach for all these techniques.

12.4.1 TRAJECTORY OPTIMIZATION

Since the SPL of a noise source diminishes when the distance to such source increases, flight paths can be designed such its impact is reduced. In this regard, two mitigation processes can be employed [284]:

1. *Arrival and Departure*. With the types of air-crafts that are able to fly sufficiently high that its noise its negligible, it is really on the arrival and departure phases of its flight when noise mitigation should be considered. To this effect, airport placement, and the design of its respective arrival and departure paths, can consider the terrain/elevation of its surroundings, as well as its demographic makeup to mitigate noise.
2. *Whole Trajectory*. As for other types of air-crafts, such as UAVs, where its flight height is not enough to mitigate its noise, the whole trajectory is to be optimized as to mitigate noise. To do this, it is required to have *a-priori* knowledge of land use by a population that is aimed to not be affected, as exemplified in Figure 12.1.

As it can be deduced, the vast majority of flight-path-design-based noise mitigation found in the literature is focused on the second process. A good example of this is the work in [293], which uses the iNoise software to simulate noise to calculate noise (following the standard ISO 9613), and using a rough estimate of land use from

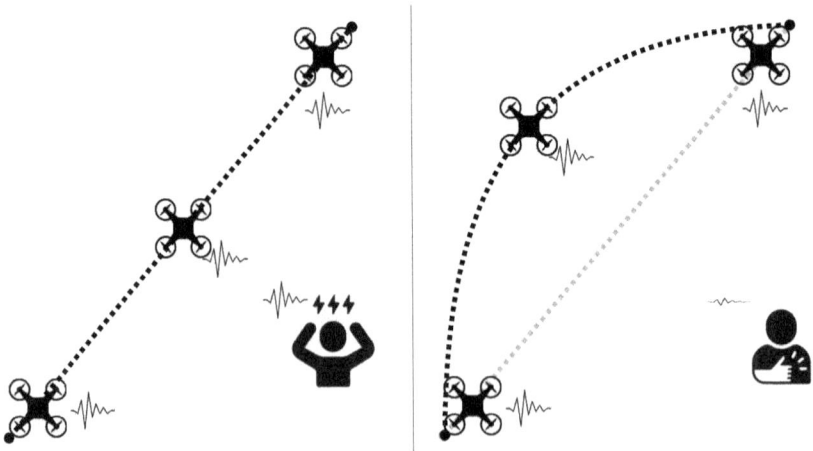

Figure 12.1 Example of trajectory optimization. Left: drone is too close to affected individual. Right: trajectory is changed so as to reduce noise impact.

an artificially chosen town, provided three flight paths. However, all three flight paths exceeded WHO's L_{DEN} value, even with a small quad-copter.

The work presented in [15] reduces the flying space to an environment, draws a grid over it, and establishes obstacles within in, as well as "quiet zones", areas that should be noise-free. Then, the A* algorithm is used to create different paths, each with a different level of "quietness". Another similar work is in [726], where the A* algorithm is modified (into what the authors refer to FairNoiseA*), where the cost of node transitioning is weighted under the amount of noise that will be generated by such transition.

The work in [570] uses agent-based modeling so that UAVs avoid paths that are currently heavily used by other UAVs and, thus, can be assumed that are noisy. Interestingly, this work uses a grid-based trajectory optimization as well, however, because of its complex-system nature, a comparison to A* is difficult.

As for the work in [58], it has an important similarity to the work [15], but uses ENVironmental Acoustic Ray-tracing Code (EnvARC) to simulate noise in a seemingly more sophisticated fashion. Some paths were proposed in a given flight space; however, no trajectory optimization was presented.

It is important to mention that an important limitation of this type of noise mitigation techniques is that the location of quiet zones need to be known *a-priori*. This can be difficult to obtain in some scenarios, such as human activity changing throughout the day. An example of this can be an industry park or a university, both of which are usually not occupied during late evening but are heavily occupied in the mornings and afternoons. Additionally, as established in [570], UAVs increase noise activity once present in an acoustic environments but decrease once they leave it. Thus, trajectory optimization requires up-to-date knowledge of the environment that can be highly dynamic and may not be available.

12.4.2   UAV AND MOTOR BLADE DESIGN

By far, this is the type of technique that is the most used for UAV noise mitigation and can be divided into two parts:

1. Modification of the whole UAV
2. Blade design

The first parts involves the modification of the structure of the UAV to reduce its noise emission. The work in [424] presents different manners of soundproofing the UAV, specifically by the use of rubber, foam pads, and aluminum foil. It was found that foam pads helped reduce the noise by a considerable margin (around 15 dB). The work in [566] proposes a UAV with the shape of flying saucer. By employing the Coanda effect [239], which states that a fluid tends to "stick" to a convex surface, the number of propellers are reduced to just one, while still creating enough thrust to lift the UAV.

However, the vast majority of this type of techniques involve the design and number of blades in UAV propellers, as exemplified in Figure 12.2.

Figure 12.2 Example of blade design. Left: drone is using "noisy" blade design. Right: drone is using "silent" feather-like blade design.

There is a considerable amount of objectives to optimize for when designing a propeller blade, such as power constraints, thrust production and, of course, noise mitigation. In the work of [222], three different optimization methods (a genetic algorithm, a simplex scheme, and steepest descent) were used to design a blade that maximizes flight and mitigates noise. The authors found that these two objectives were, in a sense, contradictory: blades that maximize flight tended tended to be the loudest. However, they also found that as the number of blades increased, the motor speed requirements to maintain lift decreased, which, in turn, decreased noise.

Biologically inspired blade designs are quite common, such as using the serrated shape of an owl feather, which the authors of [182,454,642,667] found to be effective in noise mitigation. Other biologically inspired were used, such as the shape of cicada wings and maple seeds, all of which provided similar thrust to a baseline propeller with significantly less noise.

Additionally, the authors in [182] also found that porous material is able to reduce turbulence and, thus, noise emissions.

In [596], several types of propellers were evaluated and found that "loop-type" blade provided the most noise reduction as well as the best performance in psychoactive annoyance level tests.

The authors of [453] proposed the addition of attachments they call "Gurney flaps" that add a type of serration in the bottom part of the blade, which mitigated the emitted noise at around 1 dB.

In addition to the shape of the blades, there is also the matter of the number of blades to use at each propeller. The authors in [98] found that although increasing the number of blades reduces the level of the noise (which concurs to the findings of [222]), reducing the number of blades "shifts" the noise to frequencies where human hearing is not as sensitive and, thus, diminishing the "annoyance" factor.

Additionally, the authors of [249] also found that using counter-rotate motors (with the same number of blades) are able to reduce noise while maintaining thrust. In addition, the authors propose running both motors at different speeds to "spread out" the noise into different frequency bins, reducing the annoyance factor and overall noise level.

It is important to note that the amount of noise mitigated using these blade-design-based techniques is still quite low. Of the works described in this section, the noise reduction reported was between 2 and 5 dB, with a minimum of 1 dB in [453] and up to a maximum of around 10 dB in [642]. For the latter, it should be noted that this reduction was also due to the reduction of speed in their motor when using the proposed serrated blades.

12.4.3 AUTOMATIC NOISE CANCELING

Another way to mitigate noise is by carrying out a digital signal processing technique known as automatic noise canceling. A version of this technique that is applicable to cancelling drone noise is exemplified in Figure 12.3.

As it can be seen, to carry out this technique it is necessary to, first, capture the noise that is wished to be canceled, which implies that the drone needs to carry on-board microphones to capture the noise created by its motors. Then, simultaneously, a phased version of the noise (basically, the waveform of the noise multiplied by a factor of -1) is to be reproduced close to the location where it is desired that the noise is mitigated. The result being that, by the additive nature of the acoustic interaction of two signals, a destructive interference occurs that results in the noise being canceled. There is an important limitation to this: the noise source and its phased reproduction need to be perfectly out of phase so that the cancellation occurs. This requires proper

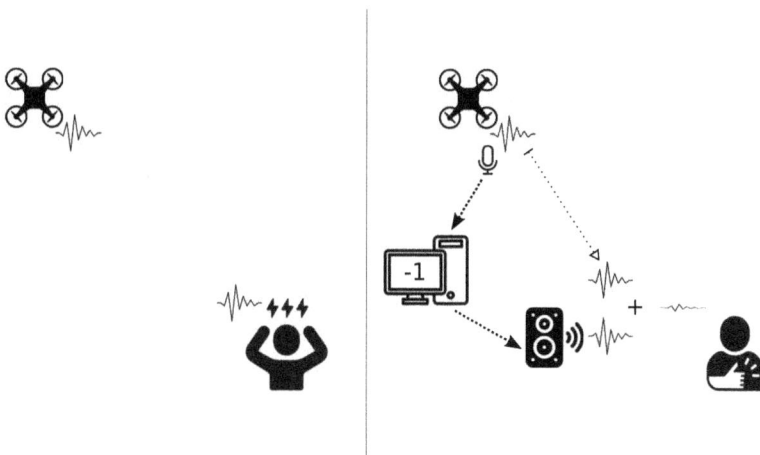

Figure 12.3 Example of automatic noise canceling. Left: no noise canceling. Right: noise canceling is carried out.

synchronization of the captured signal and the reproduced signal, as well as very low latency reproduction.

The work of [56] uses several microphones on board the UAV which capture the noise in different directions; the authors call these captured signals "secondary sources". These are modeled using spherical sector harmonics, and by using a matching process, are then canceled out. Simulated results showed between 20 and 35 dB of noise reduction.

The work in [444] actually uses the tonal nature of the UAV noise to cancel it. For each of the tones of the UAV noise, a pitch shifting technique is employed and is controlled by a closed-loop controller that monitors the output signal. The system was able to reduce the noise energy by 25%, up to 44%.

The authors of [20] proposed a noise-free recording process that uses an on-board microphone to capture the UAV noise, and a second microphone that captures the noisy target speech (which is to be filtered). The latter is expected to be mixture of the target speech and a version of the UAV noise. In the first phase of the process, an automatic noise cancellation process is carried out, with a speaker reproducing a phased version of the UAV noise in real-time. In the second phase, which seems to be carried out in an offline manner, subtracts the frequency-domain version of the recorded UAV noise from the frequency-domain version of the output of the first phase, as a type of post-filter. The final result showed a high similarity (67.5%) to the clean version of the target speech with almost a complete cancellation of the UAV noise.

As it can be seen, there are very few works that aim to cancel the UAV noise via digital signal processing techniques, which implies that it is a foundling area of research. However, it is important to note that most, if not all, of these techniques will require additional audio hardware, both on-board the UAV and on the ground (near the population sensitive to the UAV noise), to capture and reproduce audio signals. All of which may be non-viable in some scenarios.

12.5   DISCUSSION AND FINAL REMARKS

The growth of the drone market for leisure and civilian applications has raised concerns regarding public acceptance of the drone technology. These concerns include safety, health, and noise pollution issues. SPL is an important indicator used to measure drone acoustics. Exposure to high SPLs can have detrimental effects on human hearing, making it important to monitor and control noise levels. The SPL of a drone depends on several factors and ranges from 60 to over 100 dB. The lack of specific standards for drone noise underscores the need to establish means to measure its effects and develop guidelines for deployment in urban settings.

Public acceptance of drones is affected by factors such as annoyance caused by sound levels, societal safety, and environmental impacts. While public opinions on drones vary across time and countries, surveys indicate a growing trend of acceptance for urban air mobility, with between half and three-quarters of the public supporting this idea. According to [101], societal concerns related to drones include noise impact, impacts on animals and flora, visual pollution, and loss of privacy or intrusion.

Similarly, the works in [165, 245] evaluated public acceptance of civilian drones for daily-life applications.

The most pressing concerns were related to criminal activity and privacy breaches, followed by liability, safety, and damages, with noise being the least concern. Catastrophe response, research work, and civil protection had the highest acceptance rates, while parcel delivery and air taxi had low acceptance rates due to safety concerns. Personal use of drones showed a similar trend with higher approval for non-profit and social benefit applications. Noise concerns, age, and gender influenced attitudes towards drones, with advanced air mobility for passenger transport having little acceptance. The study calls for establishing regulations for noise management to ensure sustainable UAM.

The continuous presence of noisy drones flying over homes and workplaces may lead to decreased public acceptance due to safety and health concerns, similar to living or working near airports. WHO recommends avoiding exposure to levels above 45 dB, and more than 40 dB at night. To address concerns over drone noise, research groups are developing evaluation frameworks and metrics to understand the effects of drone noise on humans and animals under various scenarios. Research has also shown that different species have varying levels of sensitivity to noise at different frequencies generated by drones, emphasizing the need to reduce drone impact on target species and inform future drone use guidelines.

In order to understand the basis for noise analysis, in this chapter, we have also dedicated a section to Noise modeling, describing essential concepts such as Noise Pressure Level, Tonal, and Broadband noise found in several works touching on the topic of drone noise and measurement for the public acceptance [120, 550, 615–617].

We have also delved into noise mitigation techniques, seeking to discuss recent works on reducing or canceling the noise emitted by a drone, particularly multi-rotor drones. As mentioned in [101], noise mitigation may reduce noise annoyance, which may not be essential for one or a few drones that add noise to existing soundscapes. However, UAM envisages using a significant number of drones whose noise may cause considerable distress to human and animal life, equally or even more than conventional traffic noise.

While drone noise may cause discomfort in civilian applications, the emitted noise can be used to detect drones [81, 404, 505] when visual or radar systems may not be sufficient [732]. In contrast, developing silent or stealth drones may compromise the safety and accountability of drone operations in civilian spaces.

13 An Overview of NeRF Methods for Aerial Robotics

Luis Fernando Rosas-Ordaz
Department of Computing, Electronics and Mechatronics, Universidad de las Americas Puebla (UDLAP), Puebla 72810, Mexico.

Leticia Oyuki Rojas-Perez
Computer Science Department, Instituto Nacional de Astrofísica, Óptica y Electrónica (INAOE), Puebla 72840, Mexico.

Cesar Martinez-Torres
Department of Computing, Electronics and Mechatronics, Universidad de las Americas Puebla (UDLAP), Puebla 72810, Mexico.

Jose Martinez-Carranza
Computer Science Department, Instituto Nacional de Astrofísica, Óptica y Electrónica (INAOE), Puebla 72840, Mexico.

CONTENTS

13.1 INTRODUCTION

In recent years, the demand for the creation of synthetic images has been increasing, due to the need to generate new views or render 3D objects. One of the tools to generate synthetic views is neural networks, which receive one or several images

DOI: 10.1201/9781003385615-13

as input, generating the synthesis of novel views as output. It should be noted that according to [271] the main focus of novel view synthesis is on pixel generation and image-based rendering (IBR).

For this reason, the synthesis of novel views has become a topic with a wide range of opportunities and applications, highlighting in this overview the area of computer vision and robotics. Regarding the first one, the recent impact in augmented reality can be highlighted, which is mentioned in [475], where it is possible to recover the shape, pose, and 3D appearance of a target. Another area of great interest is facial reconstruction, addressed in [248], [668], [104] and [696], which will generate a 3D human face, through novel views and using as input one or several images.

On the other hand, regarding robotics, the impact of the synthesis of novel views focuses on robotic navigation and inspection, which is mentioned in [304], [78], [595] and [700], since the creation of novel views allows the representation of objects or environments. Another interesting application is transparent object detection, covered in [137] and [294], where novel views work to infer the geometry of a transparent object.

One of the recent methods for the synthesis of novel views is Neural Radiance Fields (NeRF) [417], which was created in 2020. This method allows the creation of novel 3D views, using one or more images as input (each image accompanied by their respective poses). It is important to mention that, due to the speed, quality of the synthetic images, and versatility in the input images and hardware, the use of NeRF has increased exponentially.

Due to this versatility, in 2021 NeRF was applied in [346], which allowed the reconstruction of spine data using ultrasound images as input. In the area of computer vision, in [187], [472] and [221], NeRF was used for facial reconstruction. Finally, in robotics, it was applied in [261] for the detection and location of transparent objects for robotic arms.

It should be noted that NeRF has great potential in aerial robotics, this is demonstrated in [13], where it is possible to synthesize simulation environments of real environments with great precision. For this reason, if NeRF continues to be used for aerial robotics, in the future a method could be developed which allows the creation of complex routes with great precision in simulation environments, in collective and photometric robotics, and even in hybrid methods.

It should be noted that the following Overview is organized as follows: in Section 1.2, the operation of NeRF will be explained and the PSNR, SSIM, LPIPS, and time metrics will be defined, which are the most used parameters in most of the reviewed methods; in Section 1.3, the NeRF methods related to computer vision and robotics will be exposed; in Section 1.4, the NeRF methods related to aerial robotics will be shown; in Section 1.5, NeRF methods in different application areas will be discussed; in Section 1.6, the methods that seek to improve NeRF will be shown; and finally, in Section 1.7 the discussion and conclusion of the authors of this overview will be presented.

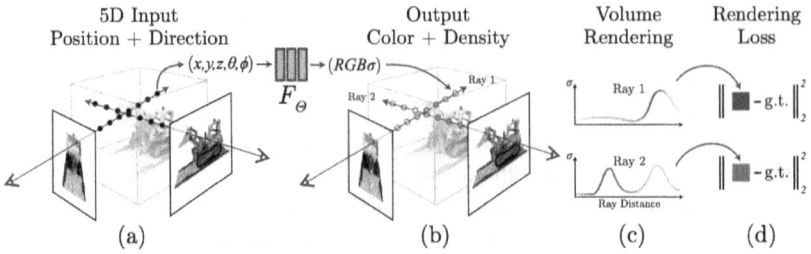

Figure 13.1 NeRF steps shown in [417] where a) 5D coordinates (x, y, z, θ, ϕ). b) NeEF output (R, G, B, σ). c) use of volume rendering techniques. d) optimization through minimizing the residual between synthesized images and the real image.

13.2 THEORETICAL FRAMEWORK

Neural Radiance Fields (NeRF) is a novel method to generate complex 3D scenes, since this neural network receives as input a set of images of the same object from different angles with their respective poses, generating synthetic images of new views as a result. This algorithm is created by [417] which uses a non-convolutional connected deep network, where it receives as input a continuous 5D coordinate (See Fig. 13.1), made up of the spatial location (x, y, z) and the viewing direction (θ, ϕ), obtaining in the output the color density (σ) and the emitted colors (R, G, B). In more detail, the steps that the NeRF follows to perform its rendering are

1. Camera rays are displaced through the scene to generate 3D point sampling.
2. Those points and their corresponding 2D viewing directions are used as inputs to the neural network to produce an output set of colors and densities.
3. Volume representation techniques are used, to accumulate those colors and densities in a 2D image.

13.2.1 COMMON METRICS

It is important to define the evaluation metrics of the synthetic images created by NeRFs, which are normally PSNR, SSIM, LPIPS (these metrics were proposed by [713]), and execution time. Next, the mentioned metrics will be defined.

Peak Signal-to-Noise Ratio
Peak signal-to-noise ratio (PSNR) is defined in [474] as a mathematical expression where the relationship between the power of the maximum possible value of a signal and the power of the distortion noise that affects the quality is made. Equation 13.1 shows the PSNR formulation.

$$PSNR = 20 \cdot \log_{10} \left(\frac{MAX_I}{\sqrt{MSE}} \right) \tag{13.1}$$

Where
MAX_I = Maximum signal value that exists in the true image;
MSE = Mean square error.

It is important to mention that, in equation 13.2 it is shown how the MSE is expressed

$$MSE = \frac{1}{mn} \sum_{i=0}^{m-1} \sum_{j=0}^{n-1} [I(i,j) - K(i,j)]^2. \tag{13.2}$$

Where
I = Matrix data of original image;
g = Matrix data of degraded image;
m = Numbers of rows of pixels of the images;
i = Index of that row;
n = Numbers of columns of pixels of the images;
j = Index of that column.

It should be noted that according to [559], the PSNR value will be small when there are differences in the pixel value between the two images.

Structural Similarity Index Measure
The Structural Similarity Index Measure (SSIM) is considered by [474] as a metric that measures the degradation of image quality caused by processing. Equation 13.3 shows the SSIM formulation

$$SSIM(x,y) = [l(x,y)]^\alpha \cdot [c(x,y)]^\beta \cdot [s(x,y)]^\gamma \tag{13.3}$$

Where
x = Original image;
y = Reconstructed image;
$l(x,y)$ = Luminance comparison function;
$c(x,y)$ = Contrast comparison function;
$s(x,y)$ = Structural comparison function.

It is important to mention that, according to [559], the maximum value of SSIM is one only if the three factors are 1, although the values can range from -1 to 1. Therefore, if the value is 1, there is perfect similarity; if it is 0 it means that there is no similarity; and finally if it is -1 it means that there is a perfect anti-correlation.

Learned Perceptual Image Patch Similarity
Learned Perceptual Image Patch Similarity (LPIPS) is defined in [474] as a metric that evaluates the distance between image patches. The lower the LPIPS value, the greater the similarity between the real image and the generated image.

Execution time

Execution time is the amount of time it takes for a neural network to fully execute the code. Execution time is usually measured in seconds (s) or hours (h). This metric is normally used to make comparisons with other methods, since this way, it is possible to parameterize which neural network can have a faster execution, which is extremely useful in code optimization.

It is important to mention that, due to the great versatility of NeRF, it has ventured into different areas of interest, resulting in specialized or improved methods. For this reason, an overview of NeRF in different areas will be shown below.

13.3 APPLICATIONS IN COMPUTER VISION AND ROBOTICS

Over the last few years, the subfields of computer vision and robotics within the field of artificial intelligence have become increasingly interconnected and dependent on one another. At the outset, NeRF methods are closer to computer vision than to robotics. However, vision sensors are more commonly used in robotic applications nowadays. For this reason, in this section, we begin by providing an overview of NeRF methods developed in the context of computer vision. Second, we revise a number of works dedicated to robotic systems involving computer vision methods that take advantage of the NeRF methodology.

13.3.1 WORKS IN COMPUTER VISION

Currently, computer vision is in need of creating and/or using more versatile and faster neural networks, since its demand and application are increasing exponentially. For this reason, since the creation of the NeRF in 2020, one of the first branches to make use of them was computer vision. This is demonstrated with the creation in 2021 of AD-NeRF [221], which is a NeRF in charge of generating high-fidelity talking head videos which adjust to the input audio sequence; HyperNeRF [472], which handles topological variations by modeling a family of shapes in a higher dimensional space, which allows a more accurate reconstruction of human faces; STaR [706], neural network which reconstructs a dynamic scene with a single moving rigid target; D-NeRF [492], a method that focuses on the dynamic domain, allowing the creation of 360° views of a moving object; FiG-NeRF [678], which allows you to create high-fidelity 3D objects, having as a complement that FiG-NeRF can make the model while the objects are separated from their respective backgrounds; S-NeRF [152], which is a NeRF that allows the detection of shadows, the synthesis of albedo and the filtering of transient objects, this by taking advantage of the effects of non-correlation; and finally, *Dynamic Neural Radiance Fields for Monocular 4D Facial Avatar Reconstruction* [187], code where photorealistic images are generated using dynamic NeRF to model the appearance and dynamics of a human face.

In the field of computer vision, the application of NeRF is very versatile, already covering areas such as stylization and sterilization of scenes, creation and edition of 3D scenes, deformation, manipulation, and reconstruction of objects, facial reconstruction, satellite photogrammetry and analysis of shadows, the solution to the problem of reflections of synthetic images, augmented reality, scanning of objects, estimation of the gaze of the human eye, and semantic portraits. Next, the advances in the aforementioned areas will be described.

1. **Scene stylization and sterilization:** A benchmark in this area is Stylized-NeRF [258], which generates stylized images of a scene since this neural network combines a 2D image stylization network and a NeRF; and NeRF-Art [655], a neural network that performs text-guided sterilization.

2. **3D scene creation and editing:** NeRFEditor [594] is a NeRF capable of editing 3D scenes, through a 360° video as input, and obtaining a high-quality 3D scene as output. On the other hand, NeRDi [151] performs a reconstruction of 2D to 3D scenes, this is done through a loss of diffusion in arbitrarily sampled views and loss of depth correlation. Also, HexPlane [90] makes a model where dynamic 3D scenes can be explicitly represented by six planes of learned features. On the other hand, RecolorNeRF [207] is a network where the scene is broken down into a set of layers of pure colors, which allows a palette to be formed, and with it, color manipulation. Finally, PaletteNeRF [320], which is a NeRF focused on 3D color decomposition.

3. **Deformation, Manipulation, and Reconstruction of objects:** NeRF-Editing [707], Control-NeRF [333], and SceneRF [91] stand out in this area. Regarding NeRF-Editing [707], the user has the ability to perform shape deformations on the content of the generated scene, which allows for generating new images from arbitrary views obtained after editing. On the other hand, Control-NeRF [333], allows manipulation of 3D content, since interpretable and controllable representations of scenes are obtained, thus allowing the synthesis of novel high-quality views. Finally, SceneRF [91], which is a self-supervised monocular scene reconstruction method.

4. **Facial reconstruction:** In this area, there is a great variety of methods, among which HeadNeRF [248], HumanNeRF [668], LipNeRF [104], and GeneFace [696] stand out. With HeadNeRF [248] you can represent and synthesize the fine-level details of the human head, such as the spaces between the teeth, wrinkles, or beard. Regarding HumanNeRF [668], it allows any frame of a monocular video of a human being to be paused, and through the NeRF to achieve new representations of the subject from arbitrary new viewpoints of the camera. LipNeRF [104] enables lip sync in talking head video generation, thus bridging the gap between GAN (neural networks ideal for good lip sync) and NeRF (optimal for posture change of the head). Finally, GeneFace [696], which is a NeRF specialized in the generation of faces in speech, allows to generate natural results which correspond to the audio.

5. **Satellite photogrammetry and shadow analysis:** The advances in this area are with ShadowNeuS [368] and Sat-NeRF [406]. With ShadowNeuS [368] shadow ray monitoring is performed, which allows optimizing samples along the ray and the location of the ray. On the other hand,

Sat-NeRF [406] is an end-to-end model for learning multi-view satellite photogrammetry in nature.

6. **Solution to the problem of reflections in synthetic images:** This solution is addressed by NeRFReN [220], which is done through a new formulation of the NeRF, where training strategies specifically designed for this case are applied, which allows the elimination of reflections in synthetic images.

7. **Augmented Reality**: Regarding this area, we have *Shape, Pose, and Appearance from a Single Image via Bootstrapped Radiance Field Inversion* [475], which is a NeRF that can recover 3D shape, pose and appearance parameterized by SDF from a single image of an object, occupying natural images and not exploring multiple views during training. This is very useful for augmented reality.

8. **Scanning of objects:** In this area, ScanNeRF [143] stands out, which is a neural network that manages to scan real objects in quantity, effortlessly, and at low cost.

9. **Eye Gaze Estimation:** Regarding this application, NeRF-Gaze [702] enables gaze estimation, as it enables the generation of dense gaze data with view consistency and accurate gaze direction.

10. **Semantic portraits:** Finally, there are compositional neural radiation fields (CNeRF) [389] which are focused on the manipulation of semantic portraits.

Next, a more detailed description of the NeRF applied in computer vision will be shown in Tables 13.1, 13.2, and 13.3.

13.3.2 WORKS IN ROBOTICS

On the other hand, the response, analysis, and processing capacity of robotic systems has to optimized, since there must be a greater speed in decision-making and they must have a better capacity for interaction in the environment in which they are found. For this reason, neural networks have begun to be integrated into robotic systems, highlighting in this case the use of NeRFs in robotics. One of the first applications that were developed in 2021 was Dex-NeRF [261], which seeks to enable the detection, localization, and inference of the geometry of transparent objects for robotic arms. Subsequently, in 2022 the NeRF development continued in the field of robotics, highlighting NARF22 [342], where a NeRF is used to carry out a completely differentiable pipeline and parameterized by configuration, allowing high-quality representations of articulated objects to be made, which are of great help for robotic perception and manipulation. On the other hand, *RGB-Only Reconstruction of Tabletop Scenes for Collision-Free Manipulator Control* [606], which is a system for collision-free control of a manipulator robot which uses only RBG views; NeRF-Supervision [698], where a representation of a scene created by a NeRF is used for training dense object descriptors for robust robotic vision systems; among others.

Table 13.1

NeRF methods applied in computer vision in 2020–2022. We highlight the custom dataset.

yr	Method	Category	Dataset	Evaluation	Metrics
2020	NeRF++ [711]	Optimization	T & T Light Field	- Model Comp.	PSNR SSIM LPIPS
2021	FiG-NeRF [678]	Creating and editing 3D scenes	Objectron real images *	- Model Comp. - Loss function - Ablation	PSNR SSIM LPIPS
	STaR [706]	Deformation, manipulation and Reconstr. of objects	Syntetic and real images*	- Model Comp. - Ablation study on optimization	PSNR SSIM LPIPS
	D-NeRF [492]	Deformation, manipulation and Reconstr. of objects	Dynamic Scenes *	- Model Comp. - Comp. of conditioning relative to time	PSNR SSIM LPIPS Time
	DNeRF for Monocular 4D Facial Avatar Rec. [187]	Facial Reconstr.	Monocular Video Sequences *	- Model Comp. - Ablation Study - Limitations Analysis	PSNR SSIM LPIPS Dist
	S-NeRF [152]	Satellite PG and shadow analysis	Satellite imagery from the WorldView-3 optical sensor	- Model Comp. - Altitude Extraction	Altitude MAE SSIM
	HyperNeRF [472]	Facial Reconstr.	Sequences of the Human Face*	- Model Comp. - Visual Comp.	PSNR SSIM LPIPS Avg
	AD-NeRF [221]	Facial Reconstr.	Short video sequence with audio track *	- Images - Intermediate Models - User study - Limitations analysis	SyncNet Scoring AU DET
2022	StylizedNeRF [258]	Sterilization and Stylization of Scenes	Forward-facing T & T	- Model Comp. - Visual Comp. - Ablation Study	Short and long-range consistency using warping error
	NeRFReN [220]	Reflections of synthetic images	FaceScape TIFace	- Model Comp. - Visual Comp. - Ablation study	PSNR SSIM LPIPS ID Score APS

PG: Photogrammetry, Avg: Average, Dist: Distance.

As mentioned, NeRF and robotics have achieved great synergy, resulting in NeRFs becoming involved in areas such as robotic navigation and inspection, detection, pose estimation, sensors, and manipulation of articulated or dense objects. Next, the advances in the aforementioned areas are described.

Table 13.2

NeRF methods applied in computer vision in 2022. We highlight the custom dataset*. We indicated the open source methods using this symbol.

yr	Method	Category	Dataset	Evaluation	Metrics
2022	Shape, Pose, and Appearance [475]	Augmented Reality	Synthetic and Real Images*	- Model Comp. - Visual Comp.	PSNR SSIM FID IoU
	NeRF-Art [655]	Sterilization and Stylization of Scenes	selfies video* H3DS LLFF	- Text - Generalization - Geometry - User Study - Ablation Study	User preference
	NeRFEditor [594]	Creating and Editing 3D Scenes	FaceScape TIFace	- Model Comp. - User Study - Ablation study	PSNR SSIM LPIPS ID Score
	NeRDi [151]	Creating and Editing 3D Scenes	DTU MVS	- Model Comp. - Visual Comp. - Ablation study	PSNR SSIM LPIPS
	ShadowNeuS [368]	Satellite PG and Shadow Analysis	DeeoShadow Synthetic and Real Images*	- Binary shadow - RBG Comp. - Ablation Study	Depth MAE
	NeRF-Editing [707]	Deformation, Manipulation and Reconstr. of objects	FVS NeRF Real Images*	- Shape Editing - Deformation Transfer - Model Comp. - Ablation Study - Limitations Analysis	PSN SSIM LPIPS FID
	SceneRF [91]	Deformation, Manipulation and Reconstr. of objects	Semantic KITTI	- Model Comp. - 3D Reconstr. - Ablation Study	PSNR SSIM LPIPS FID
	HeadNeRF [248]	Facial Reconstr.	FaceSEIP FaceScape FFHQ	- Ablation Study - Perceptual Loss - 2D Neural Rendering - Model Comp.	PSNR SSIM LPIPS Distance
	HumanNeRF [668]	Facial Reconstr.	ZJU-MoCap	- Model Comp. - Ablation Study	PSNR SSIM LPIPS
	Sat-NeRF [406]	Satellite PG and Shadow Analysis	26 Maxar WorldView-3	- Model Comp. - Ablation Study	PSNR SSIM Altitude MAE
	NeRF-Gaze [702]	Eye Gaze Estimation	ETH-XGaze GazeCapture Gaze360 MPIIFaceGaze RT-GENE	- Model Comp. - Ablation Study - Loss Function - Face Attributes	LPIPS
	PaletteNeRF [320]	Creating and Editing 3D scenes	Blender Forward-facing LLFF 360-degree Mip-NeRF360	- Model Comp. - User Study - Ablation Study	PSNR LPIPS

PG: Photogrammetry.

Table 13.3

NeRF methods applied in computer vision in 2023. We indicated the open source methods using this symbol .

yr	Method	Category	Dataset	Evaluation	Metrics
2023	ScanNeRF [143]	Object Scanning	NeRF blender LLFF DTU T & T BlendedMVG CO3D ScanNet	- Model Comp. - Visual Evaluation	PSNR Time
	Control-NeRF [333]	Deformation Manipulation and Reconstr. of objects	LLFF	- Model Comp. - Editing Comp.	PSNR SSIM LPIPS
	LipNeRF [104]	Facial Reconstr.	Talking Head Videos from Movies	- Model Comp. - Visual Evaluation	PSNR SSIM LPIPS LSE-D LSE-C
	HexPlane [90]	Creating and Editing 3D scenes	Plenoptic Video and High-resolution Multi-camera	- Model Comp. - Ablation Study - Analysis Study	PSNT D-SSIM LPIPS JOD Time
	RecolorNeRF [207]	Creating and Editing 3D scenes	Synthetic NeRF Synthetic NSVF LLFF T & T	- User study - Visual Evaluation - Ablation Study	User Preference
	GeneFace [696]	Facial Reconstr.	LRS3-TED	Model Comp. User Study Ablation Study	FID LMD Sync DIF (OOD) Sync (OOD)
	CNeRF [389]	Semantic Portraits	FFHQ Portraits Photos	- Model Comp. - Ablation Study	FID KID

1. **Robotic navigation and inspection:** E-NeRF [304], NeRF2Real [78], NeRF-Loc [595], and MIRA [700] stick out in this area. With E-NeRF [304], a volumetric scene representation in NeRF form of a fast-moving event camera is realized. On the other hand, NeRF2Real [78] is a system for applying sim2real approaches to scenes in nature, to learn vision-based whole-body robotic navigation policies. Regarding the location of objects within a NeRF scene, NeRF-Loc [595] is presented, which allows locating 3D objects for robotic purposes. Finally, Mental Imagery for Robotic Affordances (MIRA) [700] allows building artificial systems, which allow planning actions analogously on imaginary images.
2. **Detection:** Regarding object detection, GraspNeRF [137] and Evo-NeRF [294] are presented. With GraspNeRF [137] 6-DoF grasp detection is achieved for transparent and specular objects (object characteristics that are

complex for detection by a robotic system). Similarly, another NeRF specializing in transparent objects is Evo-NeRF [294], which is focused on the sequential robotic grasping of transparent objects.

3. **Poses:** Regarding this area *Neural Fields for Robotic Object Manipulation from a Single Image* [64] stands out, which seeks to make a unified and compact representation for the representation of 3D objects and the prediction of the grip pose, which can be inferred from a single image in a few seconds; and *Parallel Inversion of Neural Radiance Fields for Robust Pose Estimation* [367], which is a parallelized optimization method based on fast NeRFs, which allow estimating poses of 6-DoF objects. It should be noted that this NeRF is applied to robot manipulators and augmented reality.

4. **Sensors:** Regarding the use of sensors in robotic systems, *NeRF in the Palm of Your Hand* [722] stands out, which is an offline data scheme to improve the policies of robots that use security cameras eye-in-hand, as corrective noise is synthetically injected into visual displays using NeRF, allowing the need for depth sensors to be eliminated; *Uncertainty Guided Policy for Active Robotic 3D Reconstruction Using Neural Radiance Fields* [339], which is focused on mobile robots with a camera on the arm, where it is possible to efficiently recover the 3D shape of an object; and finally, INSPECTION-NeRF [247], which is focused on wall-climbing robots seeking to perform an inspection of surface defects, this is done by collecting RGB-D images of the surface and use of NeRF to generate the global 3D model.

Next, a more detailed description of the NeRF applied in robotics will be shown in Tables 13.4 and 13.5.

13.4   NERF FOR AERIAL ROBOTICS

Due to the successful implementation of NeRFs in robotics, efforts have begun to seek to potentiate their use in more specific and concrete areas, in this case, aerial robotics. There are several applications where robots have made use of computer vision and robotics algorithms to leverage its autonomous behaviour, for instance, for autonomous flight in tasks such as warehouse inspection [403, 405] and autonomous drone racing [427]. More recently, deep learning has been proved effective for tasks such as sense and avoid [154], autonomous object lifting [382] and development of neural pilots [523, 525].

Despite these advances, there are still very few applications using NeRF in aerial robotics, but these represent a turning point for further developments, as the given applications are of great interest and relevance. An example of this is *Vision-Only Robot Navigation in a Neural Radiance World* [13], which using NeRF, creates an interactive training environment for drones, thus allowing for accurate planning in path planning since with the NeRF the differential flatness can be considered. Another example referring to the location and/or navigation of drones is Loc-NeRF [393] which is a system for global location in real-time using neural radiation fields. It should be noted that the initial tests and data collection are carried out with a Clearpath Jackal

Table 13.4

NeRF methods applied in Robotics in 2021–2022. We highlight the custom dataset*. We indicated the open source methods using this symbol.

yr	Method	Category	Dataset	Evaluation	Metrics
2021	Dex-NeRF [261]	Detection	Synthetic Images * Images From Cannon EOS 60D* Intel RealSense*	- Synthetic and Physical Grasping	Grasp Success Rate
2022	RGB-Only Reconstr. of Tabletop Scenes for Collision-Free Manipulator Control [606]	Navigation and Inspection	HOPE	- Real tabletop dataset with ground truth - ESDF - Reconstr. - Manipulator Control	Dist. Success Rate % Collision Class.
	GraspNeRF [137]	Detection	Synthetic Multiview Grasping*	- Model Comp. - Visual Comp. of TSDFs constructed - Ablation Study	Success Rate SR RD
	NeRF2Real [78]	Navigation and Inspection	Custom	- Navigation & Obstacle Avoidance Ball Pushing	Success Time Loc. Error
	NeRF-Loc [595]	- Navigation & Inspection	NeRFLoc-Bench based on Objectron	Model Comp. Ablation Study	Average
	Neural Fields for Robotic Object Manipulation from a Single Image [64]	Poses	Variety of scenes of multiple objects	- Model Comp. - Grasping Robotics experiments	PSNR SSIM LPIPS
	Parallel Inversion of NeRF for Robust Pose Estimation [367]	Poses	NeRF Synthetic LLFF	- Model Comp. - Ablation Study	RE TE
	Uncertainty Guided Policy for Active Robotic 3D Reconstr. using NeRF [339]	Sensors	NeRF Blender	- Model Comp. - Reconstr. - Ablation Study	F-Score
	NARF22 [342]	Articulated or dense objects	Progress Tools	- Model Comp. - Image Rendering - Quality	MSE ADD Config. Error
	NeRF Supervision [698]	Articulated or dense objects	Images from iPhone 12	- Model Comp.	AEPE PCK
	MIRA [700]	Navigation and Inspection	Custom	- Model Comp. - Image Rendering - Quality - Ablation Study	Success Rate
	Evo-NeRF [294]	Detection	7 object meshes*	- Rapid Single Object Retrieval - Graspability - Ablation Study - Sequential Decluttering - Model Comp.	Success Rate Time

RE: Rotation Error, TE: Translation Error, Dist: Distance

Table 13.5

NeRF methods applied in Robotics in 2023. We indicated the open source methods using this symbol.

yr	Method	Category	Dataset	Evaluation	Metrics
2022	E-NeRF [304]	Navigation & Inspection	EDS	- Model Comp. - Image Rendering Quality	PSNR SSIM LPIPS
2023	NeRF in the Palm of Your Hand [722]	Sensors	YCB ShapeNet	- Model Comp. - Grasping Environments	Success Rates Time
2023	INSPECTION-NeRF [247]	Sensors	Custom	- Model Comp. - Qualitative Evaluation	PSNR LPIPS RSDE Time

UGV robot, but later, its use in drone simulation environments is considered, and even its application with real drones. On the other hand, there is TransNeRF [474], which is a model for efficiently synthesizing novel views of complex scenes, using only a sequence of sample images from the UAVid dataset. Similarly, another application that has a possible future application in area robotics is LATITUDE [729], which is focused on global location using NeRF for the representation of cities. The author mentions that LATITUDE [729] has potential applications for high-precision navigation in large-scale urban scenes.

On the other hand, there is Mega-NeRF [624], BungeeNeRF [677] and AsyncN-eRF [674], which are projects focused on simulating large environments (cities). For this reason, it is necessary to acquire data from different scales, therefore, to generate these shots, the use of drones is essential.

Next, a summary of the NeRF applied in aerial robotics will be shown in Table 13.6.

13.5   OTHER APPLICATIONS

As described, the applications of the Neuronal Radiance Fields in Computer Vision, Robotics, and Aerial Robotics have been very useful, since they have allowed the optimization of various processes. It is important to point out that the application of the NeRF is not only focused on these three areas, since, due to the great versatility of NeRF, it can also be diversified in different areas such as medicine, space, and underwater exploitation, large-scale rendering, elimination of noise, image processing, analysis of transparent and refractive objects, mapping, scenes from text, among others. Due to this extensive diversification of NeRF applications, advances in seventeen different areas will be described below.

1. **Medicine:** In this area, *3D Ultrasound Spine Imaging with Application of Neural Radiance Field Method* [346] stick out, which consists of applying a NeRF for the Reconstruction of the spine in 3D, and with this, to

Table 13.6

NeRF methods applied in Aerial Robotics in 2021–2022. We indicated the open source methods using this symbol.

yr	Method	Category	Dataset	Evaluation	Metrics
2021	TransNeRF [474]	Synthesis of Novel Views	UAVid	- Model Comp. - Image Rendering Quality	PSNR SSIM LPIPS
2023	Vision-Only Robot Navigation in a Neural Radiance World [13]	Navigation & Location	Variety of High-fidelity Simulated mesh Environments	- GT Comp., - Comp. to Prior Work - Performance Timing	Time Collision RE TE
	Mega-NeRF [624]	Simulation of Large Environments	Mill 19 Quad 6k UrbanScene3D	- Model Comp. - Reconstr. Quality	PSNR SSIM LPIPS
	BungeeNeRF [677]	Simulation of Large Environments	Scenes from Google Earth Studio	- Model Comp. - Reconstr. Quality	PSNR SSIM LPIPS
	AsyncNeRF [674]	Simulation of Large Environments	Asynchronous Urban Scene UrbanScene 3D	- Model Comp. Reconstr. Quality - Ablation Study	PSNR SSIM LPIPS RMSE RMSE log
	LATITUDE [729]	Navigation and Location	Urban Minimum Altitude Mill 19	- Global localization - Ablation Study of Optimization	TE RE
	Loc-NeRF [393]	Navigation and Location	LLFF	- Pose Estimation - Full System	RE TE

GT: Ground Truth, Comp: Comparison, RE: Rotation Error, TE: Translation Error.

be able to evaluate the measurement of the curvature of the spine from the reconstructed results; MedNeRF [132], which achieves the Reconstruction of computed tomography projections from few x-rays (or even the view); Ultra-NeRF [676], which enables physically enhanced implicit neural rendering (INR) for ultrasound imaging, which learns tissue properties from superimposed ultrasound scans; and *Neural Rendering for Stereo 3D Reconstruction of Deformable Tissues in Robotic Surgery* [663] which allows the Reconstruction of deformable tissue from binocular captures in robotic surgery (using a DaVinci robot) under single viewpoint settings.

2. **Space and ocean exploration:** Regarding this area, we have *3D Reconstruction of Non-cooperative Resident Space Objects using Instant NGP-accelerated NeRF and D-NeRF* [394], which consists of adapting Instant NeRF and D-NerF to map resident space objects (RSO) in orbit, this in order to identify functionality and support orbit services; WaterNeRF [558], which allows obtaining dense depth estimation, and likewise, to obtain color correction based on physics for underwater images; and finally, MaRF [203] which is a neural network which can generate a 3D scene of the

environment of the planet Mars, this in order to help in space exploration, in areas such as planetary geology, simulated navigation and analysis of shapes.

3. **Articulated pose estimation:** CLA-NeRF [623] is introduced which is a NeRF that can perform view synthesis, part segmentation, and articulated object pose estimation.

4. **Large-scale rendering:** Regarding rendering, we have Block-NeRF [603], which is a variant of Neural Radiance Fields that is focused on rendering large-scale environments.

5. **Elimination of noise:** In this area, NAN [476] stands out, which is a NeRF that manages to perform burst noise removal. This system is successful in meeting the challenges of large movements and occlusions under very high noise levels.

6. **Image processing:** Regarding this application, *Neural Radiance Fields Approach to Deep Multi-View Photometric Stereo* [288] stands out, which manages to solve the problem of Multi-View Photometric Stereo (MVSP), since this system recovers the reconstruction dense 3D of an object from images.

7. **Transparent and refractive objects:** In this area, two methods stick out, which are: LB-NeRF [183] which seeks to solve the problem of a different refractive index in the scene, through multiple points of view by introducing the effect of the refraction of light as a displacement of the straight line originating from the center of the camera; and *Sampling Neural Radiance Fields for Refractive Objects* [465], where it is possible to optimize a NeRF for refractive objects.

8. **Mapping:** For the mapping, it stands out NeRF-SLAM [529] and *Towards Open World NeRF-Based SLAM* [370], which combine SLAM with NeRF, to achieve a 3D mapping of high geometric precision.

9. **Representation of climate change:** Regarding awareness of the effects of climate change, there is ClimateNeRF [359], which allows realistic scenes of the effects of climate change, having the ability to represent scenarios with snow, flooding, and smog.

10. **Tactile Sensory Data:** *Touching a NeRF* [720] stands out in this area, which is a generative model to simulate realistic tactile sensory data using NeRF.

11. **Code Dictionary:** Neural Radiance Field Codebooks (NRC) [654] is focused on learning how to reconstruct scenes from novel views using a code dictionary of objects that are decoded through a volumetric renderer.

12. **Object detection:** In this area, NeRF-RPN [254] stands out, which aims to detect all the object-bounding boxes in a scene, thus allowing an optimized object detector.

13. **Reconstruction:** *Research on 3D reconstruction technology of large-scale substation equipment based on NeRF* [477] is presented, which is a method that greatly improves the speed and accuracy of 3D reconstruction of substation equipment at big scale

14. **Depth:** Regarding this application, we have *Neural RGB-D Surface Reconstruction* [40], which allows us to perform a surface representation using an implicit function in the NeRF framework and extend it to use depth measurements from an RGB-D basic sensor.

15. **Rendering of multiple 3D scenes:** X-NeRF [727] is a method which allows to train a 360° model with insufficient views and RGB-D images. Taking insufficient RGB-D images as input, X-NeRF first transforms them into a sparse point cloud tensor and then applies a 3D generative sparse convolutional neural network (CNN) to complete it.

16. **Audiovisual scenes:** AV-NeRF [362] stands out, which is an acoustic-aware audio generation module that integrates audio propagation in NeRF, allowing an audio generation to be associated with 3D geometry of the viewing environment.

17. **Scenes from the text:** Finally, in this area, two methods stand out, which are MAV3D [573] and *Traditional Readability Formulas Compared for English* [335]. Referring to MAV3D [573], there is a method to generate three-dimensional dynamic scenes from text descriptions. On the other hand, *Traditional Readability Formulas Compared for English* [335], performs readability of English texts for use in simplification studies of common texts and medical texts using NeRF.

Next, a summary of the NeRFs in general applications will be shown in Tables 13.7 and 13.8

13.6   IMPROVEMENTS TO THE ORIGINAL NERF

Due to the great impact and usefulness of NeRF in different research areas, the original code has begun to be improved and optimized, since doing this, allows NeRF to expand its usefulness. An example of this is the creation of NeRF++ [711] in 2020, which focused on seeking to eliminate NeRF shape radiation ambiguities and improve the fidelity of view synthesis. Afterwards, in subsequent years, a large number of improvements have emerged, which are focused on five areas, which are speed, quality, input versatility, optimization, and computational requirement. Next, the classified methods will be described.

1. **Speed:** Regarding this area, the following stand out: SteerNeRF [356], NeRF– [665], DVGO [592], DVGOv2 [593], and VaxNeRF [310]. SteerNeRF [356] speeds up the volume rendering process since it takes advantage of the fact that the view change is usually smooth and continuous. On the other hand, NeRF– [665] seeks to simplify the NeRF training process in forward-facing scenes, by eliminating the requirement for known or pre-computed camera parameters or both. With DVGO [592], it is possible to maintain the quality of the NeRF, but achieving the result in only 15 minutes and with a single GPU; it should be noted that, the authors subsequently improved the DVGO implementation using PyTorch and using the simpler dense grid representations, resulting in DVGOv2 [593]. Finally,

Table 13.7

NeRF methods applied in General Applications in 2021–2022. We highlight the custom dataset*. We indicated the open source methods using this symbol.

yr	Method	Category	Dataset	Evaluation	Metrics
2021	3D Ultrasound Spine Imaging with Application of NeRF Method [346]	Medicine	Scoliotic Patients*	- Statistical Analysis	MAD Correlation Large Discrepancy Cures
2022	WaterNeRF [558]	Space and Ocean Exploration	UWBundle Underwater Images*	- Qualitative - Color Correction - Underwater Image Quality	Angular Error UIQM SCM
	CLA-NeRF [623]	Articulated Pose Estimation	SAPIEN Shape2Motion Real Images*	- Articulated Pose Estimation - Failure Cases	PSNR SSIM LPIPS MSE
	Block-NeRF [603]	Large-Scale Rendering	Public Roads Images*	- Model Comp., - Ablation study - Limitations Analysis	PSNR SSIM LPIPS
	NAN [476]	Noise Elimination	LLFF-N	- Novel-view - Real-world results - Model Comp. - Ablation Study	PSNR SSIM LPIPS
	NeRF Approach to Deep Multi-View Photometric Stereo [288]	Image Processing	DiLiGenT-MV	- Baseline Comp. - Training and Validation - Ablation Study	PSNR LPIPS Training and Val. Loss
	LB-NeRF [183]	Transparent and Refractive Objects	Multi-view Images*	- Model Comp. - Comparative Analysis	PSNR SSIM LPIPS
	NeRF-SLAM [529]	Mapping	Cube-Diorama Replica	- Model Comp. - Reconstr. Quality - Ablation Study - Limitations Analysis	PSNR Distance
	ClimateNeRF [359]	Repr. of Climate Change	T & T Mip-NeRF360 KITTI-360	- Image Simulation - User Study -Limitations Analysis	User Preference
	Sampling NeRF for Refractive Objects [465]	Transparent and Refractive Objects	Synthetic Scenes*	- Model Comp. - Reconstr. Quality - Ablation Study	PSNR SSIM LPIPS
	MaRF [203]	Space and Ocean Exploration	Images from Curiosity Rover* Perseverance Rover* Ingenuity Helicopter*	- Model Comp. - Reconstr. Quality	PSNR
	MedNeRF [132]	Medicine	Chest Images* Knee Images*	- Model Comp. - Reconstr. Quality - Ablation Study	PSNR SSIM FID KID
	Touching a NeRF [720]	Tactile Sensory Data	YCB	- Model Comp. - Reconstr. Quality	SSIM MSE Accuracy

Table 13.8

NeRF methods applied in General Applications in 2022–2023. We highlight the custom dataset*. We indicated the open source methods using this symbol.

yr	Method	Category	Dataset	Evaluation	Metrics
2022	NeRF-RPN [254]	Object Detection	Custom	- Model Comp. - Reconstr. Quality - Ablation Study	Recall Avg. Precision
	3D Reconstr. of large-scale Substation Equipment based on NeRF [477]	Reconstr.	Custom	- Reconstr. Effect - Model Comp.	PSNR SSIM Loss Time
	Neural RGB-D Surface Reconstr. [40]	Depth	ScanNet	- Model Comp. - Reconstr. Quality - Ablation Study	Avg. positional RE IoT NC istance
	X-NeRF [727]	Rendering Multiple 3D scenes	7 RGB-D Cam*	- Model Comp. - Reconstr. Quality	PSNR SSIM LPIPS Depth
	Neural Rendering for Stereo 3D Reconstr. of Deformable Tissues in Robotic Surgery [663]	Medicine	Videos from 6 cases of in-house DaVinci Robotic Prostatectomy*	- Model Comp. - Reconstr. Quality	PSNR SSIM LPIPS
2023	3D Reconstr. using Instant NGP-accelerated NeRF and D-NeRF [394]	Space and Ocean Exploration	Images of the Target RSO*	- Model Comp.	PSNR SSIM LPIPS Time
	NeRF-Based SLAM [370]	Mapping	Replica TUM RGB-D	- Model Comp. - Reconstr. Quality	RMSE Dist.
	Neural Radiance Field Codebooks (NRC) [654]	Code Dictionary	ProcTHOR RoboTHOR CLEVR-3D NYU Depth	- Object Nav. - Depth Ordering - Ablation Study - Limitations	Success Rate SPL Depth Accuracy
	AV-NeRF [362]	Audiovisual Scenes	FAIR-PLAY Indoor Scenes*	- Audio Generation - Quality - Ablation Study	STFT ENV
	MAV3D [573]	Scenes from Text	Dynamic Scenes*	- Text-to-4D - Text-to-3D - Text-to-Video - Ablation Study	R-Precision Percentage of Human Preferences
	Ultra-NeRF [676]	Medicine	Synthetic and Phantom B-mode Images*	- Model Comp. - Reconstr. Quality	SSIM
	Traditional Readability Formulas Compared for English [335]	Scenes from Text	CCB WBT CAM CKC OSE NSL ASSET	- Model Comp.	Accuracy

RE: Rotation Error, Avg: Average, Dist: Distance

VaxNeRF [310] manages to speed up NeRFs using voxels, where only binary labels of foreground and background pixels per image are wanted.

2. **Quality:** Regarding the improvement of the quality of the synthetic images, it stands out: 4K-NeRF [664], CLONer [93], RobustNeRF [532], NeRFlow [161] and *Dynamic View Synthesis from Dynamic Monocular Video* [190]. With 4K-NeRF [664] synthesis of ultra-high resolution (4K) views is achieved. On the other hand, CLONer [93] allows high-quality modeling of large outdoor driving scenes seen from sparse views of the input sensor and building differentiable 3D Occupancy Grid Maps (OGMs). Another relevant method is RobustNeRF [532], which manages to improve the quality of synthetic images, through the removal of atypical data from a scene. On the other hand, NeRFlow [161] is presented, which allows obtaining quality synthetic images through the 4D spatial-temporal representation of a dynamic scene from a set of RGB images. Finally, there is *Dynamic View Synthesis from Dynamic Monocular Video* [190], which model is based on static NeRF invariant in time and a dynamic NeRF variable in time, which allows creating novel views in arbitrary viewpoints.

   a. **Speed and quality:** It is important to mention that a distinction will be made with FastNeRF [197], since this model is capable of generating high-fidelity photorealistic images at 200 FPS, it is also 3,000 times faster than normal NeRF.

3. **Versatility in the input** Regarding this area, the following stand out: GeCoNeRF [323], PixelNeRF [704], DietNeRF [267], BARF [364], Nerfies [471], and NoPe-NeRF [59]. Concerning the versatility of receiving few inputs, GeCoNeRF [323] is presented which performs a few-shot configuration with geometry-aware coherence regularization; PixelNeRF [704] is also presented which is a method that allows NeRF to work with one or few input images; and finally, DietNeRF [267], which is a NeRF that works with few input images. On the other hand, BARF [364] is presented, which is a package fit model (BARF) to train NeRF from imperfect (or even unknown) camera poses. Regarding the versatility of using mobile cameras, Nerfies [471] is presented, which is the first method capable of photorealistically reconstructing deformable scenes from photos and/or videos captured casually from mobile phones. Finally, NoPe-NeRF [59] is presented, which is a model which can be trained without previously calculated camera poses.

4. **Optimization:** Regarding the optimization of certain processes of the NeRF operation we have: iNeRF [699], SNeRL [565], Mip-NeRF 360 [46], NeRF-VAE [314], and Nerfstudio [604]. With iNeRF [699], you can obtain the poses of the synthetic images created, and with this, have more training data. On the other hand, with SNeRL [565] it is possible to optimize the operation of the NeRF, since in this method reinforcement learning is applied, so a convolutional encoder is used to learn the implicit conscious 3D neural representation from of multiview images. In the same way, there is

Mip-NeRF 360° [46], which would be considered as an extension of NeRF which solves unlimited scenes, since a non-linear scene parameterization, online distillation, and a regularized are used. On the other hand, regarding training optimization, NeRF-VAE [314] is presented, which is a model that allows inferring the structure of a new scene, without the need to retrain, using amortized inference. Finally, in view optimization, we have Nerf-studio [604], which is a NeRF based on a modular PyTorch framework, which has plug-and-play components to implement NeRF-based methods, and with this, obtain innovative views.

5. **Computational requirement:** Regarding the versatility of the computational requirement to implement NeRF we have *Baking Neural Radiance Fields for Real-Time View Synthesis* [238], KiloNeRF [511], MEIL-NeRF [118] and *Real-Time Neural Light Field on Mobile Devices* [92]. Regarding real-time rendering, there is *Baking Neural Radiance Fields for Real-Time View Synthesis* [238], which manages to obtain a real-time representation of a NeRF in basic hardware; and KiloNeRF [511], a method that achieves real-time rendering by using thousands of small multilayer perceptrons (MLPs) instead of a single large MLP. Regarding the memory usage requirement, MEIL-NeRF [118] is presented which corrects the problem of unlimited memory (which causes previous data to be forgotten after training with new data); this correction is done by developing a memory-efficient incremental learning algorithm. Finally, the *Real-Time Neural Light Field on Mobile Devices* [92] is presented, which is a NeRF adapted for mobile devices.

Next, a summary of the improvements to the original NeRF code will be shown in Tables 13.9 and 13.10.

13.7 DISCUSSION AND CONCLUSIONS

In this chapter, we have provided a succinct overview of the state-of-the-art technique known as NeRF (which stands for Neural Radiance Fields). This technique enables the rendering of photorealistic 3D images from 2D images using deep neural networks to model the radiance of a scene, which is obtained by learning the scene's color and light from a set of images. The network learns to generate a view of a realistic 3D model given any camera angle and position.

This overview has covered the theoretical framework behind NeRF and its application in fields key to aerial robotics, namely, computer vision and robotics. Then, we have discussed specific works in aerial robotics where NeRF have been applied. We prepared the closure of the chapter by mentioning other applications of NeRF outside of these fields and concluded our review by mentioning some works that have improved the original NeRF framework.

It is important to mention that, regarding the branch of computer vision, the trends from a general perspective focus on the use of NeRF for the creation, edition, manipulation, and reconstruction of 3D scenes; analysis of parameters in images; and augmented reality, which are areas of great interest today.

Table 13.9

NeRF methods to improve its performance in 2020–2021. We highlight the custom dataset*. We indicated the open source methods using this symbol.

yr	Method	Category	Dataset	Evaluation	Metrics
2020	NeRF++ [711]	Optim.	T & T Light Field	- Model Comp.	PSNR SSIM LPIPS
2021	PixelNeRF [704]	Input Versatility	DTU MVS	- Rendering Quality - View Reconstr. - Ablation Study	PSNR SSIM LPIPS
	FastNeRF [197]	Speed Quality	Realistic 360 Synthetic LLFF	- Rendering Quality - Cache Res. - Ablation Study	PSNR SSIM LPIPS
	iNeRF [699]	Optim.	NeRF LLFF	- Pose Estimation	RE TE
	NeRFlow [161]	Quality	Pouring Gibson Real Images*	- Model Comp. - Rendering Quality	PSNR SSIM LPIPS MSE
	Dynamic View Synthesis from Dynamic Monocular Video [190]	Quality	Dynamic Scene	- Model Comp. - View Synthesis - Ablation Study	PSNR SSIM LPIPS
	Baking NeRF for Real-Time View Synthesis [238]	Comput. Reqs.	LLFF	- Model Comp. - Ablation Study	PSNR SSIM LPIPS
	BARF [364]	Input Versatility	LLFF	- Model Comp. - Quality Results	PSNR SSIM LPIPS RE TE
	KiloNeRF [511]	Comput. Reqs.	Synthetic-NeRF Synthetic-NSVF BlendedMVS T & T	- Model Comp. - Quality Results	PSNR SSIM LPIPS Time
	Nerfies [471]	Input Versatility	Multi-view Images	- Model Comp. - Quality Results	PSNR SSIM LPIPS
	VaxNeRF [310]	Speed	NeRF-Synthetic NSVF-Synthetic	- Model Comp. - Training Speed	PSNR SSIM Time
	DietNeRF [267]	Input Versatility	Realistic Synthetic MVS DTU	- Model Comp. - Synthetic Images Quality - Ablation study	PSNR SSIM LPIPS FID KID
	NeRF-VAE [314]	Optim.	GQN CLEVR Jaytracer	- Degenerate Orientations in the GQN data - Model Comp. - Model Ablations	MSE

RE: Rotation Error, TE: Translation Error.

Table 13.10

NeRF methods to improve it's performance in 2022–2023. We highlight the custom dataset*. We indicated the open source methods using this symbol.

yr	Method	Category	Dataset	Evaluation	Metrics
2022	NeRF - - [665]	Speed	Blender Forward-Facing	- View Synthesis Quality - Damara Param. Estimation, - Breaking Point - Controlled Option Patterns	PSNR SSIM LPIPS ATE
	SteerNeRF [356]	Speed	NeRF-Synthetic T & T	- Model Comp. - Runtime Breakdown - Ablation Study	PSNR SSIM LPIPS
	4K-NeRF [664]	Quality	LLFF Synthetic-NeRF	- Model Comp. - 4K visualization Effect - Ablation Study	PSNR LPIPS NIQE
	CLONer [93]	Quality	KITTI	- Model Comp. -Depth Eval. - Ablation Study	PSNR SSIM LPIPS
	Direct Voxel Grid Optimization [592]	Speed	Synthetic-NeRF Synthetic-NSVF BlendedMVS T & T DeepVoxels	- Model Comp. - Quality - Ablation Study	PSNR SSIM LPIPS
	Improved Direct Voxel Grid Optimization for Radiance Fields Reconstr. [593]	Speed	Synthetic-NeRF T & T	- Model Comp. - Ablation Study	PSNR SSIM LPIPS Time
	MEIL-NeRF [118]	Comput. Reqs.	T & T Replica TUM-RGBD	- Model Comp. - Ablation Study	PSNR MS-SSIM
	Real-Time Neural Light Field on Mobile Devices [92]	Comput. Reqs.	Realistic Synthetic 360° Real-World forward-facing	- Model Comp. - Ablation Study - Limitations Analysis	PSNR SSIM LPIPS Latency
	NoPe-NeRF [59]	Input Versatility	T & T ScanNet	- Model Comp., - Ablation Study - Limitations Analysis	PSNR SSIM LPIPS ATE RMSE RPE
	Mip-NeRF 360 [46]	Optim.	Custom	- Model Comp. - Ablation Study - Limitations Analysis	PSNR SSIM LPIPS Time
2023	RobustNeRF [532]	Quality	Synthetic and Real Images*	- Model Comp. - Ablation study - Sensitivity Analysis	PSNR SSIM LPIPS
	GeCoNeRF [323]	Input Versatility	NeRF-Synthetic LLFF	- Model Comp. - Visual Comp. - Ablation Study	PSNR SSIM LPIPS Avg. Error
	SNeRL [565]	Optim.	Offline Datasets	- Environments Eval. - Ablation Study	Time Step
	Nerfstudio [604]	Optim.	Nerfstudio*	- Model Comp. - Protocol	PSNR SSIM LPIPS

On the other hand, NeRF trends in the field of robotics are focused on robotic navigation and inspection, detection, estimation of poses, sensors, and manipulation of articulated or dense objects. All these areas allow for better control of the robot, resulting in a greater diversification in the tasks of robotic systems, since they will have a better interaction in varied environments with greater precision.

Regarding the field of aerial robotics, we have identified two main trends: the use of NeRF for outdoors localization and map and 3D model generation of large areas. In both cases, we envisage a clear benefit of these methods in areas such as cartography, photogrammetry and autonomous navigation.

We noted that there are not many works in aerial robotics, which is somewhat expected given that NeRF is relatively new, but also because the framework was not intended for online or real-time applications. However, as discussed in Section 13.6, the latest improvement of NeRF includes the rendering in real time, which opens the door to many more applications. Furthermore, an open question is whether NeRF could be performed in less high-tech processors such as Tensor Processing Units [522, 524] or even conventional processors such that the quality of the rendering is not affected while allowing fast processing. It is an open question what applications could take advantage of a NeRF with such characteristics.

For this reason, it is interesting to see that NeRF is being used in other areas such as space and underwater exploration, where NeRFs allow the creation of simulation environments and data analysis; high precision geometric mapping, which is achieved with the combination of SLAM and NeRF; and finally, the application of NeRF in medicine is highlighted, since using NeRF improves diagnoses, resulting in better treatment or patient intervention.

Thus, from our point of view, we believe that the possible future impact of NeRFs in aerial robotics will focus on the creation of real-world simulation environments for planning complex routes with high precision flight; in the collaborative robotics (swarms), which will allow the image sampling (for the NeRF method) of a large terrain to be reduced, in photometry, in hybrid methods, and in the creation of aerial synthetic images which will eliminate certain elements.

Finally, from what is shown in this overview, the use of NeRF points to be very promising for the following years, since, in just four years since its creation, a great variety of implementations and improvements of this method have already emerged, which shows that, due to its versatility and performance, neural radiance fields are an ideal tool for the creation of synthetic images.

14 Warehouse Inspection Using Autonomous Drones and Spatial AI

Jose Martinez-Carranza*, Leticia Oyuki Rojas-Perez

Computer Science Department, Instituto Nacional de Astrofísica, Óptica y Electrónica (INAOE), Puebla 72840, Mexico.
*Corresponding Author: carranza@inaoep.mx

CONTENTS

We present and evaluate an approach to carry out warehouse inspection using small drones capable of performing autonomous flight and spatial AI. For the latter, we use a novel smart camera called OAK-D that computes depth estimation and neural network inference on its chip, with no cost to the host computer on board the drone. The neural inference is used for package scanning by detecting QR and bar codes on the colour camera of the OAK-D. We also implemented a person detector seeking to achieve a safe flight during the inspection, including when getting too close to a structure. For drone localization, we use RGB-D ORB-SLAM with grey and depth images provided by the OAK-D, which enables autonomous flight in a GPS-denied environment. Our experiments show a small drone, with a diameter of 44 *cm* and weight of less than 700 *g.*, is capable of performing the inspection autonomously, with onboard processing, and able to communicate with a GCS to visualise video and package information scanned by the drone. Moreover, our solution can be easily replicated to use two drones; hence, facilitating the distribution of the inspection task.

14.1 INTRODUCTION

E-commerce has grown significantly due to the COVID-19 pandemic, increasing the number of goods and commodities traded on the internet. Companies have seen the need of speeding up several tasks within the logistic process, including their automatisation to reduce human intervention either for safety reasons or to cope during lockdown times [166, 628]. Warehouse inventory is one of those tasks that has received wide attention in the logistic process. For the latter, Unmanned Aerial Vehicles (UAV) or drones, as commonly known, have been proposed to facilitate the automatisation of this task [403].

Based on the literature [173, 401, 628], we argue that warehouse inspection using autonomous drones has to consider the development of the following modules: 1) a localization system that enables the drone to navigate in a GPS-denied environment such as a warehouse; 2) a package scanning system, which scans the products on the warehouse shelves using bar codes, QR codes or RFIDs; 3) a sense&avoid system, which enables the drone to fly safely, seeking to avoid obstacles or people; 4) a data management system, which is used to handle the package information, this could include decoding of the package codes, either online or offline on a Ground Control Station (GCS).

Some commercial and academic works have focused on scanning and data management using drones that are piloted manually [21, 41, 159, 173]. On the other hand, localization is an essential component to be developed for autonomous solutions to succeed. Thus, some works use markers laid out on the warehouse shelves, floor and walls to carry out a visual-based localization [401, 703]; however, populating the warehouse with markers may be prone to a human error leading to incorrect localization. A more robust approach is that of using a LIDAR-based Simultaneous localization and Mapping (SLAM) technique [55], although a LIDAR sensor may not be suitable for small drones. Other approaches have proposed a model-based visual localization [491] or a combination of markers and visual SLAM [324]. A solution running fully onboard is described in [405], where a low-cost metric monocular SLAM is used for localization [427, 520], while QR scanning is performed with a second tiny camera also onboard the vehicle. This solution received the *Special Award: Best Flight Performance* in the 11$^t h$ International Micro Air Vehicle Conference and Competition (IMAV) 2019.

Motivated by the use of vision-based methods, in this work, we present a solution for warehouse inspection using small drones and a novel sensor known as OpenCV AI Kit (OAK-D) from the company Luxonis®, see Figure 14.1. The OAK-D is a

Figure 14.1   Our approach aims at enabling a small drone to perform autonomous warehouse inspection using a novel sensor, the OAK-D smart camera, capable of performing spatial artificial intelligence, such as CNN inference and depth estimation on the chip, with no computational cost to the host computer on board the drone.

smart camera capable of performing spatial artificial Intelligence on the chip with no computational cost to the host computer [522, 524]. This standalone sensor contains three cameras: a 4K colour camera and two grayscale cameras arranged as a stereo camera. It also has an Intel Movidius Myriad X processor for neural network inference. The OAK-D also computes a depth image that can be associated with any of its cameras.

Our solution uses the OAK-D and a host computer on board the drone to perform ORB-SLAM [437] for localization. Furthermore, we exploit the OAK-D's capabilities to run a YOLO v3 Tiny model [14] that has been trained to detect QR/bar codes and persons observed by the colour camera. Person detection and depth information are used in our sense&avoid module to stop the drone when a person walks in front of the drone or when it gets too close to the wall during the inspection. QR and bar codes are decoded and saved on the host computer, which communicates back to the GCS only for visualisation, displaying both the drone's position and the list of scanned QR and bar codes. In addition, our system was easy to replicate such that we could fly two drones simultaneously to distribute the inspection.

Our evaluations suggest that, in contrast to other solutions requiring heavier and larger drones, our approach offers a feasible lightweight approach. We include some illustrative examples of distributed inspection using two drones equipped with our methodology to illustrate that replication is straightforward. Finally, we consider that our work can serve as the basis for developing a professional solution that could be deployed for warehouse inspection and similar inspection tasks.

We should highlight that our approach competed in the OpenCV AI Challenge 2022 [1], winning 1st Place in the Regional Prize of the competition, region 2: South America + Central America + Caribbeans. Therefore, to present our approach, we have organised this paper as follows: Section 14.2 discusses the related work; Section 14.3 describes our system; experiments and evaluation are discussed in Section 14.4; finally, our conclusions are drawn in Section 14.5.

14.2 RELATED WORK

The work in [173] provides an excellent review of the most representative works regarding warehouse inspection using drones until 2019. Therefore, in this section, we will describe the latest developments regarding warehouse inspection using drones.

We begin by mentioning those solutions where the drone is manually piloted by a human. In these works, the main contribution is that of package scanning. For instance, the work in [174] describes a system based on block-chain to receive the inventory data registered from sensing RFID tags in the warehouse with a drone, which are validated and made available in their system for consumption. The work in [282] describes a Virtual Reality application for human pilots to train their flight skills in warehouse facilities.

Regarding vision-based systems, the work in [142] uses the drone to capture images of the warehouse, which are later processed offline with a CNN to recognise alphanumeric characters and bar codes. Similarly, the work in [295] uses vision to detect pallets on the images captured with a drone. It is important to mention that any of these works have a sense&avoid systems, perhaps because they rely on the ability of the pilot to avoid crashes.

Regarding autonomous solutions, some works use tags or markers in the environment to localise the drone. The more advanced ones use SLAM either with LIDAR or vision. For instance, the work in [703] uses QR placed on the floor, which are detected with a CNN, and then used to calculate the position of the drone in the facility. QR codes are also used to detect packages in the warehouse. The work in localization [401] uses whycon markers placed on the products to enable drone localization and also QR codes to label the products, which are also detected on the image captured by the drone. The work in [134] uses a bottom camera looking at the ground, seeking to detect direction lines that are processed on the GCS to calculate the drone's position. This solution also detects and decodes QR codes on the products. An interesting approach combining two robotic platforms is presented in [281]. In this work, an underground vehicle (UGV) uses a LIDAR-based SLAM to localise itself. This UGV has Infra-Red (IR) markers on the top of it, which are observed with an IR downward camera onboard the drone. In this manner, the drone can use the UGV's position to localise itself. This approach also performs QR detection using deep learning to scan the products. In a similar fashion, the work in [217] uses a UGV to carry the drone to the desired shelve, then the drone takes off, flying vertically and scanning bar codes that are transmitted to the GCS. The drone also uses a downward camera to track the UGV, using it as a reference for localization. The UGV uses LIDAR to perform localization among the racks. In a simpler approach, the work in [113] uses 2D barcodes, efficiently detected with an IR camera, utilised to estimate the drone's position in the warehouse. In this work, path planning is performed to define the drone's flight path.

A more sophisticated approach is that of [55], where LIDAR is used to perform SLAM with a the drone's speed of 7.8 m/s, although an initial map has to be built by flying the drone manually before autonomous flight. For package scanning, this work uses April tags and RFIDs on the packages. A solution with redundancy in the localization is presented in [324], which uses a combination of LIDAR and IMU to perform SLAM and three onboard cameras: a forward camera to recognise AprilTag markers, an upward camera to perform visual SLAM on the ceiling, and a downward camera to perform lane recognition. Although this solution seems quite robust, this approach employs a relative large drone, measuring 74×71 *cm* in the horizontal plane and weighing 3.2 *kg*.

Finally, it is worth mentioning that the works in [55, 217, 324] have also implemented a sense&avoid system, which makes them more robust and safe to be deployed in warehouses with personnel around.

14.3   SYSTEM OVERVIEW

In this section, we describe each one of the modules designed and implemented in our approach. We should highlight that all the computations are processed on board the drone, either processed by the OAK-D or on the host computer.

14.3.1   DRONES AND HARDWARE

We used two Bebop 2.0 Power Edition (PE) drones from the Parrot Company. The Bebop 2.0 PE offers a Software Development Kit for communication with the drone using a compute program. This drone has an altimeter and an embedded bottom camera that uses optic flow to maintain its position during hovering, an Inertial Measurement Unit and a forward camera. We did not use the IMU since it becomes inaccurate, especially close to metallic structures. We did not use the forward camera either since we considered the OAK-D to be enough.

The Bebop 2.0 PE can not run custom programs on board. Instead, we mounted a host computer, the Intel Stick Computer m3, which communicates directly with the drone to send flight commands. By using the Bebop's local wireless network, the host computer also communicates with the GCS, an Alienware 15 r3 Laptop. The drone's position and a selected camera view are transmitted from the host computer to the GCS. The drone can be piloted manually with the GCS if needed.

To place both devices steadily, we used a 3D printed mount that also replaced the OAK-D's back-case to save weight. In our experience, the Bebop 2.0 PE has a payload of 200 g.; beyond that, the drone becomes unstable to fly. The host computer, the OAK-D and the mount added up a final weight of 187 *g*. The Bebop 2.0 PE has

Figure 14.2 To the left, the mount design to carry the OAK-D sensor and the host computer, the Intel Stick Computer m3. To the right, the mount placed on the Bebop 2.0 PE.

a weight of 500 g. Thus, when adding the mount with the devices, the drone has a total weight of 687 g., see Figure 14.2. For more details on the mount design for the Bebop 2.0 PE, see [524].

14.3.2 ROS-BASED COMMUNICATION SYSTEM

To develop and implement our approach, we used the Robotic Operating System (ROS) [498]. This allowed us to implement independent task-oriented node packages that could communicate through the ROS network protocol via publishers and subscribers. In this manner, we were able to run nodes on the Intel Stick Computer to run ORB-SLAM, the flight controller, and to publish low-resolution colour images (from the OAK-D). The OAK-D smart camera also has its internal neural processor capable of running Convolutional Neural Networks (CNN). This is a great advantage since it is similar to having a GPU processor but without having to occupy the significant space and weight that an embedded GPU would require. On the other hand, depth estimation and object detection tasks become inexpensive in computational terms to the host computer. For the network communication, we used the local network created by the Bebop 2.0 PE.

A general block diagram of the ROS nodes implemented in our system is shown in Figure 14.3. This diagram indicates which nodes are run on the onboard computer, the Intel Stick Computer (green boxes), and which ones run on the GCS (orange box). In the same figure, we can see the ROS node that communicates with the OAK-D camera to receive the colour, OAK-D's right image of the stereo rig, depth image, and the object detections (persons, QR and bar codes). Green arrows indicate data published by the nodes. Red arrows indicate data consumed by other nodes.

14.3.3 DRONE LOCALIZATION USING ORB-SLAM

For a genuine autonomous flight in indoor environments, the drone must know its position within such an environment where GPS is not available. This is a well-known problem in robotics, for which several solutions have been created. From the

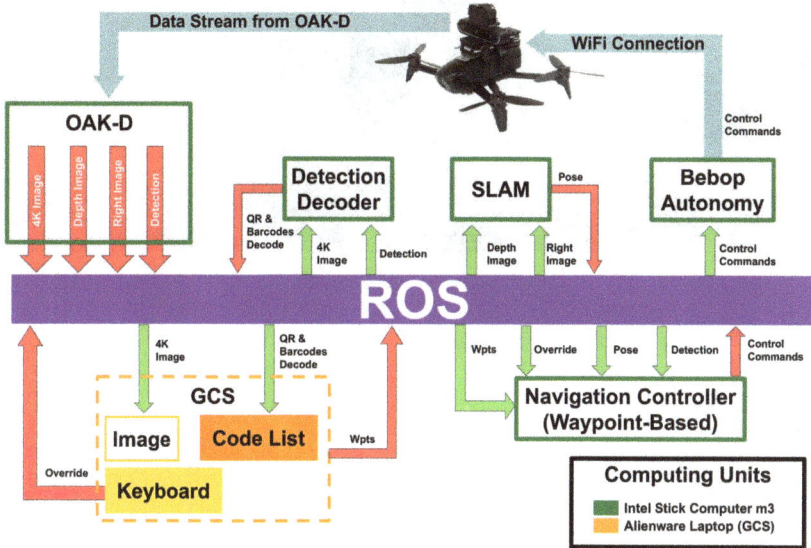

Figure 14.3   Communication system based on the Robotic Operating System. Nodes running on the Intel Stick are marked in green. Nodes running on the Ground Control Station (GCS) are marked in Orange. The GCS is used for visualisation only.

traditional Simultaneous Localization and Mapping (SLAM) using LIDAR [163] to Visual Odometry (VO) [549] and visual SLAM [140].

To use the OAK-D for visual SLAM and after some experimental testing, we chose to use the right image of the stereo rig and the depth image (aligned to the right image) to be processed by ORB-SLAM [437] in its RGB-D version. Camera images were captured at 640×400 at a frequency of $30\ Hz$. Regarding the right camera, we also used the camera mode "NEGATIVE", which inverted the grey values in the image. We noted that this effect helps visual features to be more persistent. Figure 14.4 shows an example of a map created with ORB-SLAM using the OAK-D imagery.

We do not use special marks, external sensors or any other calibration in the environment. Note that in Figure 14.1 and Figure 14.4 random blue tape can be seen on the grey columns in the corridor or on the floor for the following reasons: 1) For indoor flight, the Bebop 2.0 PE has a downward camera that performs optical flow to maintain stable flight during hovering; without texture, the drone would drift; 2) Optical flow and ORB-SLAM require texture to work, the test facility lacks texture on the floor and some of the columns, especially if the drone flies too close to them.

14.3.4   CODE AND PERSON DETECTION WITH YOLO V3 TINY

Our detection network based on YOLO v3 Tiny was trained to recognise four classes: background, person, QR and bar codes. A total of 8,270 images with a resolution of

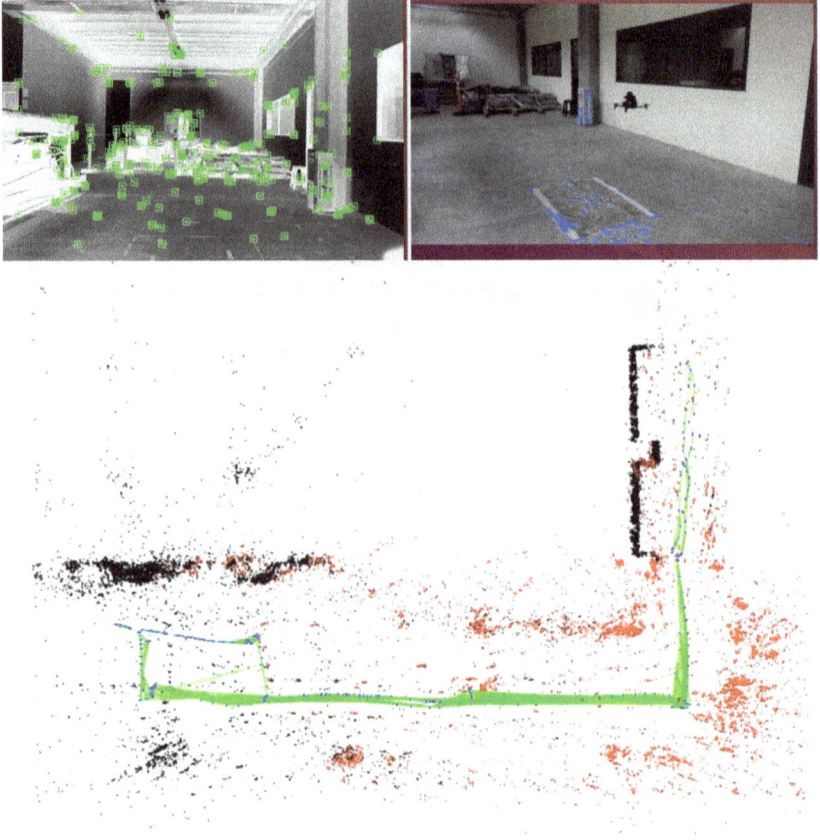

Figure 14.4 First row: ORB-SLAM features as green dots on the right image of the OAK-D, using the effect mode "NEGATIVE"; to the right, an external view of the drone in the test facility. Second row: top view of the 3D map and the drone's trajectory in green.

416×416 were collected, containing different examples of the classes mentioned above. For the background, we take images from the environment. For QR codes and bar codes, some images were taken from the internet, not only from the test area, even blurred images. Finally, we include images of people with and without mouth covers, full-body view, back view, side view, and groups of people to detect a person. Figure 14.5 shows examples of the dataset collected to train and evaluate our network.

Subsequently, data augmentation was performed by rotating the images 180 degrees. We obtained 16540 images with 25,662 labels, distributed as follows: 5,518 labels to the background (class 0), 6,530 labels to QR code (class 1), 4,986 labels to bar code (class 2) and 8,628 labels to person (class 3). We used Google Colab to train our network, using the following custom configuration: 8000 epochs, batch size of 64, a learning rate of 0.001 and leaky ReLu as an activation function. From the

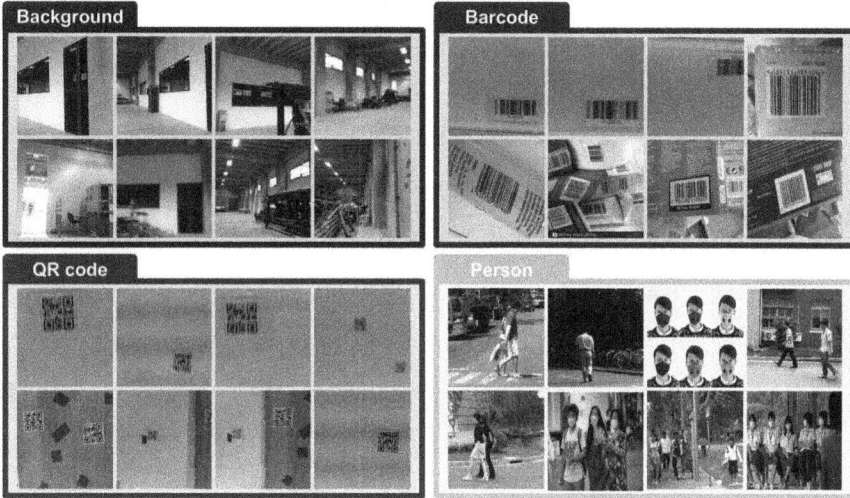

Figure 14.5 Examples of the dataset collected to detect background, person, QR and bar codes.

16540 images, 80% were used for training and 20% for validation. From the evaluation, we obtained a precision of 94%, a recall of 80% and an F1-score of 86%. Once trained, the model was uploaded to the OAK-D chip using the instructions provided by the sensor maker. For real-time detection, we used the depth image to measure the distance of detected people w.r.t. the OAK-D. Image regions of detected QR and Bar codes are processed by the host computer to perform decoding using the open-source library Zbar [76].

14.3.5 CONTROLLER FOR AUTONOMOUS FLIGHT

We implemented a proportional controller that enables the drone to fly forward/backwards (navigation mode) or sideways to the left/right (inspection mode), depending on the type of waypoints set by the user via the GCS.

The Bebop's SDK offers a control communication with the Bebop's inner controller via four control signals: roll s_ϕ (lateral motion), pitch s_θ (moving forward/backwards), yaw s_ψ (rotation to the left/right) and altitude s_h (elevation). Therefore, our controller defines these control signals using the waypoints and the current position of the drone, which is assumed to be the position of the OAK-D smart camera, estimated with ORB-SLAM and represented in the right-hand coordinate system.

Assuming that the drone flies on a horizontal plane, we operate with vectors obtained from the translation $\mathbf{t}$ and rotation matrix $\mathbf{R}$ estimated with ORB-SLAM. Using a unit vector $\mathbf{v} = [1,0,0]$, a heading vector is set as $\mathbf{h} = \mathbf{Rv}$. A departing waypoint $\mathbf{w}_s$ and the next waypoint $\mathbf{w}_g$ are used to define the direction vector $\mathbf{d} = \mathbf{w}_g - \mathbf{w}_s$, with its corresponding rotation matrix representation $\mathbf{R}_d = Rot(\mathbf{d})$, where $Rot(\cdot)$ is a function that calculates such matrix. Finally, we also compute the drone's position relative to $\mathbf{w}_s$:

$$\mathbf{r} = \mathbf{R}_d^\top (\mathbf{t} - \mathbf{w}_s) \tag{14.1}$$

If the waypoints are set to navigation mode, then the control signals are calculated as follows:

$$s_\theta = K_{p_\theta}(\|\mathbf{d}\| - r_x) \tag{14.2}$$

$$s_\phi = K_{p_\phi}(-r_y) \tag{14.3}$$

$$\mathbf{n} = \mathbf{d} \times \mathbf{h} \tag{14.4}$$

$$s_\psi = K_{p_\psi} sign(\mathbf{n}) acos\left(\frac{\mathbf{d} \cdot \mathbf{h}}{\|\mathbf{d}\|\|\mathbf{h}\|}\right) \tag{14.5}$$

$$s_h = K_{p_h}(w_{gz} - r_z) \tag{14.6}$$

where *sign* is defined as:

$$sign(\mathbf{n}) = \begin{cases} 1: & n_z \geq 0 \\ -1: & n_z < 0 \end{cases} \tag{14.7}$$

and $\mathbf{w}_g = [w_{gx}, w_{gy}, w_{gz}]$, $\mathbf{r} = [r_x, r_y, r_z]$, $\mathbf{n} = [n_x, n_y, n_z]$.

Note that similar calculations can be arranged for the case when waypoints are set for navigation mode but with backwards flight.

If the waypoints are set to inspection mode then:

$$s_\theta = K_{p_\theta}(-r_y) \tag{14.8}$$

$$s_\phi = K_{p_\phi}(\|\mathbf{d}\| - r_y) \tag{14.9}$$

and equations (4-7) can also be used, except that $\mathbf{d}$ becomes the perpendicular vector to $\mathbf{w}_g - \mathbf{w}_s$ on the plane $X - Y$ of the global coordinate system. The gains $K_{p_\theta}, K_{p_\phi}, K_{p_\psi}, K_{p_h}$ were tuned exmpirically.

Our controller was implemented within the navigation controller node that received the pose estimation, published by ORB-SLAM, see Figure 14.3. Pose estimation ran at $10 - 12\ Hz$, which allow the drone to fly at a speed of $0.25\ m/s$.

14.3.6 SENSE-&-AVOID SYSTEM

For Sense-and-Avoid, we considered two scenarios during autonomous flight. First, the drone performs an autonomous frontal flight, and during this mode, a person or persons may walk in front of the drone. In this case, the drone has to stop and enter into hovering mode until the person or persons moves away, getting out of the viewpoint of the OAK-D's colour camera. If a person is detected, the depth image

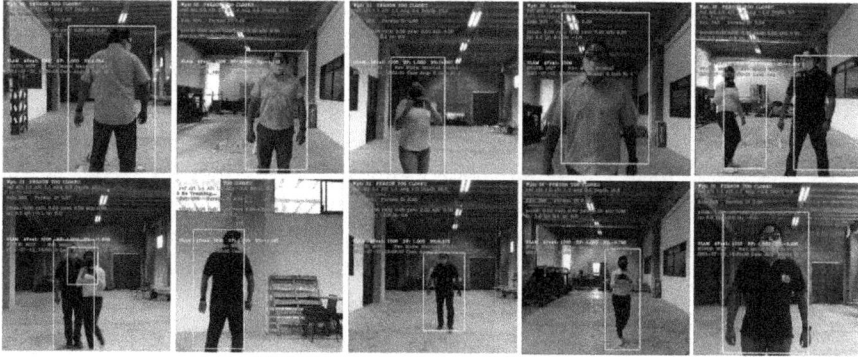

Figure 14.6 Examples of person detection, including when the person is seen from the back and with more than one person in the view.

is used to calculate their distance w.r.t. the drone. If this distance is between 0.5 and 7 metres, then the drone will enter into hovering mode. Once the person gets away beyond 7 metres or out of the field of view, the autonomous flight will resume, commanding the drone to fly towards the next waypoint. Examples of person detection are shown in Figure 14.6. Note that more than one person can be detected at the same time, both genders, different clothing and with different backgrounds.

The second scenario considers the inspection mode, where the drone performs autonomous flight with lateral motion, To avoid the drone getting too close to the wall, the depth image is used to sense how close to the structure the drone is. If the latter is less than 0.5 metres, the inspection controller stops and an avoidance manoeuvre commands the drone to fly backwards, in the direction of the next waypoint until the average depth distance is larger than 0.8 metres.

14.4   EXPERIMENTAL FRAMEWORK

In this section, we present our experimental framework, designed to test our system fully integrated with the OAK-D smart camera. To validate our approach, we carried out several flights for 3 scenarios: 1) Inspection of a corridor and other locations in the facility; 2) distributed inspection using 2 drones; 3) corridor-only inspection with 2 drones; Experiments with two drones aim at demonstrating that we managed to replicate our system to be used with a second drone, which illustrates that warehouse inspection could be carried out more rapidly and efficiently using several drones. During all test flights, the drone takes off from an initial location called *home*. Except for the corridor inspection with two drones, the drones take off and navigate towards the locations to be inspected, returning home after concluding the inspection. A video illustrating these experiments can be watched at https://youtu.be/kUTHPsFlEU0.

Note that take-off and landing are performed manually for safety reasons using the GCS; however, once in the air, the user can switch to autonomous flight mode, being able to interrupt the autonomous flight to take over and manually control the drone if needed or to perform the landing. An experiment is considered successful if the QR and bar codes are detected and decoded for at least 95 % of the total of codes to be registered. In addition, the flight is also considered successful if the sense&avoid system is triggered and works as expected. ORB-SLAM is started before taking off.

For these experiments, the drone follows a trajectory defined by the user using the GCS interface. This path can be stored in a *yaml* settings file. The path can be edited in the file and changed in the GCS. Currently, during inspection mode, our system allows the user to define two waypoints from which the system can calculate a set of intermediate waypoints with a separation distance defined by the user. It is important to mention that for every flight, the drone maps the scene from scratch. However, the map could be stored to be reused and enhanced in further inspections as long as the facility does not change too much.

When using two drones, we used two twin Laptops to launch their corresponding GCS. This was necessary because each drone runs its own ROS master on their Intel Stick computer. As a consequence, a GCS could connect only to one drone; hence, the need for using two separate GCSs. However, in our future work, we will consider one GCS for several drones.

14.4.1   ORB-SLAM EVALUATION USING THE OAK-D

We conducted an evaluation of ORB-SLAM running on the IntelStick Computer on board the drone and using right and depth images provided by the OAK-D. The 6 Degree of Freedom (DoF) camera pose estimated by ORB-SLAM was compared against that obtained with the VICON® motion-captured system. Two categories of experiments were conducted: 1) Random flight with the camera facing forward (FF); 2) Random flight with variable camera orientation. For the former, the drone was flown manually following a random trajectory, but the drone always kept an orientation in yaw of 0 degrees; this meant that the drone flew in lateral, forward and backward motion without rotation changes. This type of flight is similar to the flights performed during inspection of the wall in the corridor. The drone's speed was 0.5 m/s. For the second category, the drone change heading randomly. The drone's speed was between 0.5 m/s to 1.0 m/s, seeking to assess the quality of the pose estimation under a faster flight.

Trajectory comparisons of the ORB-SLAM's output (in green) against the VICON (in red) are shown in Figure 14.7. In qualitative terms, ORB-SLAM resembles the trajectory captured by the VICON system, meaning that the OAK-D's depth estimation provides quality metric information. For a quantitative evaluation, we calculated the error between the VICON and the ORB-SLAM trajectories. The VICON system ran at 100 Hz, whereas the ORB-SLAM ran at $10 - 12$ Hz. We synchronised both trajectories by using the time stamp. The results of this evaluation are shown in Table 14.1. Note that the percent error (average error divided by total distance) is less than 0.5%, the expected percentage for state of the art visual SLAM systems [629].

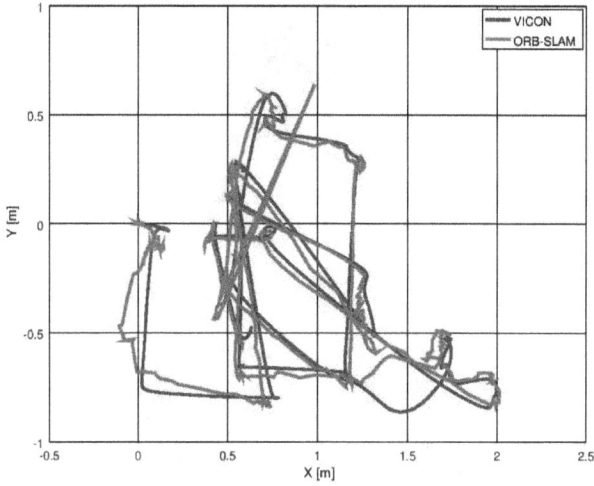

(a) Random Facing Forward

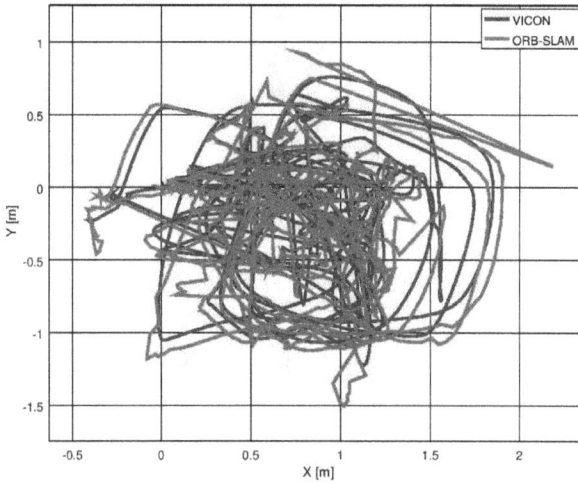

(b) Full random

Figure 14.7 Top view of the experiments carried out to evaluate the accuracy of the pose estimation obtained with ORB-SLAM using grey and depth images from the OAK-D and compared against the trajectory measured with the VICON system.

14.4.2 TEST FACILITY

Due to the COVID-19 pandemic, finding a warehouse to test and evaluate our proposal was not possible. However, we used a test facility that resembles an industrial warehouse with a corridor that could be used to emulate a shelve with packages.

(a) Pictures of the test scenario

(b) QR and bar code laid out on the corridor

Figure 14.8   Test facility used in the experiments.

Figure 14.8(a) shows photographs of the place. On the left wall of the corridor, we laid QR and bar codes out, using tape to draw shapes of boxes, only to emulate an expected appearance of warehouse shelves containing products. Dimensions of the QR and bar codes are also indicated in Figure 14.8(b). The code sizes were chosen as

Table 14.1

Average error and percentage error of ORB-SLAM w.r.t the VICON system.

Trajectory	Avg Error [m]	Std Dev. [m]	Total Dist. [m]	Error %
Random FF	0.05	0.06	14.88	0.3
Random	0.07	0.11	71.68	0.1

the smallest size to be detected by the CNN running on the OAK-D and considering the minimum safe distance the drone could fly w.r.t the wall.

Note that the corridor poses a difficult challenge for a drone to fly across it. The corridor has a usable space of 20 metres in horizontal length and a height of 3 metres. Some columns that stand out from the left wall in the corridor make a narrow entrance from the main hall. Each column has a diameter of 0.55 metres, with the entrance having a width of 1.45 metres; a narrow space to be flown by a drone with a diagonal diameter of 0.44 metres. Assuming that the drone maintains its position exactly at half of the free space between the column and the right wall, the drone has only half a metre around it to move freely. For this reason, the minimum safe distance is 0.5 metres.

14.4.3   INSPECTION OF CORRIDOR AND OTHER LOCATIONS

The user uses the GCS to trigger the drone's take-off, located at home, while also activating the autonomous flight. The flight path is already loaded into the flight controller. The drone navigates toward a corridor in the test facility. Once in the corridor, the drone performs a turn to face the wall and begins a lateral motion, stopping at every waypoint set by the system, in order to allow enough time for the OAK-D to detect and decode the detected QR and bar codes.

After inspecting the corridor, the drone navigates towards other locations in the back of the warehouse. In these locations, QR codes are laid out on the top of two columns at the height of 3 metres. After registering these codes, the drone returns home and once there, the user activates automatic landing. During the navigation, a person walks in front of the drone to test the person detection system. Figure 14.9 shows a screenshot of the GCS during one of the experiments.

We carried out ten runs for this scenario, which were considered successful. In very few occasions, the QR/bar decoder did not work properly. We attribute this to the open-source decoder used in this work. However, the OAK-D did an excellent job detecting QR/bar codes and persons. A top view with the ten trajectories obtained with ORB-SLAM is shown in Figure 14.10. The consistency in the drone's motion can be appreciated in these trajectories. On average, the drone flew a distance of **98.9 metres,** with an average flight time of **6 minutes 30 seconds,** from taking off to landing. This means that the drone flew on average at a speed of **0.25 m/s.**

14.4.4   DISTRIBUTED INSPECTION WITH 2 DRONES

To demonstrate that we are able to replicate our solution, we performed experiments where we used two drones to distribute the warehouse inspection, see Figure 14.11. Both drones use the OAK-D and the Intel Stick as the onboard computer with a twin ROS-based system. The only difference is the flight trajectory to be flown.

We carried out two cases of distributed inspection: 1) One drone flies towards the corridor, the second drone flies to the back of the facility to inspect the codes on the top of the columns; 2) the two drones fly towards the corridor, the first one inspect the top of the corridor, the second one inspects the bottom of the corridor.

Since we also performed ten runs for each one of these cases, the inspection of the corridor was shorter (half of the corridor). This was done merely to save testing time

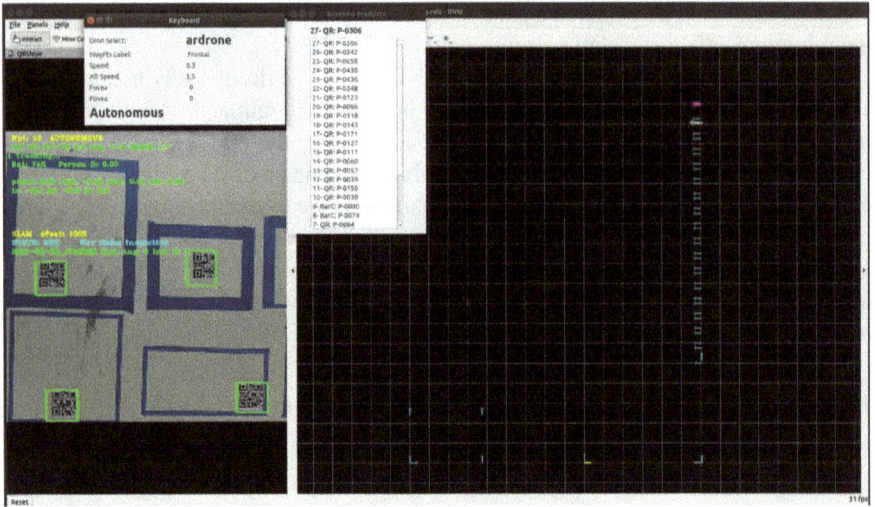

Figure 14.9 GCS showing the OAK-D's colour camera. Waypoints are depicted as cyan arrows and the drone's position as a white arrow, keyboard and the list of scanned packages are also shown. A video illustrating these experiments can be watched at `https://youtu.be/kUTHPsF1EU0`.

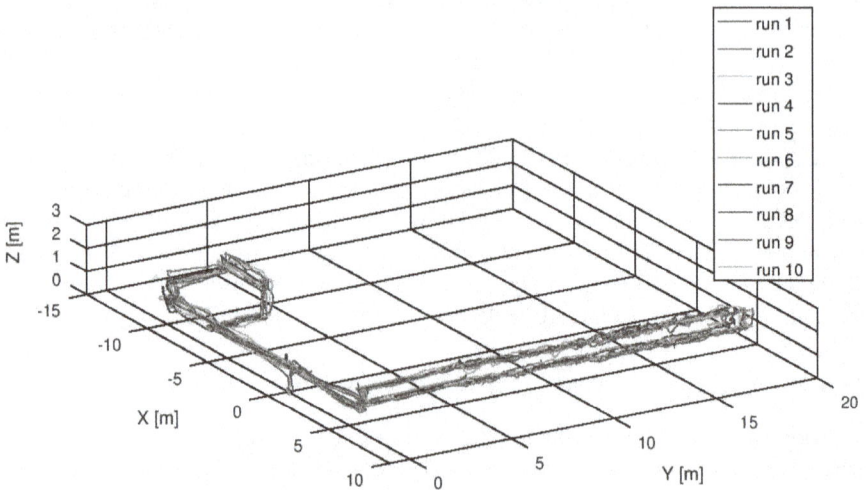

Figure 14.10 Trajectories estimated with RGB-D ORB-SLAM using the OAK-D for 10 runs of the inspection of corridor and other locations in the test facility using one drone. Home is located in the coordinate $(0,0,0)$.

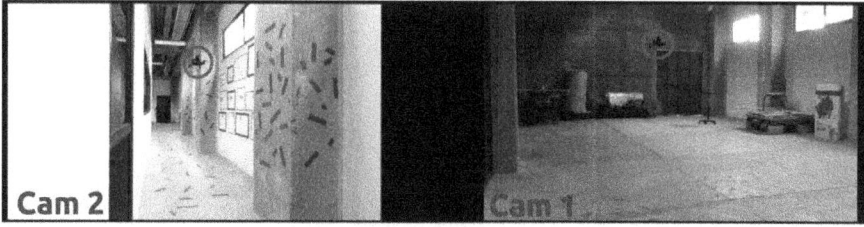

(a) External views of two drones inspecting different places in the test facility.

(b) External views of the drones inspecting the same corridor.

Figure 14.11 Snapshots of experiments with two drones, highlighted in a red circle, performing a distributed inspection. A video illustrating these experiments can be watched at https://youtu.be/kUTHPsF1EU0.

since the access to the facility was limited. However, the inspections were successful as illustrated in the video https://youtu.be/kUTHPsF1EU0.

Table 14.2 shows a summary of the average traversed distance during the autonomous flight for each inspection scenario, its duration time, total number of codes to be scanned and number of QR and bar codes scanned successfully.

Table 14.2

Average results for the inspections performed with our approach.

	Distance [m]	Time [min:sec]	Total # codes	#QR codes	#Bar codes
Inspection of Corridor Other Locations	98.9	6:30	68	56	10
Distributed Inspection with 2 Drones	77.5	3:00	37	26	10
Inspection of Corridor-only with 2 Drones	25.0	2:00	33	22	10

14.5   CONCLUSIONS

We have presented an approach that has the potential of being used for warehouse inspection or similar inspections tasks using autonomous drones equipped with a novel sensor called OAK-D, capable of performing spatial artificial intelligence tasks such as depth estimation and CNN inference on the chip without cost to the host computer on board the drone

This work has been motivated by the problem of warehouse inspection using drones, mainly focusing on lightweight autonomous drones. We assessed the use of a novel sensor, a smart camera known as OAK-D, which provides an RGB camera and a pair of stereo monochromatic images, plus the capability of running neural inference on board the sensor. We have described a methodology that exploits the OAK-D's capabilities to enable several compute-intensive tasks running on board the drone, for instance, QR/Bar code detection, person and 3D structure detection, and visual Simultaneous Localization and Mapping, used for localization of the drone without using GPS and to achieve autonomous flight.

Our experiments show that our fully integrated framework enables a small drone to perform visual SLAM, package scanning by detecting and decoding QR and bar codes, sense&avoid for person detection and structure closeness, and effective communication with a GCS, seeking to enable an effective and safer flight operation. All this processing was carried out on board the drone, an essential feature to achieve genuine autonomous performance. Compared to other solutions employing larger and heavier drones, equipped with several sensors, ours offers a lightweight solution that can be easily replicated, thus enabling distribution of the inspection.

Our evaluations suggest that, in contrast to other solutions requiring heavier and larger drones, our approach offers a feasible lightweight approach. We include some illustrative examples of distributed inspection using two drones equipped with our methodology to illustrate that replication is straightforward. Finally, we consider that our work can serve as the basis for developing a professional solution that could be deployed for warehouse inspection and similar inspection tasks.

In our future work, we will explore lighter visual-based localization methods that can take advantage of the OAK-D neural inference, such as in [124, 522]. We will also experiment with newer versions of the OAK-D, which promise a smaller size with lighter weight, seeking to test a neural localization approach similar to [522] as much as testing with a neural pilot that can pilot the drone without requiring an explicit map or 3D representation of the scene [523, 525]

15 Cognitive Dynamic Systems for Cyber-Physical Engineering

Cesar Torres Huitzil
Computer Science Department, Tecnologico de Monterrey, Puebla campus, México

CONTENTS

15.1 INTRODUCTION

Computing has become ubiquitous in daily life. Thanks to silicon technologies have evolved towards nanoscale dimensions, computing devices become deeply embedded, invisible, in the fabric of everyday life. Embedded and mobile computing systems interact both with humans and a large number of other alike systems due to an increasingly connected society and industry. Physical objects augmented with computing, communication, and sensing capabilities collect massive data, ranging from raw physical variables, e.g., temperature, up to high-level information, e.g., traffic

DOI: 10.1201/9781003385615-15

level and crowd concentration in cities [481]. Under such connection-centric scenario, systems are composed of devices that work together as part of a larger complex system in order to satisfy a global need or multiple ones. On the other hand, the increasing silicon complexity on a chip, derived from process scaling and the introduction of new materials, devices and interconnect infrastructures, demands for increased functionality, lower cost, and shorter time to market by higher levels of abstraction for the design process management. Furthermore, nanoscale electronic devices will become inherently unreliable and unpredictable due to process variation, noise, soft errors, and other non-idealities in nanometer process technologies, i.e., the number of defects in physical substrates is becoming a major constraint that affects the design of computing devices. It is arguable that advances in manufacturing technology alone cannot address this problem cost-effectively, where statistical behavior is the primary attribute of device and circuit fabric [561] [619].

Computing devices are intended to be truly connection-centric artifacts, globally interconnected via the Internet, either always or intermittently online, with other physical systems. At a high level of abstraction, a device employs a combination of software and hardware components that tightly work in a reactive and time-constrained environment. Normally, software provides functional features and flexibility, and hardware is used to provide performance. Both hardware and software supply computing, communication, sensing capabilities, defining a three-dimensional space where a device/system can operate, as shown conceptually in Figure 15.1. Furthermore, in some situations, devices, individually or as whole, must be able not only to *observe*, but to *interact* with the physical world to *control* it. For instance, autonomous vehicles, such as self-driving cars, operate in dynamic environments, and face complex decision-making challenges. They must perceive and interpret their surroundings using sensors, plan and navigate routes, and make real-time decisions to ensure safe and efficient operation. The interaction with other vehicles, pedestrians, and infrastructure adds to the complexity of their dynamic behavior. Yet, in the task of formation control of agents, e.g., swarm robots, the mutual goals of the agents are to maintain a desired spatial configuration or to perform a desired behavior, such as translating as a group or maintaining the center of mass of the group [175]. Thus, an another key dimension can be added to Figure 15.1 to conform a *computation, communication, control* and *sensing* (C^3S) space. Note that each device functionality can be considered as a point in this high-dimensional space where each axis operates at different spatial and temporal scales.

The huge amount of data generated by such interacting devices need to be timely processed to extract useful information. This can be achieved by moving the decision-making capabilities to the devices and thus allowing them to take actions locally [514]. However, in real-world scenarios, where computing devices are deployed, often give rise to situations neither anticipated by system designers nor addressed properly at run-time by the software/hardware. Thus, such complex systems should operate with a high degree of autonomy in a loosely supervised manner in complex and uncertain environments, and must be safety critical since humans-in-the-loop are involved. The dynamic nature of the environment, where the physical system-environment interaction generates high dimensionality and rich content on

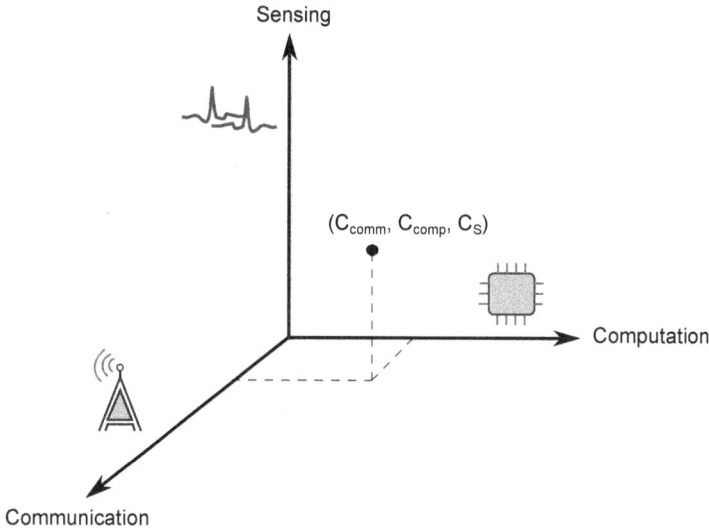

Figure 15.1   Visualization of the communication, computation, and sensing space for a device, with the application/device functionality represented as a point $(C_{comm}, C_{comp}, C_s)$ within this space.

different time scales, must be properly considered. In this context, if the environment is stable, representations should remain stable and if the environment suddenly changes, representations must dynamically adapt themselves and stabilize again onto the new environment. The emergent functionality of such large-scale system is already considered no longer fully manageable or understandable a priori, both by the myriad of interacting components and the huge amount of generated data streams. Using learning-based decision-making components is useful to address this challenge, but it also makes them hard to analyze or verify.

To effectively tackle the challenges at hand, it is crucial to consider exploring alternative models of computation. This is especially important in the field of engineering intelligent systems that must operate with a high degree of autonomy in complex and uncertain environments, while being only loosely supervised. Thus, researchers and practitioners can expand the range of tools and techniques available for creating intelligent systems that are more robust, efficient, and adaptive. These alternative models of computation may incorporate successful approaches such as neural networks, deep learning, or probabilistic reasoning, among others. In this chapter, the Cognitive Dynamic Systems (CDS) paradigm, proposed by Simon Haykin based on the Joaquin Fuster's paradigm of cognition, is presented as engineering framework that can enable the design of such systems [561]. The principles of human cognition still remain to be elucidated; nevertheless, neuroscience has made progress toward understanding the information processing mechanisms in the brain, which is considered as hybrid analog/digital, event driven, distributed, fault tolerant, and massively parallel structures that make extensive use of adaptation, self-organization, and learning [647]. In spite of computer engineering and neuroscience research objectives with respect to computation and cognition may differ, the focus is on the basic underlying principles, so that they can be exploited to build systems that roughly mimic the cognitive capabilities of the brain [578].

The main purpose of this chapter is not to present novel cognitive architectures or algorithms. Instead, its aim is to introduce the core CDS architecture, which is firmly grounded in cognitive studies and artificial intelligence research. The purpose of this introduction is to establish a framework that can be utilized for the engineering of complex systems, including autonomous vehicles and robotic swarms. The CDS architecture serves as a fundamental building block for constructing complex systems by providing a set of principles and guidelines, which are grounded in cognitive science, and are meant to facilitate the development of systems that exhibit intelligent behavior and are adaptive, flexible, and robust. The goal of CDS is to provide systems with a sense of autonomy and the ability to learn from their surrounding environment through experience over time based on the some principles of human cognition.

The rest of this chapter is organized as follows. In Section 15.2, some relevant concepts related to complex dynamic systems, Internet of Things (IoT) and cyber-physical systems (CPS) are briefly described. Section 15.3 presents in detail the CDS architecture and its main components. Central aspects to cognitive control are provided in Section 15.4. Finally, concluding remarks are presented in Section 15.5.

15.2 COMPLEX DYNAMIC SYSTEMS

15.2.1 BACKGROUND

Dynamic systems are complex systems that exhibit dynamic behavior and undergo changes in their state over time, i.e., the future behavior of a dynamic system not only depends on the current stimulus but also on its past evolution. The state of a dynamic system is the minimum amount of information required to define its actual condition at any given time. This state evolves over time, and any change in the state represents the behavior of the system. The behavior of a dynamic system can be described by its *state trajectory* [234], which captures the evolution of the system's state over time. It's important to note that the state of a dynamic system is often not directly measurable. Instead, it is indirectly observed through a set of observables or noisy measurements. These observations provide partial and often imperfect information about the *system's state*, making it challenging to accurately predict the system's future behavior. Therefore, developing effective methods for inferring the system's state from these observations is a crucial aspect of studying dynamic systems.

Dynamic systems can be classified as either *active* or *passive*. An active dynamic system takes actions and explores its environment by continuously interacting with it. The set of all possible actions is called the *action space*, which, similar to the state space, plays a fundamental role in determining the tasks that the dynamic system can perform. By searching over the action space, an active dynamic system can find an optimal policy to perform optimal control, which aims to steer the system toward a desired state while minimizing a cost function related to the control task [234]. This is achieved by designing a control law that determines the best action to take at each time step. Overall, the complexity of an action space has key implications for the type of tasks that a system can perform.

On the other hand, a passive dynamic system mainly focuses on modeling its environment and using the model for classification and state tracking purposes. Passive systems do not take any actions to influence their environment. Instead, they rely on the observations of the environment to build a model that accurately describes its behavior. Passive dynamic systems can be used for various tasks, such as predicting future states of the environment, identifying patterns in the environment, or estimating the parameters of the model. While active dynamic systems have the advantage of being able to interact with their environment and perform optimal control, passive ones have the benefit of being able to accurately model the environment and use the model for various prediction and estimation tasks.

Finally, an active dynamic system is considered to be of a *large scale kind* due to a combination of three factors, as described in [230]:

> High dimensionality of the environment state space, for instance, due to a large number of inputs with different spatial and temporal scales.
> High computational complexity of the predictive and nonlinear model used in tracking the state of the environment.
> High search complexity of the action space due to a huge number of actions, countless number of actions would be necessary to account for all possible states.

Modern complex dynamic systems are characterized by high interconnectivity and interdependence, both physically and through various information and communication network constraints. Examples of such systems include the Internet of Things (IoT) and CPS, which enable the merging of physical and digital worlds to offer diverse and enhanced services in some domains. These systems have become essential components of our modern world, and their continuous development and improvement are vital for the efficient functioning of many industries.

15.2.2 IOT AND CPS

An IoT system is composed of several physically separated, perceptual, and communicating devices that provide observation and data measurement from the physical world, thanks to a communication network where virtually any object equipped with a radio interface can be connected, whereby, most of the time, the communication does not need a human in the loop. IoT technology facilitates sensor-based monitoring through a cooperative operating environment that provide the necessary observability for spatially separated systems. However, with the rapid deployment of IoT devices and applications, networks are exploding in terms of the number of devices but also in complexity, as well as a large amount of sensory data is introduced into the network, complemented by relevant Web-based and social data of various modalities [562]. Many IoT applications are widely distributed, some have stringent real-time requirements, and in all cases, it is necessary to maintain trustworthy communication and adaptability to dynamic environments [515].

Like IoT, CPS integrate sensor devices to interpret data from the physical world even if data is noisy, incomplete, remote, etc., but they are not necessarily connection centric. CPS go beyond the IoT by directly interacting with the physical world using data analytics and control mechanisms. CPS are considered active systems, and some examples of their applications include autonomous vehicles, distributed and cloud-controlled robotics, tele-health systems, and smart energy grids [280]. CPS are distributed and hybrid real-time dynamic systems operating, at different time and space scales, in highly dynamic and unpredictable environments, where the systems will learn from their internal and external environments, and adapt their behavior in response to the continuous change in the context of operation, variable workloads, physical infrastructures, and network topologies [280]. As timing is imposed by the external environment, a deep understanding of dynamics and control is essential in the CPS *perception-action loop*. CPS integrate of computational (cyber) and physical components with a feedback loop. In the loop, the actuators may receive commands from control units via wireless communication channels and operate based on the received instructions [150].

IoT/CPS can potentially provide rich spatial and temporal coverage of most of the physical signals acquired by the available embedded sensors at different spatial and time scales. Figure 15.2 shows a conceptual view of a device/system and a three-dimensional space that highlight the three dimensions, sensory, spatial and temporal that can be exploited in such device. These dimensions are the focus of attention for smart devices when used as ubiquitous sensing platforms in the IoT framework. Variation across sensory, spatial and time dimensions at different scales will require detection and appropriate adaptation of device applications [431]. These aspects are typically determined by the phenomena being monitored in some application domain.

The distributed data and processing required by IoT/CPS applications may span different portions of the physical space, sense different aspects, and be triggered at different instants in time. Interestingly, space and time are orthogonal, and applications might cover all combinations of these two dimensions. For instance, in sense-only and static applications, devices acquire data streams through time and most of the processing occurs without considering space at all, but only within a very localized space of interest, as it is indicated by the rectangle projected in the time-sensory plane in Figure 15.2. This is by far the most common case in current standalone applications deployed in smart devices in the IoT/CPS. On the other hand, the rectangle projected in the time-space plane comprises some other kind of domain-related applications where the main phenomena of interest span a considerable geographical area through time, such as tracking of unmanned aerial vehicles, thus introducing an even greater degree of dynamism due to mobility. Finally, the box labeled as spatio-temporal stream in Figure 15.2, is used to illustrate richer IoT-based applications that require continuous monitoring of the environment and might involve several collaborative static/mobile devices that span over a large geographical space of the environment.

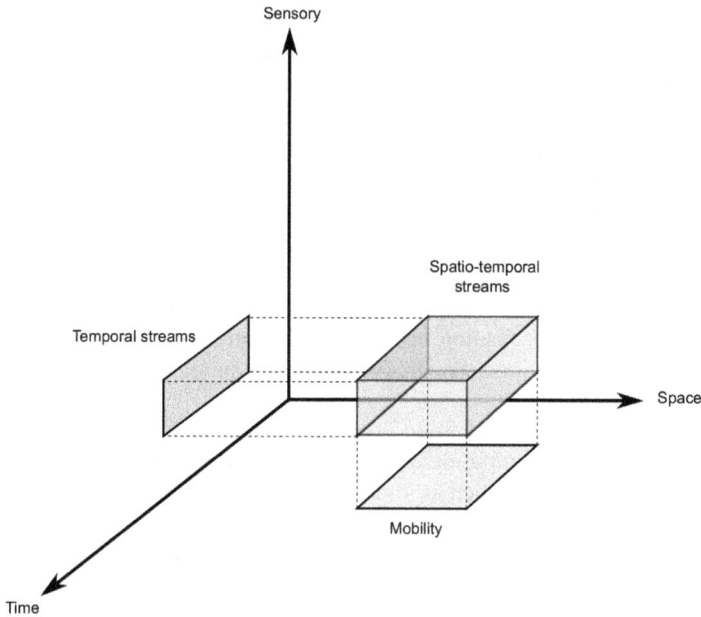

Figure 15.2 Conceptual representation of the time-spatial-sensory dimensions in context-aware applications, where IoT/CPS technologies have the potential to provide abundant spatial and temporal data streams, enabling richer sensory experiences.

15.2.3 IOT/CPS CHALLENGES

Through the integration and extensive collaboration of *computation, communication, control* and *sensing* ($C^3 S$) technology, IoT/CPS can achieve real-time sensing, dynamic control, and information processing services for a wide range of potential applications. However, the widespread deployment of IoT/CPS devices needs to address some important computational challenges and physical constraints for next-generation platforms. As these systems become increasingly complex and interconnected, the need for effective management and control becomes of paramount importance. Robust and reliable communication networks, efficient algorithms, and sophisticated analytics tools are essential to ensure optimal performance and mitigate potential risks. Before any further discussions on potential solutions, it is necessary to highlight some of the vital requirements of these complex systems:

The increasing amount of data generated by IoT/CPS requires distributed and intelligent data processing on edge devices, which must also adhere to other constraints such as energy efficiency. Centralized processing through cloud computing can cause delays due to long-distance transmission [251]. To analyze complex physical signals, devices require more learning and less explicit programming. However, current computing with embedded learning were not initially designed to interact dynamically with the physical world in a natural way across different spatio-temporal scales. Therefore, devices should be able to adapt, self-organize, and exchange capabilities dynamically to meet system needs and environmental dynamics [515].

Coping with error-prone devices and reliability from unreliable/faulty components. It is important to consider fault tolerance and graceful degradation as miniaturized objects deployed in dynamic and unpredictable environments raise significant challenges not properly addressed with currently dominant passive fault-avoidance approaches [619].

Dependability and security. In the area of systems-of-systems, scalability, modularity, and composability pose technical difficulties. Ensuring correctness and system verification becomes even more challenging when timing constraints and continuous operation are also required [557] [192].

In order to address some of these open challenges, particularly the two former, major breakthroughs in computing are expected to benefit from unconventional and new models of computation such as brain-inspired computing. The challenge is then to find not only high-performance and energy-efficient, but also intelligent fault-tolerant computing solutions. New computational models are needed to break away the traditional von Neumann architecture and to rethink some fundamental concepts of computing to redefine how we design a device/system and how to instruct it what to do. The power of data and learning machines has led to the popularity of data-driven nonlinear design paradigms, such as deep learning for the exploitation of data. Artificial intelligence algorithms and data are both crucial for meeting the computational challenge in IoT/CPS design, but neither can deliver alone all the autonomous data-driven self-decision-making capabilities required in a resource-constrained environment.

The development of cognitive computing plays a crucial role in meeting key technical challenges in IoT/CPS, including the generation of vast sensory data, efficient computing/storage, and integration of diverse data sources and types [715]. Due to the physical size and complexity of the dynamical model, there is a compelling need for decentralized analysis and control design of large-scale systems. The incorporation of measurements during the modeling and controlling stage has become increasingly necessary with the emergence of data-driven applications in various fields of applied sciences. In this regard, model-based approaches can be replaced by data-driven strategies that directly construct controllers from experimental data. Moreover, many potential IoT/CPS devices require some level of machine learning or cognitive capability to achieve optimal effectiveness [201].

15.3   CDS PRINCIPLES AND ARCHITECTURE

15.3.1   BIOLOGICAL INSPIRATION

Computing has undoubtedly become an integral part of modern life, but it still lags behind in performing some of the most fundamental tasks that biological systems, particularly humans, execute effortlessly and efficiently. Tasks such as perception, motor control, and language processing pose significant challenges for conventional computers, which, even in the best-case scenario, consume a high amount of power and space resources to match the brain's ability to learn and adapt. Moreover, the disparity between conventional computing technologies and biological nervous systems is even more evident when it comes to autonomous real-time interactions with the environment, particularly in the presence of noisy and uncontrolled sensory input, such as sensory processing and motor control. Nervous systems carry out robust and reliable computation using hybrid analog/digital unreliable processing elements. They emphasize distributed, event-driven, collective, and massively parallel mechanisms, making extensive use of adaptation, self-organization, and learning [111].

The field of computational neuroscience offers a potential solution to bridge the gap between existing fixed computing systems and dynamic self-organized substrates. By defining neural models that exhibit properties such as unsupervised learning, self-adaptation, self-organization, and fault tolerance, more efficient computing in embedded and autonomous systems can be achieved. Recent advancements in machine learning and artificial intelligence provide an opportunity to connect these seemingly divergent perspectives. However, the challenge still remains in transitioning from designing reactive artificial neural modules to building complete neuro-inspired systems with cognitive abilities [111]. To address these challenges, this chapter introduces the CDS paradigm as a framework for engineering complex cyber-physical systems. The CDS, proposed by Simon Haykin in [231], inspired by ideas drawn from neuroscience and cognition in the human brain, is based on Joaquin Fuster's studies on the prefrontal cortex [186]. Fuster's research suggests that the prefrontal cortex endows the rest of the cortex with the ability to anticipate and adapt to changes in the environment. The executive functions of the prefrontal cortex, including top-down attention, working memory, preparatory set, planning, and decision-making, allow for restructuring of past experience to organize new adaptive structures for goal-directed behavior.

15.3.2   CDS ARCHITECTURE

CDS draws inspiration from the structure and function of the human brain [235], which does not solely rely on a modular paradigm where different brain areas function independently to carry out specific tasks. Instead, it operates as a large-scale network where various areas work together to perform cognitive functions. From an engineering perspective, a CDS consists of three main components:

A perceptor o perception components in order to perceive the environment using appropriate sensors.
Memory in order to learn from experience, anticipate the outcome of actions and adapt to the changing circumstances, and
An actuator or action execution component in order to purposely act to achieve goals.

The actuator and perceptor are connected via a feedback channel to establish a global perception-action loop. This loop enables the CDS to continually interact with

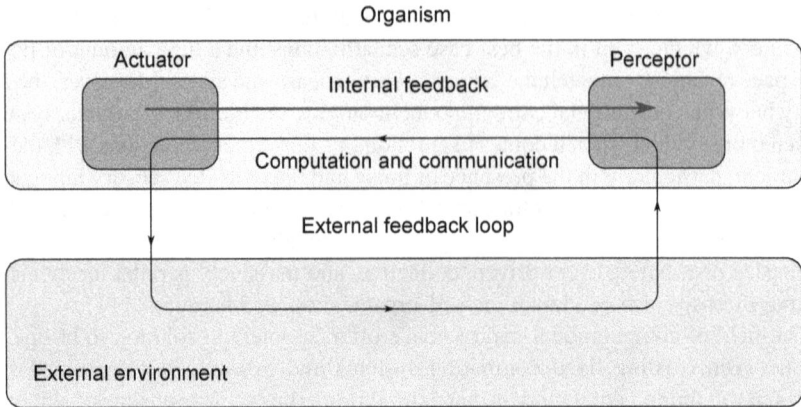

Figure 15.3 External perception-action loop showing how an organism perceives and processes environmental information, computes actions, and acts on the environment through the actuator, adapted from [349]. Internal feedback, relevant for cognitive mechanisms, flows in the opposite direction of the external feedback loop. Memory component is not included for simplicity.

the environment, as depicted in Figure 15.3. A feedback link delivers the extracted relevant information about the environment and system from the perceptor to the executive. The memory component, hierarchical and distributed in nature, is not explicitly illustrated in Figure 15.3, since its main purpose is to emphasize the CDS's interaction with the environment.

15.3.3 PERCEPTION-ACTION LOOP

The perception-action loop is the core in the organism-environment interaction in the CDS. It is inspired from the models of sensorimotor control [186] that assume that sensory information from the environment naturally leads to actions, which then act back on the environment, creating a single, unidirectional perception-action loop as shown in Figure 15.3. The environment plays an important role since it can heavily influence how behaviors emerge in different settings.

Within this loop, the perceptor plays the role of extracting useful information from noisy sensory measurements and transmitting signals from sensors to the actuator. In the context of IoT/CPS, the perceptor senses the environment, whether physical or digital, by processing incoming stimuli or measurements, also known as observables. Based on the information acquired about the environment, the actuator computes potential actions and then acts on the environment to complete the cycle [170]. Specifically, the actuator performs cognitive control, which is considered a key function of the actuator as it acts on the environment on a cycle-by-cycle basis [185]. Due to its relevance, further details on cognitive control will be explained in Section 15.4.

The single perception-action loop model, as already described, neglects a well-known and ubiquitous feature of sensorimotor processing in the brain [349]: *internal feedback.* This feedback includes all signals that do not flow from sensing towards action but rather in the opposite direction, as indicated by the red arrow in Figure 15.3. Such feedback emphasizes that perceptor and actuator subsystems are not indeed isolated. However, the single-loop model has endured since it provides a conceptual framework and a set of tools from control theory that allows subsystems to be treated in isolation. Nonetheless, it is essential to acknowledge that internal feedback plays a significant role in sensorimotor processing and should be considered when developing more complex models.

In the context of the CDS paradigm, internal feedback is viewed as a potential mechanism to achieve rapid and accurate control despite the spatial and temporal limitations of the brain components and communication systems. Recent studies have classified this feedback into two distinct categories [349]:

> Counter directional internal feedback is in the opposite direction of the single-loop model; these signals flow from action toward sensing.
> Lateral internal feedback consists of recurrent connections within and between areas with access to both bottom up and top down information.

A clearer visual description of internal feedback, internal and lateral loops, can be observed in Figure 15.4. This figure also explicitly introduces memory as a distributed and hierarchical component within the CDS architecture. Memory serves as a vital connector between system components, storing multi-modal information on various levels of abstraction. The goal of providing a CDS with memory is to allow for the capture and storage of long- and short-term information. In addition to this role, memory also serves other crucial functions, such as attention and support associations of knowledge.

15.3.4   MULTILAYERED MEMORY

Human memory is one of the most complex systems in nature as it receives a huge amount of sensory data to be processed in a highly distributed way. Memory, a key element in any cognitive architecture, not only connects CDS components and store multi-modal information on different levels of abstraction but provide mechanisms for data abstraction of semantics [480]. From an information processing perspective, memory serves the purpose of learning from the environment, storing and continuously updating stored knowledge in response to environmental changes, and predicting the consequences of actions taken and/or selections made by the system as a whole. In addition, memory must facilitate the association of knowledge, via the internal feedback loops, as the system requires an understanding of how perception and action are interconnected.

The CDS framework has been inspired by the prefrontal cortex of the brain, and its architecture is designed to mimic the structural expansion of this region. The framework incorporates memory as an integral component, in addition to the perceptor and actuator layers. These three components are coupled with each other, as illustrated in Figure 15.4. Perceptual memory enables the perceptor to interpret observables,

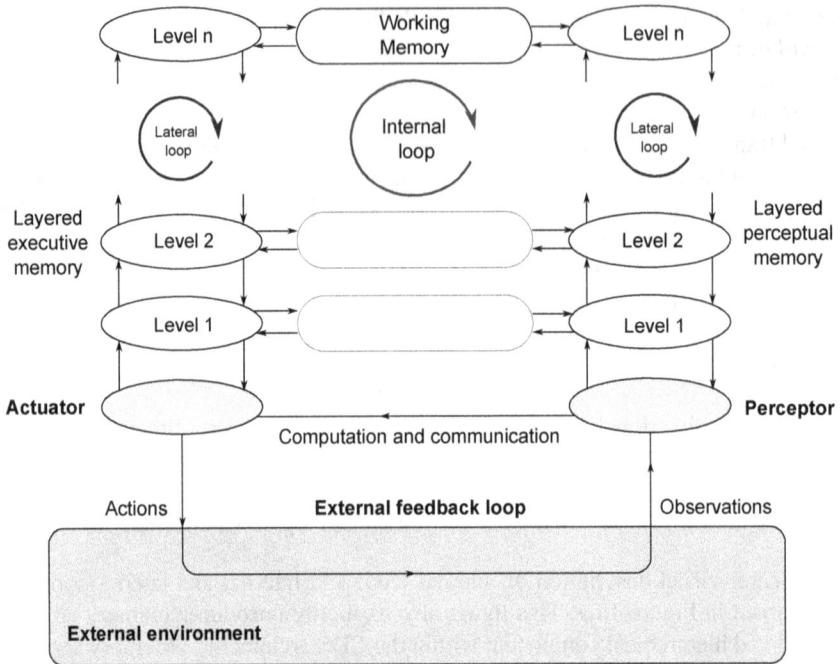

Figure 15.4 The hierarchical organization of memory in the CDS with the different internal feedback loops distributed in the architecture, adapted from [231].

recognize their distinctive features, and categorize the learned features statistically. The structure of perceptual memory is desirably hierarchical and consists of several layers so as to perform perceptual abstractions of incoming stimuli that represent the essence of an object, event, or experience. Executive memory, on the other hand, tracks past actions and their effectiveness in achieving the pursued task [170]. The executive memory associates an action to each incoming measurement from the perceptor, which can be used as a reference for future actions. It is worth noting that the actuator, through its memory, prioritizes exploration and exploitation to optimize resource utilization in the action-space.

The working memory is also an important component of the CDS framework to reciprocally couple the executive and perceptual memories. The working memory holds information for a limited duration and can be controlled by attention mechanisms. This makes it crucial for reasoning, learning, problem-solving, and other cognitive processes. Working memory allows for the manipulation of stored information, and not refers only to the short-term storage of information. Feedback between areas, at the highest cortical level, are essential to the proper functioning of working memory.

15.3.5 CDS ATTENTION AND INTELLIGENCE

Attention and intelligence are fundamental building blocks of cognition in the CDS framework, and their manifestation is conceived as algorithmic mechanisms, emerging from distributed interactions with memory, perception, and action components [235]. As a consequence, attention and intelligence are not assigned to specific physical locations unlike other components of the CDS architecture. This perspective aligns with the inherent distributed processing in the brain, which associates some cognitive functions with dynamic interactions among distributed areas.

Attention has been been a crucial aspect of study of visual sensory processing. The human visual system possesses the ability to selectively focus on certain parts of a scene for faster and efficient perception. This biological mechanism is known as human attention. Deep learning models have incorporated attention mechanisms that concentrate on the most relevant aspects of the input signal for further processing. This is commonly referred to as machine/neural/artificial attention [329]. In the CDS framework, attention can be added to both the perceptual and executive sides of the perception-action cycle, similar to memory. Attention prioritizes the allocation of available resources, enabling the focus of information gathering and processing on critical strategic components. Perceptual attention addresses the information overflow issue, while executive attention implements a version of the principle of minimum disturbance [170]. Executive attention also exploits the explore-exploit tradeoff to facilitate strategies for the learning and planning of cognitive actions for future cycles. Attention is facilitated by the iterative local feedback loops, illustrated by the blue circles in Figure 15.4, whereas the attentional internal feedback that automatically selects and represents the most salient information for actions [349], is represented by the red circle. The CDS multilayer structure results in an increased number of local and global feedback loops within the perception-action cycle, which enhances attention and intelligence.

Intelligence is a complex and multifaceted concept that is difficult to define. It encompasses a vast amount of knowledge, as well as the ability to manipulate that knowledge to draw conclusions about novel situations, acquire new information, and apply complex strategies [631]. In various animal species, including humans, decision-making is a crucial cognitive function necessary for survival in the natural world. It is also a fundamental ability for intelligence in artificial agents. To better understand the computational mechanisms underlying human-like decision-making, the CDS framework regards intelligence as the most powerful principle in human cognition. It is built on some other fundamental elements of the CDS: perception-action cycle, internal feedback loops, memory, and attention. In this context, one of the primary objectives of intelligence in the CDS is to achieve optimal control over the environment in the most effective and efficient manner possible. The CDS framework also incorporates the predictive pre-adaptive characteristic, which originated in cognitive neuroscience and has yet to be fully utilized in engineering [170]. Multiple levels of feedback within the CDS facilitate intelligence by enabling the system to strive for optimal decision-making even in the face of inevitable uncertainties in the environment.

15.4 COGNITIVE CONTROL

Brain architectures have evolved to support highly efficient and robust sensorimotor control, despite the fact that biological components communicate and compute more slowly and noisily than engineered components. Therefore, understanding these architectures better is crucial for building systems that are as scalable, efficient, and robust as biological organisms [583] [544] [369]. Moreover, the prevalence of large-scale distributed systems emphasizes the need for scalable control strategies that only require local communication.

The term cognitive control originated in psychology and neuropsychology. Within the CDS architecture, cognitive control serves to adapt the flow of information from the perceptual part of the system to its executive part, reducing the *information gap* and *risk*, which might be defined as the expected loss associated with a decision [232]. From an engineering perspective, the brain's property of predictive adaptation to future environment changes is of key practical importance, particularly when systems encounter unexpected adverse events or obstacles while pursuing goals or performing tasks. These events are commonly referred to as risks [233].

15.4.1 COGNITIVE CONTROL STRUCTURE

Cognitive control is a fundamental aspect of the human brain that enables individuals to adjust their behavior in response to changing environmental demands. Fuster's paradigm provides a useful framework for understanding cognitive control and its interaction with other functional blocks of a CDS architecture. Figure 15.5 shows the cognitive controller alongside other components of the CDS, highlighting the integral role that cognitive control plays in regulating behavior. The actuator observes the environment and the external system indirectly through the perceptor, thanks to the feedback link connecting the perceptor to the controller.

It is important to note that cognitive controller is not a replacement for physical state-control controllers, but rather an addition to the classical control paradigm. By relying on the actuator, cognitive control is able to decrease the information gap between an individual and their environment through a variety of cognitive actions. These actions can take many different forms, such as attentional control, inhibitory control, and working memory, all of which are vital for efficient cognitive processing [232]. Cognitive controller provides the CDS with the means for effective decision-making through the concepts of learning and planning, often resulting in improved utilization of computational resources. Based on the results of the planning stage, the cognitive controller is responsible for deciding on which cognitive actions to actually execute by updating the policy in a manner to reduce the cost of the predefined value-to-go function.

At any given time, the cognitive controller looks at the current state and determines which actions are executable in that state. Actions define operations that a reasoning engine could perform to change the state of its environment [123]. All the possible resulting states are then evaluated to see which is optimal. Rather than preprogramming this list of actions, a learning engine is responsible for

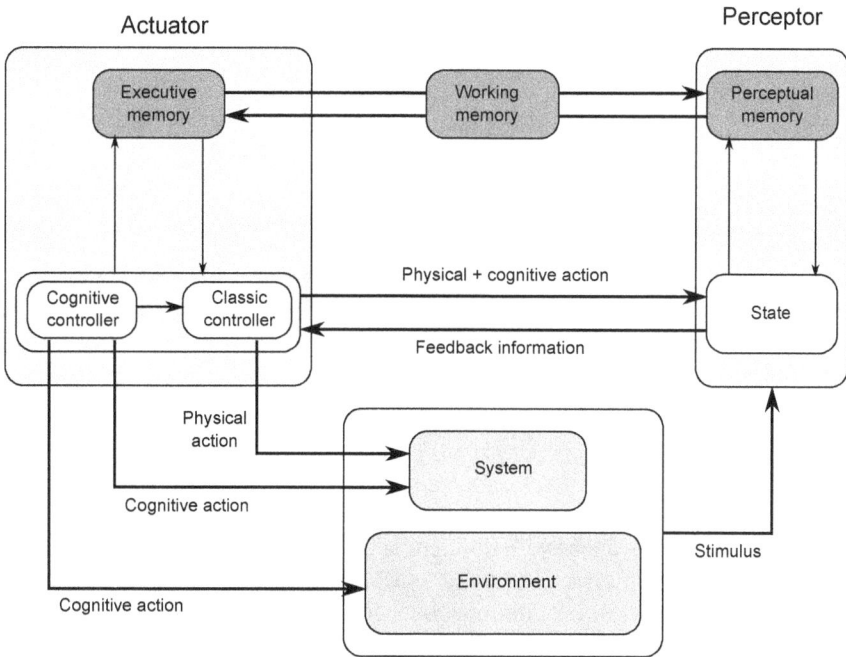

Figure 15.5 Simplified block diagram of the CDS architecture with the cognitive controller, which aims to reduce the information gap through different actions, adapted from [232]. Cognitive actions can influence different parts of the CDS.

augmenting the list of actions available to the controller that allow it to adapt to a changing environment.

To gain a deeper understanding of these principles, Section 15.4.2 explores some cognitive control concepts in greater detail, so as to gain a better understanding of the mechanisms underlying cognitive control and its potential applications in various domains.

15.4.2   INFORMATION FLOW AND INFORMATION GAP

Information is the useful or relevant portion of the data. Relevance finds a meaning only in the context of a perceptual task aimed for performing decision making or control. The information gap concept in CDS is a theoretical framework proposed by Simon Haykin to explain how humans interact with their environment, learn from it, and adapt to it. According to this concept, the interaction between an agent (e.g., a human or a machine) and its environment can be viewed as a process of reducing the information gap between the agent and the environment by iteratively acquiring new information about the environment, processing this information to extract relevant features, and using these features to guide its actions.

The information gap is the difference between what the agent knows about the environment and what it needs to know in order to achieve its goal. This gap arises

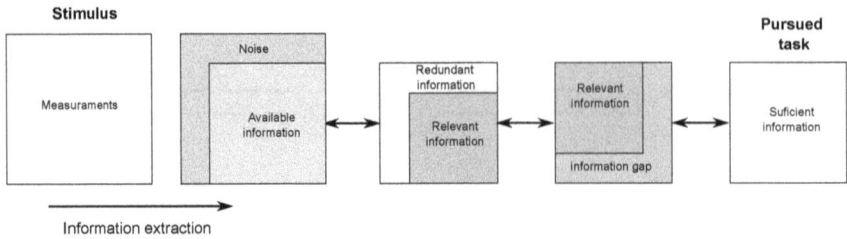

Figure 15.6 Illustration of the information gap concept, which is difference between what the agent knows about the environment and what it needs to know in order to achieve its goal.

due to different factors such as the complexity, uncertainty, and ambiguity of the environment, as well as the limitations of the agent's sensors, actuators, and cognitive processes. The information gap concept is closely related to the notion of active perception: an active process of selecting and interpreting relevant information based on the agent's goals and prior knowledge. The agent may need to explore the environment actively, experiment with different actions, and learn from its mistakes to acquire the sufficient information needed to achieve its goals.

Regarding the critical role of information, complex systems can significantly benefit from a mechanism that controls the directed flow of information in a way to decrease a properly defined task-specific information gap. Decreasing the information gap will reduce the risk involved in achieving a satisfactory level of performance. Figure 15.6 summarizes the concept of information gap and its relationship of some other kinds of information involved in the perception-action loop:

> *Available information* refers to the information that is currently accessible to the agent through its sensors or other means of perception. This information may be incomplete, noisy, or ambiguous, and may not be sufficient for the agent to achieve its goals.
>
> *Redundant information* refers to information that is not necessary or informative, as it can be derived from other available information. In other words, redundant information provides no new or additional knowledge beyond what is already available.
>
> *Relevant information* is information that is necessary and informative for the task at hand. It provides new and useful knowledge that is not already known or derived from other available information.
>
> *Sufficient information* is the minimal amount of information that is needed to achieve a certain goal or perform a task. It is the information that is both necessary and sufficient for the task at hand and it may be hidden or implicit, and may not be immediately available to the agent.

Recently, the literature has shown an increasing interest in research to provide physical systems with the ability to operate effectively despite uncertainties and effectively manage actions that are sensitive to risk.

15.4.3   COGNITIVE AND CONTROL ACTIONS

The role of perception is to extract the available information out of noisy sensory measurements. Recall that the difference between sufficient information for performing a task and relevant information forms the information gap [232]. The theory of Haykin and Fuster proposes that cognitive actions can be used to reduce the information gap between the perceptual and executive parts of the brain, leading to better decision making, based on the idea that the brain uses cognitive control to regulate the flow of information between these two parts of the brain.

In response to information extracted through the perceptor, the actuator performs actions in order to continually enhance this information in subsequent cycles. These actions could be called cognitive actions, which are not being applied to change the state of the environment, but to mitigate the level of uncertainty [232]; they are usually applied to the environment to indirectly affect the system's perception. Thus, cognitive actions might be applied:

To the environment in order to indirectly affect the perception process.
To the system itself in order to reconfigure the sensors and/or actuators.
To the system, but with the goal of decreasing the information gap, with or without other goals, as a part of state-control actions or physical actions.

The cognitive controller is responsible for the decision-making for such cognitive actions in the executive component. Nevertheless, all these different types of cognitive actions do not necessarily exist in a given problem. A real-world problem might include only one of the aforementioned types of cognitive actions, even without the system's state controller.

15.5   CONCLUDING REMARKS

To address the challenges of developing large-scale dynamic systems such as IoT and CPS, it is imperative to leverage innovative computing paradigms that support composability, interoperability, and intelligence at various levels. Brain-inspired computing models have emerged as promising frameworks for engineering cyber-physical systems, given their recent success in artificial intelligence and machine learning based applications.

This chapter provided an in-depth overview of CDS, an engineering framework for developing IoT/CPS systems. CDS is a highly capable information-processing machine that can potentially create intelligent and autonomous devices/systems that interact with their environment. The hierarchical memory organization enables higher levels to abstract the behavior of lower levels, naturally expanding the system's spatial-temporal scope. Additionally, it accounts for multi-modal representations, which are essential for supporting memory associations and achieving objectives through emergent behavior. In the CDS architecture, cognitive control helps to adjust the information flow between the perceptual and executive parts of the system, minimizing the information gap and mitigating risk.

By exploiting feedback at multiple levels, the CDS architecture enables optimal decision-making in the face of environmental uncertainties, potentially achieving desired behaviors without explicit programming for all possible scenarios. Furthermore, the inherent mechanisms in the CDS to control the directed flow of information reduces the task-specific information gap and minimizes the risk associated with achieving satisfactory performance. For instance, i) attentional mechanisms focus on specific sensory inputs while ignoring others, can help to enhance the signal-to-noise ratio of sensory information, ii) the working memory, which temporarily holds information in mind while performing cognitive tasks, can help to bridge the information gap between the perceptual and executive parts, and iii) learning mechanisms to adapt the processing of sensory information based on prior experience, allowing the CDS to better predict and respond to future stimuli.

To fully leverage the potential of the CDS architecture, the key task will be to map out its target applications across different domains and make its components more concrete. One promising approach to achieve this is to use advanced machine learning techniques, such as deep learning methods like transformers, reinforcement learning, and other related techniques. By doing so, we can improve our understanding of how the CDS architecture can be adapted to suit different use cases and domains, and explore new opportunities for improvement. This mapping may require a multidisciplinary effort, involving collaborations between experts in CDS and machine learning, as well as specialists in various fields. Furthermore, it is important to conduct rigorous testing and validation to ensure the accuracy and reliability of the resulting models. By successfully mapping out the potential applications of the CDS architecture and making its components concrete, we can unlock new possibilities for intelligent and autonomous systems in a wide range of domains.

16 EEG-Based Motor and Imaginary Movement Classification: ML Approach

Francisco Javier Ramírez-Arias, Juan Miguel Colores-Vargas
UABC, Facultad de Ciencias de la Ingeniería y Tecnología, Blvd. Universitario No. 1000, Valle de las Palmas, 21500, Baja California, Mexico

Jován Oseas Mérida-Rubio
UABC, Facultad de Ciencias de la Ingeniería y Tecnología, Blvd. Universitario No. 1000, Valle de las Palmas, 21500, Baja California, Mexico

Enrique Efrén García-Guerrero, Everardo Inzunza-González
UABC, Facultad de Ingeniería Arquitectura y Diseño, Carretera Ensenada-Tijuana No. 3917, Ensenada, 22860, Baja California, México

CONTENTS

DOI: 10.1201/9781003385615-16

16.1   INTRODUCTION

The central nervous system comprises the brain and spinal cord, with the brain being a vital component [548]. The human brain is made up of approximately 100 billion neurons that work together to generate thoughts, emotions, and actions [341]. The brain is divided into two hemispheres, with the right hemisphere controlling the left side of the body and the left hemisphere regulating the right side [94]. Various neuroimaging techniques are used to measure and record brain activity, including electrocorticography (ECoG) [683], functional near-infrared spectroscopy (fNIRS) [708], and functional magnetic resonance (fMRI) [60]. Among these techniques, electroencephalography is the most widely accepted by the scientific and private sectors for research purposes in fields such as neuroscience, robotics, home automation, the Internet of Things, and education [178, 650]. EEG, or Electroencephalography, is a non-invasive way of measuring the electrical activity generated in the brain during different mental processes [502]. This is done by placing electrodes on the scalp and collecting signals with different frequencies and amplitudes that reflect a person's mental state [611]. The frequency ranges of these signals are classified into delta (0-4 Hz), theta (4-7 Hz), alpha (8-12 Hz), beta (12-30 Hz), and gamma (30-100 Hz) [387], which are useful for identifying different clinical problems such as schizophrenia [456], epilepsy [10], and brain tumors [440]. Additionally, EEG can be used to identify the central nervous system roots of motor disabilities in neural disorders [29]. Compared to other neuroimaging techniques, EEG offers advantages such as portability, safety, cost-effectiveness, simple equipment, and temporal resolution [536].

The EEG neuroimaging method is the most commonly used approach to develop brain-computer interfaces (BCIs) in both academic and private sectors [463]. BCIs have historically been used clinically to understand motor and cognitive impairment and offer an advantage over electromyography (EMG) pattern recognition in amputees due to the lack of neuromuscular signals [446, 609]. BCIs provide direct communication and control channels between the user's brain and computers without the need for muscular activity [473, 640]. These systems are classified as exogenous or endogenous based on whether external stimuli are required for the brain to generate a particular response or if users need to regulate their brain rhythms through training [434]. Despite their differences, most BCI models have four key stages: signal acquisition, information preprocessing, feature extraction, and classification [77, 434]. Signal acquisition involves placing electrodes on the scalp to obtain analog signals that are digitized using analog-digital converters [625]. Preprocessing is then used to remove noise and artifacts: noise induced by the electrical line; the background noise of the brain; various artifacts that the EEG signals present as a result of some muscular activity such as eye movement, facial muscle activity, among other things [215]. Feature extraction is a crucial step that involves obtaining relevant features from the EEG signals, which are then used as input for classification algorithms [659]. Classification is carried out by different algorithms, including Linear discriminant analysis (LDA), Support Vector Machine (SVM), k-nearest neighbors (KNN) [506], decision tree (D.T.) [250], naive Bayes (N.B.), and artificial deep neural networks (ANNs) [196]. At present, various fields such as science, engineering, and research are utilizing brain computer interfaces (BCIs) to develop applications that provide solutions to complex problems [47, 365]. Advancements in high-density electronics, data acquisition systems, intelligent systems, and neural networks have made this possible. BCIs can be used in six application scenarios: replace, restore, augment, enhance, supplement, and research tools [77]. In [253, 635], the authors identified device control, user status monitoring, assessment, training and education, gaming and entertainment, cognitive enhancement, safety, and security as current and future areas of BCI application. Machine learning (ML) is commonly used in intelligent systems, which can learn from training data to automate the analytical model generation process [38-42]. Deep learning (DL) is a paradigm within ML that uses artificial neural networks (ANNs) [139]. ML algorithms typically classify EEG signals related to motor and imaginary movements [506, 612], while DL is useful for high-dimensional data processing [53]. However, for low-dimensional data input, machine learning algorithms may achieve superior results that are more interpretable than deep neural network results [51, 250].

The authors of [27] used ANNs and SVM to classify EEG signals associated with the right and left hands based on power, mean, and energy as features. In the research [70, 487], ML algorithms such as SVM, LDA, and ANNs were used to control a wheelchair and a robotic arm based on the mean, energy, maximum value, minimum value, and dominant frequency of EEG signals. Other studies have used ML and DL algorithms to classify EEG signals for various applications including pattern analysis [710], group membership classification [272], and brain computer interfaces apps [5, 57]. However, there are still challenges such as real-time processing of EEG signals and optimization of ML algorithms for implementation on embedded systems or edge computing devices [430]. The authors emphasize the need for

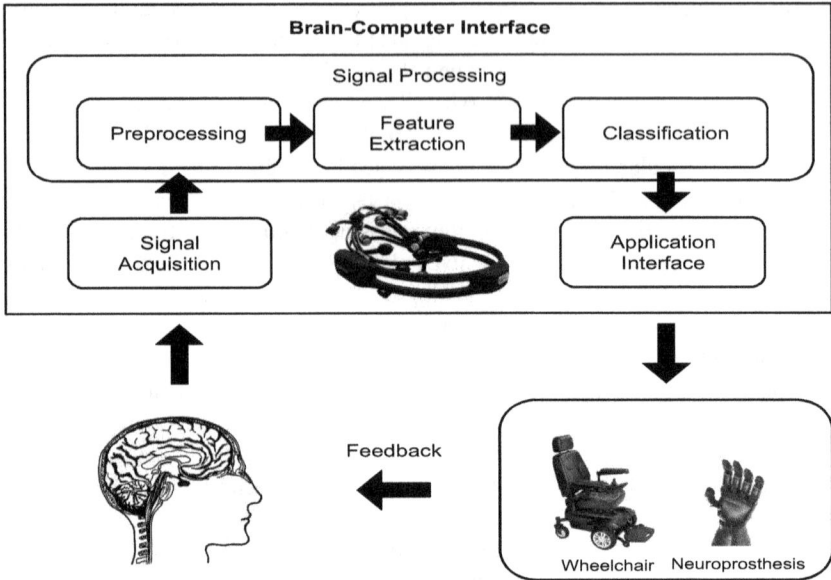

Figure 16.1 Basic structure of Brain–Computer Interface.

continued research and development of reliable and efficient systems for EEG signal classification, especially for programming human-like movements [8, 433].

Figure 16.1 shows the basic structure of brain-computer interfaces. It shows the different stages involved in developing applications based on this type of technology. The first is the acquisition of brain signals through a headband with several electrodes. The second stage is the pre-processing of the EEG signals through some filter. Feature extraction is the third stage. This stage seeks to obtain the most representative signals' parameters relevant for the classification stage. In the classification stage, the aim is to label the feature vectors obtained from the signal analysis in their respective class to encode them in some commands later that the particular application can use. In some applications, feedback is often available.

The purpose of this paper is to assess the performance of nine different ML algorithms in classifying EEG signals for identifying movement patterns in order to control a robotic hand prosthesis. The dataset used contains over 1500 EEG recordings from 109 subjects, with each recording being 1–2 minutes in length [205]. In this study, 20 subjects were randomly selected for training, validation, and testing of the proposed method. The goal is to aid in the development of robotic limb prosthetics, which can recognize patterns in EEG signals with complex dynamics using ML algorithms. It is hypothesized that ML algorithms will outperform standard methods in signal classification tasks. The novelty of this study is that it provides a methodology for classifying EEG signals by training several ML algorithms and employing processing, analysis, and feature extraction techniques in the time domain of various

EEG signal epoch related to motor and imaginary tasks. This methodology can be translated into commands for controlling mechatronic systems such as wheelchairs, robotic prostheses, and mobile robots.

16.2   MATERIALS AND METHODS

Signal acquisition, preprocessing, feature extraction, and classification are the four main stages of an EEG signal classification system. The part where signals are gathered is done by electrodes placed on the scalp's surface. The user's brain activity can be picked up and sent to the data acquisition system with these electrodes. After the EEG signal is acquired, it is necessary to make specific improvements or modifications, which are carried out through different signal preprocessing techniques in the time, frequency, or time-frequency domains. Reducing line noise and muscle artifacts are some of the main goals of preprocessing. The feature extraction stage uses various digital signal processing techniques to derive the feature vector. These vectors aim to train machine learning algorithms to classify EEG signals. The result obtained from these algorithms is the proper identification to which the feature vectors belong. Figure 16.2 illustrates the proposed framework for developing this investigation. We can see that two development environments were used; LabVIEW and Matlab. LabVIEW was used for reading EEG signals, channel selection, signal segment extraction, noise reduction and EEG band extraction, signal analysis, and generation of feature vectors with their respective labels. MATLAB was used for the training, validation, and testing of the different classification algorithms and to obtain their respective performance metrics.

16.2.1   HARDWARE AND SOFTWARE

The hardware required for carrying out these experiments fulfills the following requirements: Operating system Microsoft Windows 10 Pro, OptiPlex 3070 system model, x64-based computer, Intel Core i5-9500 processor, 3 GHz, 6 Cores, and 6 Logical Processors, Memory (RAM): NVIDIA GeForce GT 1030 GDDR5 2GB PCI-Express x16, 16.0 GB DDR4 2666 MHz (2x 8 GB). LabVIEW 2015 was used to read the EEG signals, choose the electrodes, segment the signals, perform preprocessing and analysis, extract features, and prepare the dataset. The Biomedical Toolkit and Signal Express libraries, which are a part of the LabVIEW development environment, were used. The various ML algorithms, a Statistics and ML Toolbox package component, were trained and tested using MATLAB 2021a.

16.2.2   EEG INPUT DATA

The dataset created by Schalk and colleagues at Nervous System Disorders Laboratory was utilized for EEG signal classification and is freely accessible on Physionet [205]. Subjects carried out 14 tasks while 64 electrodes on the BCI2000 device captured and recorded EEG signals [551]. The data are in EDF+ format [289], containing 64 EEG signals, each displayed at a rate of 160 samples per second, and an

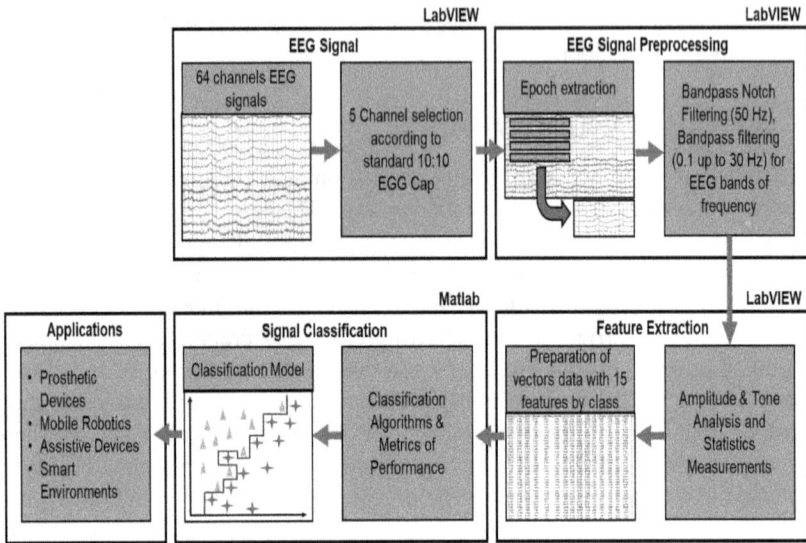

Figure 16.2 Proposed framework for EEG signal classification.

annotation channel, which refers to the actions performed during the task. The data consist of over 1,500 EEG recordings from 109 individuals, each lasting between one and two minutes. The protocol for the Schalk agreement experiment is described in Table 16.1.

16.2.3 EEG CHANNEL SELECTION

Due to the EDF format of the EEG signals, the Biomedical Toolkit was used to integrate them. The LabVIEW software 2015 version was used as the development platform. In Figure 16.3 , the designated electrodes are depicted. These electrodes present neuronal activity associated with the execution of the left-hand and right-hand movements (contained in electrodes C3, C4, and CZ [445]) and the neuronal activity associated with the movement of both feet (contained in electrodes C1 and C2 [229]); because the different EEG channels tend to represent redundant information, as stated in [576], electrodes C3, C1, CZ, C2, and C4 were chosen for this study. The selected electrodes were placed around the center of the cranium, within the motor cortex region; these electrodes are the least affected by various anomalies [338], allowing for the reliable extraction of features.

Table 16.1

Tasks to train ML algorithms to classify EEG signals.

Task	Real Motion	Imaginary Motion	T0	T1	T2
1	Open Eyes	-	Relaxing	-	-
2	Close Eyes	-	Relaxing	-	-
3	Fits	-	Relaxing	Left	Right
4	-	Fits	Relaxing	Left	Right
5	Fits/Feet	-	Relaxing	Fist	Feet
6	-	Fist/Feet	Relaxing	Fist	Feet
7	Fits	-	Relaxing	Left	Right
8	-	Fits	Relaxing	Left	Right
9	Fits/Feet	-	Relaxing	Fist	Feet
10	-	Fist/Feet	Relaxing	Fist	Feet
11	Fits	-	Relaxing	Left	Right
12	-	Fits	Relaxing	Left	Right
13	Fits/Feet	-	Relaxing	Fist	Feet
14	-	Fist/Feet	Relaxing	Fist	Feet

16.2.4   EEG BAND SEPARATION

Distinguishable frequency bands, such as Delta (1-4 Hz), Theta (4-8 Hz), Alpha (8-12 Hz), Beta (12-30 Hz), and Gamma (30–50 Hz), are frequently separated into an EEG signal for examination. (30–50 Hz). This separation was performed by employing third-order bandpass Butterworth IIR filters with various cut-off frequencies, as indicated in Table 16.2. The use of these filters is part of the digital signal processing stage.

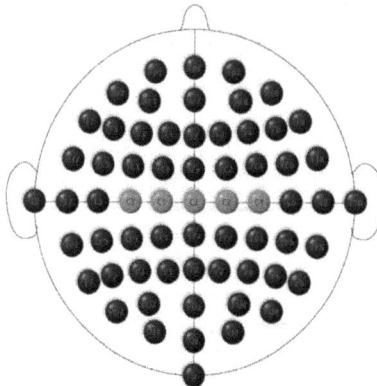

Figure 16.3 Channel used in the analysis of EEG signal classification.

Table 16.2

Bandpass filters use cut-off frequencies to extract EEG signals.

Band of EEG Signal	Low Cut-Off Frequency	High Cut-Off Frequency
Delta-δ	1.0 Hz	4.0 Hz
Theta-θ	4.0 Hz	8.0 Hz
Alpha-α	8.0 Hz	12.0 Hz
Beta-β	12.0 Hz	30.0 Hz
Gamma-γ	30.0 Hz	50.0 Hz

16.2.5 FEATURE EXTRACTION

A broad range of digital signal processing techniques can be performed to obtain the features associated with the EEG rhythm, we used the equations 16.1–16.8, to obtain the features. Measurements of tone, amplitude, and level and statistical analyses were made in these analysis methodologies. Figure 16.4 shows the electrodes used to extract the epochs of the EEG signals to extract and obtain the vector of features.

16.2.6 SIGNAL ANALYSIS

Tone measurements. The tone measurements carried out in the EEG signals epochs are the following: amplitude, frequency, and phase.
Level measurements. The level measurements implemented in the EEG signal epoch are the following: peak to peak, negative peak, and positive peak.
Statistical features. The statistical measurements applied to the different signal epochs are the following:

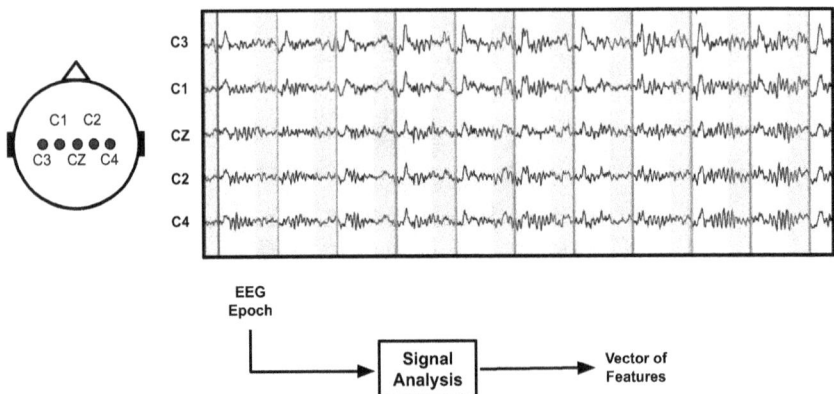

Figure 16.4 Feature extraction of EEG signal to obtain the vector of features.

Median [68]

$$Median = \begin{cases} \frac{(N+1)}{2}, & \text{when N is odd} \\ \frac{N}{2} + \frac{(N+1)}{2}, & \text{when N is ever} \end{cases} \tag{16.1}$$

Mode is the number that occurs most frequently in the set [736].
Mean [582]

$$\widetilde{x} = \frac{1}{N}\sum_{i=1}^{N} x_i \tag{16.2}$$

Root Mean Square (RMS) [736]

$$RMS = \sqrt{\frac{1}{N}\sum_{i=1}^{N} x_i^2} \tag{16.3}$$

Standard Deviation [582]

$$S = \sqrt{\frac{1}{N}\sum_{i=1}^{N} (x_i - \widetilde{x})^2} \tag{16.4}$$

Summation [736]

$$\sum_{i=1}^{N} x_i \tag{16.5}$$

Variance [582]

$$S^2 = \frac{1}{N}\sum_{i=1}^{N} (x_i - \widetilde{x})^2 \tag{16.6}$$

where $\widetilde{x}$ is the mean
Kurtosis [582]

$$Kurtosis = \sum_{i=1}^{N} \frac{(x_i - \widetilde{x})^4}{(N-1)s^4} \tag{16.7}$$

Skewness [582]

$$Skewness = \sum_{i=1}^{N} \frac{(x_i - \widetilde{x})^3}{(N-1)s^3} \tag{16.8}$$

16.2.7 TRAINING FOR MACHINE LEARNING ALGORITHMS

In this work, we selected five ML algorithms to evaluate their performance for classifying EEG signals related to real and imaginary motor movements related to the right hand, left hand, fists, feet, and the state of relaxation. The five selected algorithms are N.B., KNN, D.T., LDA, and SVM. The statistical and machine learning toolbox of MATLAB includes several tools that may be used for both the

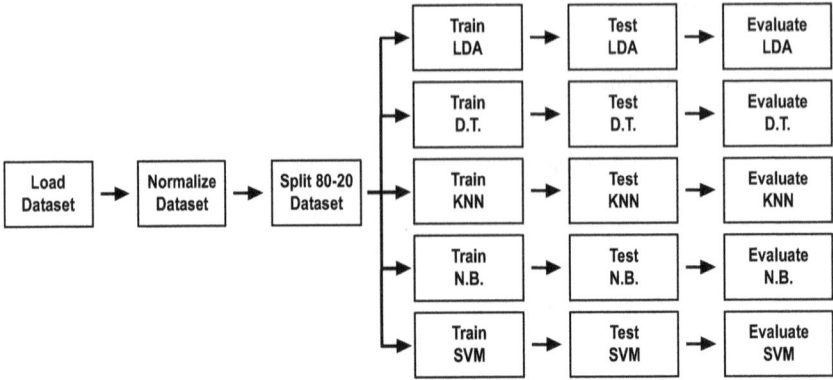

Figure 16.5 Block diagram for training, testing and evaluating the ML algorithms.

pre- and post-processing of data. These ML methods are a component of that toolbox. The block diagram for the training, testing, and evaluation of the chosen machine-learning algorithms is shown in Figure 16.5. The chosen dataset, available in [205], consists of 1500 EEG signal recordings from 109 participants that last between one and two minutes and have a sampling rate of 360 samples per second. Twenty participants were randomly chosen to train, test, and validate the proposed approach for this study. Normalizing the data between 0 and 1 is the second step to obtain better results. The dataset was then divided at random into 20 for testing and 80 for training in the following stage. Then, the ML model is trained. The next stage is to acquire the performance metrics of the ML models (for example, using the confusion matrix), i.e., the performance metrics to evaluate the ML algorithms, such as the area under the curve (AUC) and accuracy, among others.

16.3 DATASET

Each one of the different datasets generated is composed as follows: The data vector consists of 15 features in the time domain, which correspond to 5 electrodes placed at positions C3, C1, CZ, C2, and C4, which are related to motor and imaginary movements, and these belong to one of the five classes, among which they are; "Relaxation," "Right hand," "Left hand," "Fists" and "Feet." Figure 16.6 shows how the data vector is composed. This vector structure allows for better results in the classification of EEG signals. We can observe that the characteristics extracted from the EEG signals belong to a specific frequency band assigned to a specific class.

Table 16.3 presents the different characteristics of the datasets generated and used in the development of this study. We can see that 12 datasets were generated, of which six are related to motor movements and another six are related to imaginary movements. The delta and alpha bands are used in dataset 1 for the relaxation and right-hand classes. dataset 2 employs the delta and theta bands for the relaxation

Electrode Positions

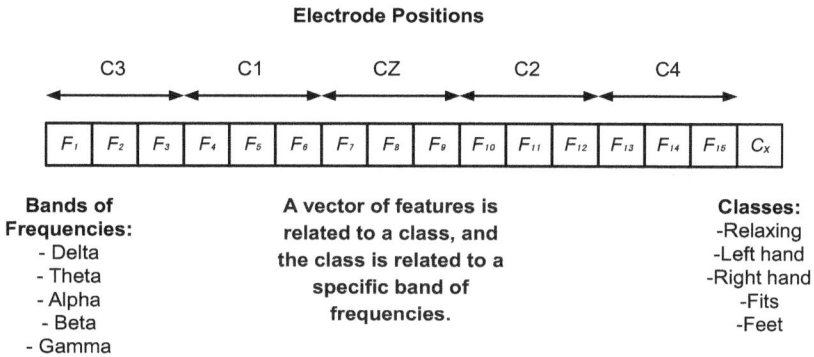

Figure 16.6 Structure of data vector.

and left-hand classes. For dataset 3, we use the delta and beta bands and fists and relaxation classes. In these three datasets, the classification is considered binary. For dataset 4, the delta, theta, and alpha bands are used, which refer to the relaxation classes for the left and right hands, respectively. The delta, beta, and gamma bands are assigned to the relaxation classes, fists, and feet, respectively, for dataset 5. In dataset 6, theta, alpha, beta, and gamma bands are used, assigned to the classes left hand, right hand, fists, and feet. It is worth noting that the classification is considered multiclass for sets 4, 5, and 6.

Table 16.3

Parameters of the datasets used in the experiments

Dataset	Samples	Samples by Class	Classes	Balanced
1	925	480, 445	2	Yes
2	933	480, 453	2	Yes
3	909	472, 437	2	Yes
4	1378	480, 453, 445	3	Yes
5	1356	472, 438, 446	3	Yes
6	1790	454, 445, 445, 446	4	Yes
7	935	480, 455	2	Yes
8	925	480, 445	2	Yes
9	931	480, 451	2	Yes
10	1380	480, 455, 445	3	Yes
11	1380	480, 451, 449	3	Yes
12	1800	455, 445, 451, 449	4	Yes

EEG								Type of Signals
Motor				Imaginary				Type of EEG Signals
Binary			Multiclass	Binary			Multiclass	Type of Classification
-Relaxing -Left hand	-Relaxing -Right hand	-Relaxing	-Relaxing -Right hand -Left hand	-Relaxing -Fits -Feet	-Right hand -Left hand -Fits -Feet	-Relaxing -Left hand	-Relaxing -Right hand	-Relaxing -Fits	-Relaxing -Right hand -Left hand	-Relaxing -Fits -Feet	-Right hand -Left hand -Fits -Feet	Dataset Classes
LDA, D.T., KNN, N.B., SVM				LDA, D.T., KNN, N.B., SVM				Models training
Metrics				Metrics				Metrics performed by model

Figure 16.7   General view of the experiments for EEG signal classification.

16.3.1 EXPERIMENT PROPOSAL

Figure 16.7 shows the plan for the different experiments that will be done to develop the study. We can see that the plan includes experiments with both real and imagined movements. Within the motor and imaginary experiments approach, we find the classification of EEG signals related to 2, 3, and 4 classes, so the classification of motor and imaginary signals is of the binary and imaginary type. datasets had to be made for either the binary or multiclass classification of the EEG signals. Six datasets were made for the binary classification, and six datasets were made for the multiclass classification. Five machine learning algorithms could be trained, checked, and tested on these datasets. They obtained a total of 60 ML models.

16.4 RESULTS

Table 16.4 shows the different experiments carried out during the development of this study. Twelve experiments were carried out, for which it was necessary to generate a different dataset for each of these experiments. Six of these experiments use EEG signals of motor movements, which relate to the movement of the right hand, left hand, fists, feet, and relaxation. Experiments 1 to 3 are binary, while experiments 4 to 6 are called multiclass. Experiments 4 and 5 contain three classes, while experiment 6 contains four. The remaining six experiments are related to EEG signals of imaginary movements of the right hand, left hand, fists, feet, and relaxation. Experiments 7 to 9 are binary, while experiments 10 through 12 are multiclass. Experiments 10 and 11 contain three classes, while experiment 12 contains four. To evaluate the performance of machine learning algorithms, we use the following commonly used performance metrics: accuracy, error, recall, specificity, precision, and F1-score. These performance metrics are derived from the correlation matrix of each model generated. Performance metrics such as the area under the curve, Cohen's Kappa coefficient, Matthews correlation coefficient, and model loss are computed. In the following

Table 16.4

Parameters of the experiments performed.

Experiment	Movement	Band of Frequencies	Number of Classes
1	Motor	δ, α	2
2	Motor	δ, θ	2
3	Motor	δ, β	2
4	Motor	δ, β, γ	3
5	Motor	δ, θ, α	3
6	Motor	$\theta, \alpha, \beta, \gamma$	4
7	Imaginary	δ, α	2
8	Imaginary	δ, θ	2
9	Imaginary	δ, β	2
10	Imaginary	δ, β, γ	3
11	Imaginary	δ, θ, α	3
12	Imaginary	$\theta, \alpha, \beta, \gamma$	4

sections, we will look at how the results of each experiment that was done to classify EEG signals, which are linked to real and imaginary movements, were analyzed.

16.4.1   RESULTS OF EXPERIMENT 01: MOTOR - 2 CLASSES

Table 16.5 shows the scores obtained in each performance metric of the five algorithms selected in this study. N.B. achieved the highest accuracy score of 0.9842 and the lowest error metric score of 0.0158. LDA, D.T., KNN, and SVM models had error metric scores above 0.02. LDA, KNN, and N.B. obtained the highest recall scores of 1.0000, while D.T. and SVM had the lowest scores of 0.9783 and 0.9815 respectively.

16.4.2   RESULTS OF EXPERIMENT 02: MOTOR - 2 CLASSES

Table 16.6 shows the results obtained from the experiment 2. The N.B. model achieved the highest accuracy (0.9844), precision (0.9896), F1-Score (0.9845), and specificity (0.9895) among the models. KNN had the best recall (0.9897), while SVM had the lowest recall (0.9691). N.B. also had the highest Cohen Kappa Coefficient and Matthews correlation coefficient. D.T. consistently scored the lowest in various metrics.

16.4.3   RESULTS OF EXPERIMENT 03: MOTOR - 2 CLASSES

Table 16.7 presents the results obtained in experiment 3. The most important details are that D.T., KNN, and N.B. had the best accuracy scores, while LDA and SVM had the lowest scores. KNN and N.B. had the lowest loss, while LDA and SVM had

Table 16.5
Experiment 01: Score parameters of EEG classification.

Metric	LDA	D.T.	KNN	N.B.	SVM
Accuracy	0.9791	0.9686	0.9791	0.9842	0.9737
Error	0.0209	0.0314	0.0209	0.0158	0.0263
Recall	1.0000	0.9783	1.0000	1.0000	0.9815
Specificity	0.9583	0.9596	0.9596	0.9694	0.9634
Precision	0.9596	0.9574	0.9583	0.9684	0.9725
F1-Score	0.9794	0.9677	0.9787	0.9840	0.9770
AUC	1.0000	0.9690	0.9800	1.0000	0.9990
Coehn's Kappa Coefficient	0.9581	0.9371	0.9581	0.9684	0.9463
Matthews Correlation Coefficient	0.9590	0.9373	0.9590	0.9689	0.9463
Loss	0.0199	0.0306	0.0191	0.0145	0.0275

the highest loss. D.T. and SVM had the best specificity scores, F1-Score parameters, AUC scores, Cohen s Kapp coefficients, and Matthews correlation coefficients.

16.4.4 RESULTS OF EXPERIMENT 04: MOTOR - 3 CLASSES

Table 16.8 shows the results of experiment 4. The D.T. algorithm obtained the best score in accuracy, recall, specificity, precision, F1-Score, average AUC, Cohen's Kappa coefficient, Matthews correlation coefficient, and ML model with the lowest loss. All algorithms obtained scores higher than 0.98, with the highest score being the SVM model.

Table 16.6
Experiment 02: Score parameters of EEG classification.

Metric	LDA	D.T.	KNN	N.B.	SVM
Accuracy	0.9738	0.9688	0.9792	0.9844	0.9688
Error	0.0262	0.0312	0.0208	0.0156	0.0312
Recall	0.9792	0.9720	0.9897	0.9794	0.9691
Specificity	0.9684	0.9647	0.9684	0.9895	0.9684
Precision	0.9691	0.9720	0.9697	0.9896	0.9691
F1-Score	0.9741	0.9720	0.9796	0.9845	0.9691
AUC	0.9790	0.9680	0.9790	1.0000	0.9990
Coehn's Kappa Coefficient	0.9476	0.9367	0.9583	0.9688	0.9375
Matthews Correlation Coefficient	0.9477	0.9367	0.9585	0.9688	0.9375
Loss	0.0260	0.0316	0.0206	0.0157	0.0312

Table 16.7

Experiment 03: Score parameters of EEG classification.

Metric	LDA	D.T.	KNN	N.B.	SVM
Accuracy	0.9681	0.9840	0.9840	0.9840	0.9681
Error	0.0319	0.0160	0.0160	0.0160	0.0319
Recall	0.9691	0.9691	1.0000	1.0000	0.9691
Specificity	0.9670	1.0000	0.9670	0.9670	0.9670
Precision	0.9691	1.0000	0.9700	0.9700	0.9691
F1-Score	0.9691	0.9843	0.9848	0.9848	0.9691
AUC	0.9700	0.9850	0.9840	0.9990	0.9990
Coehn's Kappa Coefficient	0.9361	0.9681	0.9680	0.9680	0.9361
Matthews Correlation Coefficient	0.9361	0.9686	0.9685	0.9685	0.9361
Loss	0.0319	0.0161	0.0158	0.0158	0.0319

16.4.5 RESULTS OF EXPERIMENT 05: MOTOR - 3 CLASSES

The results of experiment 5 are in Table 16.9. The D.T. model obtained the best score for accuracy, error, recall, specificity, precision, F1-Score, average AUC, Cohen's Kappa coefficient, Matthews correlation coefficient, and loss. The N.B. model obtained the lowest score, with 0.9557. The D.T. model obtained the highest score, with 0.0447.

Table 16.8

Experiment 04: Score parameters of EEG classification.

Metric	LDA	D.T.	KNN	N.B.	SVM
Accuracy	0.9825	0.9964	0.9855	0.9674	0.9928
Error	0.0175	0.0036	0.0145	0.0320	0.0072
Recall	0.9820	0.9963	0.9851	0.9675	0.9935
Specificity	0.9910	0.9981	0.9925	0.9835	0.9965
Precision	0.9840	0.9967	0.9875	0.9686	0.9923
F1-Score	0.9827	0.9965	0.9857	0.9680	0.9928
AUC	0.9837	0.9981	0.9888	0.9979	0.9999
Coehn's Kappa Coefficient	0.9605	0.9918	0.9674	0.9266	0.9837
Matthews Correlation Coefficient	0.9742	0.9947	0.9785	0.9515	0.9893
Loss	0.0186	0.0036	0.0150	0.0323	0.0063

Table 16.9

Experiment 05: Score parameters of EEG classification.

Metric	LDA	D.T.	KNN	N.B.	SVM
Accuracy	0.9893	0.9963	0.9926	0.9557	0.9786
Error	0.0107	0.0037	0.0074	0.0443	0.0214
Recall	0.9899	0.9964	0.9931	0.9542	0.9789
Specificity	0.9947	0.9982	0.9965	0.9782	0.9895
Precision	0.9894	0.9963	0.9918	0.9550	0.9792
F1-Score	0.9895	0.9963	0.9948	0.9971	0.9883
AUC	0.9849	0.9973	0.9948	0.9971	0.9883
Coehn's Kappa Coefficient	0.9759	0.9917	0.9834	0.9004	0.9518
Matthews Correlation Coefficient	0.9842	0.9945	0.9888	0.9328	0.9684
Loss	0.0105	0.0034	0.0065	0.0447	0.0214

16.4.6   RESULTS OF EXPERIMENT 06: MOTOR - 4 CLASSES

Table 16.10 presents the results of experiment 6. The classification type for this experiment was multiclass.The D.T. and KNN models achieved the highest accuracy, error, recall, specificity, precision, F1-Score, average AUC, Cohen's Kappa coefficient, Matthews correlation coefficient, and loss parameters. The D.T. obtained the lowest score of 0.0088, while the KNN model presented the highest loss with a score of 0.1687.

Table 16.10

Experiment 06: Score parameters of EEG classification.

Metric	LDA	D.T.	KNN	N.B.	SVM
Accuracy	0.9893	0.9963	0.9926	0.9557	0.9786
Error	0.0107	0.0037	0.0074	0.0443	0.0214
Recall	0.9899	0.9964	0.9931	0.9542	0.9789
Specificity	0.9947	0.9982	0.9965	0.9782	0.9895
Precision	0.9894	0.9963	0.9918	0.9550	0.9792
F1-Score	0.9895	0.9963	0.9948	0.9971	0.9883
AUC	0.9849	0.9973	0.9948	0.9971	0.9883
Coehn's Kappa Coefficient	0.9759	0.9917	0.9834	0.9004	0.9518
Matthews Correlation Coefficient	0.9842	0.9945	0.9888	0.9328	0.9684
Loss	0.0105	0.0034	0.0065	0.0447	0.0214

Table 16.11
Experiment 07: Score parameters of EEG classification.

Metric	LDA	D.T.	KNN	N.B.	SVM
Accuracy	0.9585	1.0000	0.9845	0.9843	1.0000
Error	0.0415	0.0000	0.0155	0.0157	0.0000
Recall	0.9500	1.0000	1.0000	1.0000	1.0000
Specificity	0.9677	1.0000	0.9674	0.9663	1.0000
Precision	0.9694	1.0000	0.9712	0.9714	1.0000
F1-Score	0.9596	0.9963	0.9948	0.9971	0.9883
AUC	0.9849	0.9973	0.9948	0.9971	0.9883
Coehn's Kappa Coefficient	0.9759	0.9917	0.9834	0.9004	0.9518
Matthews Correlation Coefficient	0.9842	0.9945	0.9888	0.9328	0.9684
Loss	0.0105	0.0034	0.0065	0.0447	0.0214

16.4.7   RESULTS OF EXPERIMENT 07: IMAGINARY 2 - CLASSES

Table 16.11 displays the outcomes of Experiment 7. D.T. and SVM models scored highest in performance metrics. The LDA model scored the lowest on the accuracy, error, recall, specificity, precision, F1-score, average AUC, Cohen's Kappa coefficient, Matthews correlation coefficient, and loss. The D.T. and SVM models scored the best on the accuracy, error, recall, precision, F1-score, average AUC, Cohen's Kappa coefficient, Matthews correlation coefficient, and loss.

16.4.8   RESULTS OF EXPERIMENT 08: IMAGINARY 2 - CLASSES

Table 16.12 shows the results of the experiment 8. The KNN model obtained the best score in accuracy, error, recall, specificity, precision, F1-Score, average AUC, Cohen's Kappa coefficient, Matthews correlation coefficient, and loss. All models obtained a score greater than 0.95, with the KNN model obtaining the best score of 0.9948 and the N.B. model with 0.9585.

16.4.9   RESULTS OF EXPERIMENT 09: IMAGINARY 2 - CLASSES

Table 16.13 displays the performance metrics of ML algorithms utilized to develop Experiment 9. The KNN model had the best score with 0.9948, while the LDA model had the lowest score with 0.9740. All models obtained scores greater than 0.94, with the KNN model having the best score with 0.9896.

16.4.10   RESULTS OF EXPERIMENT 10: IMAGINARY 3 - CLASSES

In Experiment 10, the performance metrics derived from the various ML algorithms are displayed in Table 16.14. The SVM model obtained the best performance metrics.

Table 16.12

Experiment 08: Score parameters of EEG classification.

Metric	LDA	D.T.	KNN	N.B.	SVM
Accuracy	0.9689	0.9741	0.9948	0.9585	0.9689
Error	0.0311	0.0259	0.0052	0.0415	0.0311
Recall	0.9706	0.9796	0.9894	0.9515	0.9681
Specificity	0.9670	0.9684	1.0000	0.9667	0.9697
Precision	0.9706	0.9697	1.0000	0.9703	0.9681
F1-Score	0.9706	0.9746	0.9947	0.9608	0.9681
AUC	0.9703	0.9740	0.9947	0.9969	0.9984
Coehn's Kappa Coefficient	0.9376	0.9482	0.9896	0.9168	0.9378
Matthews Correlation Coefficient	0.9376	0.9482	0.9897	0.9170	0.9378
Loss	0.0312	0.0258	0.0055	0.0411	0.0311

The LDA and SVM models received 1.0000, while the KNN model received 0.9867.

16.4.11   RESULTS OF EXPERIMENT 11: IMAGINARY 3 - CLASSES

Results of the experiment 11 are display in Table 16.15. The SVM model obtained the best score in the different performance metrics. The LDA and SVM models received 1.0000, while the KNN model received 0.9867. Cohen's Kappa coefficient was 0.9918, Matthews's correlation coefficient was 0.9947, and the loss parameter was 0.0036.

Table 16.13

Experiment 09: Score parameters of EEG classification.

Metric	LDA	D.T.	KNN	N.B.	SVM
Accuracy	0.9740	0.9844	0.9948	0.9792	0.9844
Error	0.0260	0.0156	0.0052	0.0208	0.0156
Recall	0.9811	0.9688	0.9899	0.9901	0.9703
Specificity	0.9651	1.0000	1.0000	0.9670	1.0000
Precision	0.9720	1.0000	1.0000	0.9709	1.0000
F1-Score	0.9765	0.9841	0.9949	0.9804	0.9849
AUC	0.9814	0.9844	0.9949	0.9956	0.9983
Coehn's Kappa Coefficient	0.9473	0.9688	0.9896	0.9582	0.9687
Matthews Correlation Coefficient	0.9473	0.9692	0.9896	0.9584	0.9692
Loss	0.0268	0.0162	0.0055	0.0211	0.0152

Table 16.14

Experiment 10: Score parameters of EEG classification.

Metric	LDA	D.T.	KNN	N.B.	SVM
Accuracy	0.9928	0.9928	0.9819	0.9710	0.9964
Error	0.0072	0.0072	0.0181	0.0290	0.0036
Recall	0.9927	0.9927	0.9823	0.9706	0.9963
Specificity	0.9962	0.9962	0.9911	0.9854	0.9981
Precision	0.9934	0.9934	0.9815	0.9711	0.9967
F1-Score	0.9930	0.9930	0.9817	0.9708	0.9965
AUC	1.0000	0.9906	0.9867	0.9933	1.0000
Coehn's Kappa Coefficient	0.9837	0.9837	0.9592	0.9348	0.9918
Matthews Correlation Coefficient	0.9894	0.9834	0.9729	0.9564	0.9947
Loss	0.0072	0.0072	0.0176	0.0296	0.0036

16.4.12   RESULTS OF EXPERIMENT 12: IMAGINARY - 4 CLASSES

Table 16.16 shows the results of experiment 12. The D.T. and KNN models had the best and lowest scores in accuracy, error, recall, specificity, precision, F1-score, average AUC, Cohen's Kappa coefficient, Matthews correlation coefficient, and loss metric. The D.T. model had the lowest score with 0.0058, while the LDA model had the most significant loss with 0.1061.

Table 16.15

Experiment 11: Score parameters of EEG classification.

Metric	LDA	D.T.	KNN	N.B.	SVM
Accuracy	1.0000	0.9964	0.9928	0.9239	0.9964
Error	0.0000	0.0036	0.0072	0.0761	0.0036
Recall	1.0000	0.9966	0.9927	0.9258	0.9966
Specificity	1.0000	0.9982	0.9964	0.9613	0.9983
Precision	1.0000	0.9964	0.9930	0.9265	0.9959
F1-Score	1.0000	0.9965	0.9928	0.9261	0.9962
AUC	1.0000	0.9974	0.9945	0.9838	1.0000
Coehn's Kappa Coefficient	1.0000	0.9918	0.9837	0.8288	0.9918
Matthews Correlation Coefficient	1.0000	0.9947	0.9892	0.8875	0.9945
Loss	0.0000	0.0032	0.0074	0.0720	0.0032

Table 16.16

Experiment 12: Score parameters of EEG classification.

Metric	LDA	D.T.	KNN	N.B.	SVM
Accuracy	0.8917	0.9944	0.7917	0.9306	0.9833
Error	0.1083	0.0056	0.2083	0.0694	0.0167
Recall	0.8938	0.9942	0.8060	0.9312	0.9846
Specificity	0.9635	0.9981	0.9300	0.9769	0.9945
Precision	0.9091	0.9945	0.8026	0.9319	0.9833
F1-Score	0.8967	0.9944	0.8010	0.9306	0.9836
AUC	0.9811	0.9962	0.8680	0.9909	0.9981
Coehn's Kappa Coefficient	0.7111	0.9852	0.4444	0.8148	0.9556
Matthews Correlation Coefficient	0.8644	0.9925	0.7336	0.9083	0.9783
Loss	0.1061	0.0058	0.1909	0.0687	0.0152

16.5   MACRO RESULTS

Figure 16.8 shows the macro results of the training, validation, and testing of the machine-learning algorithms used in this study. Accuracy and loss, two metrics considered the most pertinent in classification, were used to acquire macro results. We observe that the macro score obtained from all experiments presents an accuracy of 0.9705 and a loss of 0.0338. The macroscores obtained from the EEG signals

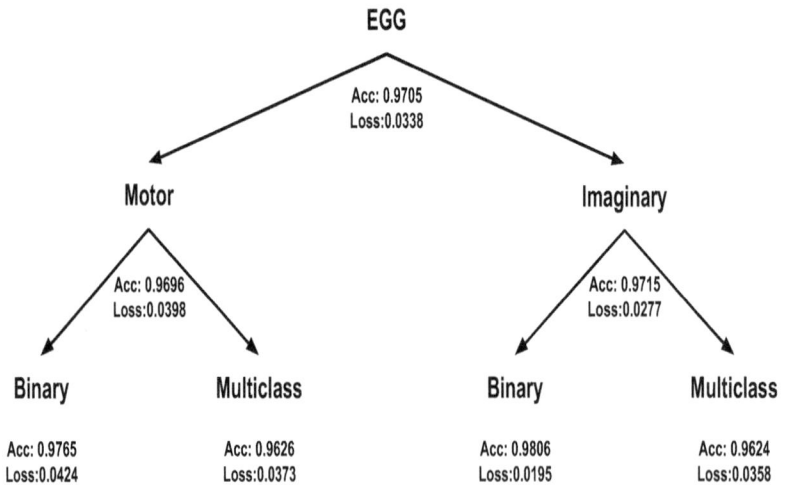

Figure 16.8   Macro results of the EEG experiments.

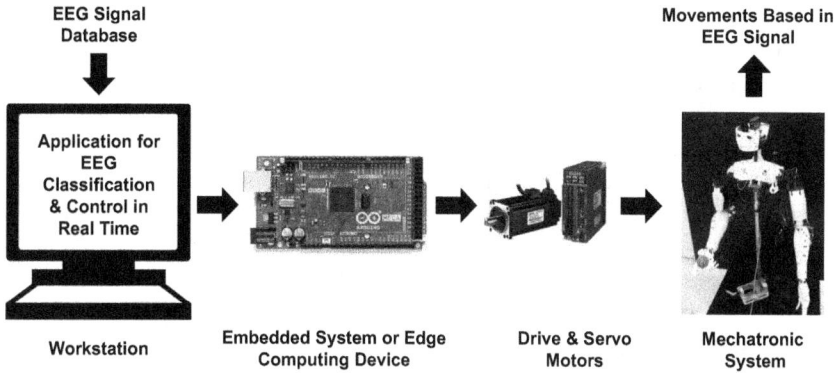

Figure 16.9   A hypothetical application scenario for a mechatronic control system.

related to actual motor movements were 0.9696 for accuracy, while the loss score was 0.0398. In contrast to the macro results obtained from EEG signals related to imaginary movements, we observed that the accuracy obtained was 0.9715, while the score for loss was 0.0277. The results obtained from binary motor movements involving only two classes were 0.9765 for accuracy and 0.0424 for loss. In the case of multiclass motor movements, the accuracy score was 0.9626, and the loss score was 0.0373. Regarding the results involving imaginary binary movements, the accuracy score obtained was 0.9806, while the loss was 0.0195. The accuracy score obtained for the multiclass imaginary movements was 0.9624, and the loss was 0.0358.

16.6   PROPOSED SCENARIO FOR CLASSIFICATION OF EEG SIGNAL

According to other studies [130, 507], the ML algorithms described in this research study could be used in high-performance embedded systems or edge computing devices. They are the primary command structure coordinating with the BCI to gather EEG inputs. In the configuration proposal, the embedded system decodes the control commands into digital signals captured by the servo drivers that allow the movements to be carried out. The workstation handles the digital signal processing step, feature extraction, and control commands. This configuration of EEG signals allows the mechatronic system to test different machine learning models and develop a wide range of applications, among which could be found as a service robot interpreting those that allow users with motor disabilities to interact with the environment in some way.

Figure 10 depicts a hypothetical diagram of the potential mechatronic control system. The development of intelligent service robots that integrate the system's many components starts with this idea. Future attributes to be developed include cheap cost, compact size, portability, low power consumption, and dependable connection with the BCI.

16.6.1   LIMITATIONS OF THE RESEARCH

One of the main limitations of this research is the need to have a brain-computer interface that allows us to capture, process, and classify the EEG signals of some of the research group members. We solve this by using outstanding databases used in other investigations [Ref]. It is worth mentioning that even with the BCI, it presents some minor issues, such as the electrodes must use saline solution when capturing the EEG signals, users and preferably have short hair for better contact of the electrodes with the surface of the scalp, as well as the design of the BCI, it is preferred that it be ergonomic so that it does not cause any discomfort to the user.

16.7   CONCLUSIONS

In this study, we present a methodology for classifying EEG signals related to movements and imaginaries of 20 randomly selected users from a public database. The classification of EEG signals is related to movements of the right hand, left hand, fists, feet, and relaxation. As a result of the digital processing of the EEG signals, 12 custom datasets were created to train machine learning algorithms. In total, 60 machine learning algorithms were operational, validated, tested, and their performance metrics were found. Macrometrics of the different experiments were obtained based on accuracy and loss metrics. The models that achieved the best results were related to the EEG signal classification experiments related to binary imaginary movements with an accuracy of 0.9806 and a loss of 0.0165. The EEG signal classification experiments that obtained the lowest accuracy and loss scores were multiclass imaginary movements, with scores of 0.9624 and 0.0358, respectively. In global terms, we consider that the experiments carried out present an accuracy score of 0.9705 and a loss score of 0.0338, both considered outstanding.

With the proposed method, it is also possible to estimate quantifiable information about both real and imagined movements. This can be done by extracting features in the time domain and measuring how well ML algorithms work. As mentioned earlier, the proposed method allows us to generate different datasets employing custom software that was developed and customized for EEG signal analysis and that can be upgraded with additional features, among which are those in the frequency and time-frequency domains.

References

1. Opencv ai competition 2021: An international ai competition celebrating opencv's 20th anniversary. `https://opencv.org/opencv-ai-competition-2021/`. Accessed on June 1, 2023.

2. Machine learning un python, 2022. `https://www.scikit-learn.org`.

3. Huamán Palma. R. A. Diseño de un sistema de ampliación de cobertura celular para las bandas de frecuencia en 850 MHz y 1900 MHz. 2018.

4. Rhoads A. and K. F. Au. Pacbio sequencing and its applications. *Genomics, Proteomics and Bioinformatics*, 13:278–289, 2015.

5. Bardia Abbasi and Daniel M. Goldenholz. Machine learning applications in epilepsy. *Epilepsia*, 60(10):2037–2047, 2019.

6. Sherif I. Abdelmaksoud, Musa Mailah, and Ayman M. Abdallah. Control strategies and novel techniques for autonomous rotorcraft unmanned aerial vehicles: A review. *IEEE Access*, 8:195142–195169, 2020.

7. Gh Abdi, F. Samadzadegan, and F. Kurz. Pose estimation of unmanned aerial vehicles based on a vision-aided multi-sensor fusion. *International Archives of the Photogrammetry, Remote Sensing & Spatial Information Sciences*, 41, 2016.

8. Sarah N. Abdulkader, Ayman Atia, and Mostafa-Sami M. Mostafa. Brain computer interfacing: Applications and challenges. *Egyptian Informatics Journal*, 16(2):213–230, 2015.

9. I. Abraham, S. Ren, and R. E. Siferd. Logistic function based memristor model with circuit application. *IEEE Access*, 7:166451–166462, 2019.

10. U. Rajendra Acharya, S. Vinitha Sree, G. Swapna, Roshan Joy Martis, and Jasjit S. Suri. Automated eeg analysis of epilepsy: A review. *Knowledge-Based Systems*, 45:147–165, 2013.

11. Amit Adam, Ehud Rivlin, Ilan Shimshoni, and Daviv Reinitz. Robust real-time unusual event detection using multiple fixed-location monitors. *IEEE transactions on pattern analysis and machine intelligence*, 30(3):555–560, 2008.

12. I. A. B. Adames, J. Das, and S. Bhanja. Survey of emerging technology based physical unclonable funtions. In *Int. Great Lakes Symposium on VLSI (GLSVLSI)*, pages 317–322. IEEE, 2016.

13. Michal Adamkiewicz, Timothy Chen, Adam Caccavale, Rachel Gardner, Preston Culbertson, Jeannette Bohg, and Mac Schwager. Vision-Only Robot Navigation in a Neural Radiance World. *IEEE Robotics and Automation Letters*, 7(2):4606–4613, April 2022. Conference Name: IEEE Robotics and Automation Letters.

14. Pranav Adarsh, Pratibha Rathi, and Manoj Kumar. Yolo v3-tiny: Object detection and recognition using one stage improved model. In *2020 6th International Conference on Advanced Computing and Communication Systems (ICACCS)*, pages 687–694, 2020.

15. Revant Adlakha, Wansong Liu, Souma Chowdhury, Minghui Zheng, and Mostafa Nouh. Integration of acoustic compliance and noise mitigation in path planning for drones in human–robot collaborative environments. *Journal of Vibration and Control*, 2023; Vol. 29, No. 19-20, PP. 4757–4771. doi:10.1177/10775463221124049.

16. R. Agarwal, D. Kumar, G. Sandhu, D. Agarwal, and G. Chabra. Diagnostic values of electrocardiogram in chronic obstructive pulmonary disease (copd). *Lung India*, 25:78–81, 2008.

17. D. S. Aguila-Torres, J. A. Galaviz-Aguilar, and J. R. Cárdenas-Valdez. Reliable comparison for power amplifiers nonlinear behavioral modeling based on regression trees and random forest. In *2022 IEEE International Symposium on Circuits and Systems (ISCAS)*, pages 1527–1530, 2022.

18. O. A. Aguirre-Castro, E. E. García-Guerrero, O. R. López-Bonilla, E. Tlelo-Cuautle, D. López-Mancilla, J. R. Cárdenas-Valdez, J. E. Olguín-Tiznado, and E. Inzunza-González. Evaluation of underwater image enhancement algorithms based on retinex and its implementation on embedded systems. *Neurocomputing*, 494:148–159, 7 2022.

19. Oscar Adrian Aguirre-Castro, Everardo Inzunza-González, Enrique Efrén García-Guerrero, Esteban Tlelo-Cuautle, Oscar Roberto López-Bonilla, Jesús Everardo Olguín-Tiznado, and José Ricardo Cárdenas-Valdez. Design and construction of an rov for underwater exploration. *Sensors (Switzerland)*, 19, 12 2019.

20. Hyohoon Ahn, Duc Tai Le, Thien Binh Dang, Siwon Kim, and Hyunseung Choo. Hybrid noise reduction for audio captured by drones. In *Proceedings of the 12th International Conference on Ubiquitous Information Management and Communication*, pages 1–4, 2018.

21. Airvant. Dron: Inventario: Inteligencia artificial: Almacén. `https://www.airvant.es/`. Accessed on June 1, 2023.

22. Alfonso Alcántara, Jesús Capitán, Rita Cunha, and Aníbal Ollero. Optimal trajectory planning for cinematography with multiple Unmanned Aerial Vehicles. *Robotics and Autonomous Systems*, 140:103778, June 2021.

23. S. Ali, S. Khan, A. Khan, and Amine B. Memristor fabrication through printing technologies: A review. *IEEE Access*, 2021.

24. Mansour Alkmim, João Cardenuto, Elisa Tengan, Thomas Dietzen, Toon Van Waterschoot, Jacques Cuenca, Laurent De Ryck, and Wim Desmet. Drone noise directivity and psychoacoustic evaluation using a hemispherical microphone array. *The Journal of the Acoustical Society of America*, 152(5):2735–2745, 2022.

25. Sam Allison, He Bai, and Balaji Jayaraman. Estimating wind velocity with a neural network using quadcopter trajectories. In *AIAA Scitech 2019 Forum*, page 1596, 2019.

26. Md Zahangir Alom, Tarek M. Taha, Christopher Yakopcic, Stefan Westberg, Paheding Sidike, Mst Shamima Nasrin, Brian C. Van Esesn, Abdul A. S. Awwal, and Vijayan K. Asari. The history began from alexnet: A comprehensive survey on deep learning approaches. *arXiv preprint arXiv:1803.01164*, 2018.

27. Mohammad H. Alomari, Aya Samaha, and Khaled AlKamha. Automated classification of l/r hand movement eeg signals using advanced feature extraction and machine learning. *arXiv preprint arXiv:1312.2877*, 2013.

28. K. Alqudaihi, N. Aslam, I. Khan, A. Almuhaideb, S. Alsunaidi, N. Ibrahim, F. Alhaidari, F. Shaikh, Y. Alsenbel, D. Alalharith, H. Alharthi, W. Alghamdi, and M. Alshahrani. Cough sound detection and diagnosis using artificial intelligence techniques: Challenges and opportunities. *IEEE Access*, 9:102327–102344, 2021.

29. Fahd A. Alturki, Khalil AlSharabi, Akram M. Abdurraqeeb, and Majid Aljalal. Eeg signal analysis for diagnosing neurological disorders using discrete wavelet transform and intelligent techniques. *Sensors*, 20(9), 2020.

30. Karim Amer, Mohamed Samy, Mahmoud Shaker, and Mohamed ElHelw. Deep convolutional neural network based autonomous drone navigation. In *Thirteenth International Conference on Machine Vision*, volume 11605, pages 16–24. SPIE, 2021.

31. J. Amoh and K. Odame. Technologies for developing ambulatory cough monitoring devices. *Critical Reviews? in Biomedical Engineering*, 41:457–468, 2013.

32. Vincent Andrearczyk and Paul F. Whelan. Using filter banks in convolutional neural networks for texture classification. *Pattern Recognition Letters*, 84:63–69, 2016.

33. S. Andrews. *Babraham Bioinformatics* https://www.bioinformatics.babraham.ac.uk/ projects/fastqc/ Fastqc, 2010.

34. M. T. Arafin, C. Dunbar, G. Qu, N. McDonald, and L. Yan. A survey on memristor modeling and security applications. In *Sixteenth Int. Symp. on Quality Electronic Design*, pages 440–447. IEEE, 2015.

35. Edoardo Arnaudo, Fabio Cermelli, Antonio Tavera, Claudio Rossi, and Barbara Caputo. A contrastive distillation approach for incremental semantic segmentation in aerial images. In *Image Analysis and Processing–ICIAP 2022: 21st International Conference, Lecce, Italy, May 23–27, 2022, Proceedings, Part II*, pages 742–754. Springer, 2022.

36. P. Arno, F. Launay, J. M. Fournier, and J. C. Grasset. A simple method based on AM-AM, AM-PM measurements and CDMA signal statistics for RF power amplifier characterization. *The 2004 47th Midwest Symposium on Circuits and Systems, MWSCAS '04, Hiroshima, Japan*, 2004.

37. K. Athanasopoulou, M. A. Boti, P. G. Adamopoulos, P. C. Skourou, and A. Scorilas. Third-generation sequencing: The spearhead towards the radical transformation of modern genomics. *Life*, 12:1–22, 2022.

38. Sayantan Auddy, Jakob Hollenstein, Matteo Saveriano, Antonio Rodríguez-Sánchez, and Justus Piater. Continual learning from demonstration of robotic skills. *arXiv preprint arXiv:2202.06843*, 2022.

39. D. Aytan-Aktug, P. T. L. C. Clausen, V. Bortolaia, F. M. Aarestrup, and O. Lund. Prediction of acquired antimicrobial resistance for multiple bacterial species using neural networks. *mSystems*, 5:e00774–19, 2020.

40. Dejan Azinović, Ricardo Martin-Brualla, Dan B. Goldman, Matthias Nießner, and Justus Thies. Neural RGB-D Surface Reconstruction. In *2022 IEEE/CVF Conference on Computer Vision and Pattern Recognition (CVPR)*, pages 6280–6291, June 2022. ISSN: 2575-7075.

41. Sung Moon Bae, Kwan Hee Han, Chun Nam Cha, and Hwa Yong Lee. Development of inventory checking system based on uav and rfid in open storage yard. In *2016 International Conference on Information Science and Security (ICISS)*, pages 1–2. IEEE, 2016.

42. Fei Bai, Minghui Pang, Qiuming Zhu, Hanpeng Li, Xiaomin Chen, Weizhi Zhong, and Naeem Ahmed. A 3GPP-based height-dependent LoS probability model for A2G communications. In *2021 13th International Conference on Wireless Communications and Signal Processing (WCSP)*, pages 1–5, 2021.

43. Vassileios Balntas, Shuda Li, and Victor Prisacariu. Relocnet: Continuous metric learning relocalisation using neural nets. In *Proceedings of the European Conference on Computer Vision (ECCV)*, pages 751–767, 2018.

44. Peng Bao, Zonghai Chen, Jikai Wang, Deyun Dai, and Hao Zhao. Lifelong vehicle trajectory prediction framework based on generative replay. *arXiv preprint arXiv:2111.07511*, 2021.

45. Richard Barker. How many cctv cameras are in london?, 2022.

46. Jonathan T. Barron, Ben Mildenhall, Dor Verbin, Pratul P. Srinivasan, and Peter Hedman. Mip-NeRF 360: Unbounded Anti-Aliased Neural Radiance Fields. In *2022 IEEE/CVF Conference on Computer Vision and Pattern Recognition (CVPR)*, pages 5460–5469, June 2022. ISSN: 2575-7075.

47. Carla Barros, Carlos A. Silva, and Ana P. Pinheiro. Advanced eeg-based learning approaches to predict schizophrenia: Promises and pitfalls. *Artificial Intelligence in Medicine*, 114:102039, 2021.

48. Yakoub Bazi and Farid Melgani. Convolutional svm networks for object detection in uav imagery. *Ieee transactions on geoscience and remote sensing*, 56(6):3107–3118, 2018.

49. J. A. Becerra, M. J. Madero-Ayora, J. J. Reina-Tosina, C. Crespo-Cadenas, J. García-Frías, and G. Arce. A Doubly Orthogonal Matching Pursuit Algorithm for Sparse Predistortion of Power Amplifiers. *IEEE Microwave and Wireless Components Letters*, 28(8):726–728, 2018.

50. J. A. Becerra, A. Pérez-Hérnandez, M. J. Madero-Ayora, and C. Crespo-Cadenas. A Reduced-Complexity Direct Learning Architecture for Digital Predistortion Through Iterative Pseudoinverse Calculation. *IEEE Microwave and Wireless Components Letters*, 31(8):933–936, 2021.

51. Nejra Beganovic, Jasmin Kevric, and Dejan Jokic. Identification of diagnostic-related features applicable to eeg signal analysis. In *Annual Conference of the PHM Society*, volume 10, 2018.

52. Dinara Bekkozhayeva, Mohammadmehdi Saberioon, and Petr Cisar. Automatic individual non-invasive photo-identification of fish (sumatra barb puntigrus tetrazona) using visible patterns on a body. *Aquaculture International*, 29:1481–1493, 8 2021.

53. Yoshua Bengio, Yann Lecun, and Geoffrey Hinton. Deep learning for ai. *Communications of the ACM*, 64(7):58–65, 2021.

54. Diego Benjumea, Alfonso Alcántara, Agustin Ramos, Arturo Torres-Gonzalez, Pedro Sánchez-Cuevas, Jesus Capitan, Guillermo Heredia, and Anibal Ollero. Localization system for lightweight unmanned aerial vehicles in inspection tasks. *Sensors*, 21(17):5937, 2021.

55. Marius Beul, David Droeschel, Matthias Nieuwenhuisen, Jan Quenzel, Sebastian Houben, and Sven Behnke. Fast autonomous flight in warehouses for inventory applications. *IEEE Robotics and Automation Letters*, 3(4):3121–3128, 2018.

56. Hanwen Bi, Fei Ma, Thushara D Abhayapala, and Prasanga N Samarasinghe. Spherical sector harmonics based directional drone noise reduction. In *2022 International Workshop on Acoustic Signal Enhancement (IWAENC)*, pages 1–5. IEEE, 2022.

57. Luzheng Bi, Xin-An Fan, and Yili Liu. Eeg-based brain-controlled mobile robots: a survey. *IEEE transactions on human-machine systems*, 43(2):161–176, 2013.

58. Haoyu Bian, Qichen Tan, Siyang Zhong, and Xin Zhang. Assessment of uam and drone noise impact on the environment based on virtual flights. *Aerospace Science and Technology*, 118:106996, 2021.

59. Wenjing Bian, Zirui Wang, Kejie Li, Jia-Wang Bian, and Victor Adrian Prisacariu. NoPe-NeRF: Optimising Neural Radiance Field with No Pose Prior, December 2022. arXiv:2212.07388 [cs].

60. Osman Tayfun Bişkin, Cemre Candemir, Ali Saffet Gonul, and Mustafa Alper Selver. Diverse task classification from activation patterns of functional neuro-images using feature fusion module. *Sensors*, 23(7), 2023.

61. Z. Biolek, D. Biolek, and V. Biolkova. SPICE model of memristor with nonlinear dopant drift. *Radioengineering*, 18(2), 2009.

62. Hunter Blanton, Connor Greenwell, Scott Workman, and Nathan Jacobs. Extending absolute pose regression to multiple scenes. In *Proceedings of the IEEE/CVF Conference on Computer Vision and Pattern Recognition Workshops*, pages 38–39, 2020.

63. Hunter Blanton, Scott Workman, and Nathan Jacobs. A structure-aware method for direct pose estimation. In *Proceedings of the IEEE/CVF Winter Conference on Applications of Computer Vision*, pages 2019–2028, 2022.

64. Valts Blukis, Taeyeop Lee, Jonathan Tremblay, Bowen Wen, In So Kweon, Kuk-Jin Yoon, Dieter Fox, and Stan Birchfield. Neural Fields for Robotic Object Manipulation from a Single Image, October 2022.

65. M. Boesch, F. Rassouli, F. Baty, A. Schwärzler, S. Widmer, P. Tinschert, I. Shih, D. Cleres, F. Barata, E. Fleisch, and M. Brutsche. Smartphone-based cough monitoring as a near real-time digital pneumonia biomarker. *ERJ Open Research*, pages 518–2022, 2023.

66. D. Bolognini, A. Sanders, J. Korbel, A. Magi, V. Benes, and T. Rausch. Visor: a versatile haplotype-aware structural variant simulator for short- and long-read sequencing. *Bioinformatics*, 36:1267–1269, 2020.

67. J. J. Bongiorno. Real-frequency stability criteria for linear time-varying systems. *IEEE Proceedings*, 52(7):832–841, 1964.

68. Poomipat Boonyakitanont, Apiwat Lek-uthai, Krisnachai Chomtho, and Jitkomut Songsiri. A review of feature extraction and performance evaluation in epileptic seizure detection using eeg. *Biomedical Signal Processing and Control*, 57:101702, 2020.

69. G. Botha, G. Theron, M. Warren, M. Klopper, K. Dheda, P. Van Helden, and T. Niesler. Detection of tuberculosis by automatic cough sound analysis. *Physiol Meas*, 39:45005, 2018.

70. Rihab Bousseta, I. El Ouakouak, M. Gharbi, and F. Regragui. Eeg based brain computer interface for controlling a robot arm movement through thought. *Irbm*, 39(2):129–135, 2018.

71. R. N. Braithwaite. A Self-Generating Coefficient List for Machine Learning in RF Power Amplifiers using Adaptive Predistortion. *European Microwave Conference*, pages 1229–1232, 2006.

72. D. Branigan, S. Deborggraeve, M. Kohli, E. Maclean, L. Mckenna, and M. Ruhwald. Pipeline report-2021 tuberculosis diagnostics tuberculosis diagnostics. Technical report, Treatment Action Group, 2021.

73. L. Breiman. Classification and Regression Trees. *Routledge*, 2017.

74. Guillaume Bresson, Yu Li, Cyril Joly, and Fabien Moutarde. Urban localization with street views using a convolutional neural network for end-to-end camera pose regression. 2019.

75. A. Brihuega, M. Abdelaziz, L. Anttila, Y. Li, A. Zhu, and M. Valkama. Mixture of Experts Approach for Piecewise Modeling and Linearization of RF Power Amplifiers. *IEEE Transactions on Microwave Theory and Techniques*, 70(1):380–391, 2022.

76. Jeff Brown. Zbar bar code reader. `http://zbar.sourceforge.net/`. Accessed on June 1, 2023.

77. Clemens Brunner, Niels Birbaumer, Benjamin Blankertz, Christoph Guger, Andrea Kübler, Donatella Mattia, José del R. Millán, Felip Miralles, Anton Nijholt, Eloy Opisso, Nick Ramsey, Patric Salomon, and Gernot R. Müller-Putz. Bnci horizon 2020: towards a roadmap for the bci community. *Brain-Computer Interfaces*, 2(1):1–10, 2015.

78. Arunkumar Byravan, Jan Humplik, Leonard Hasenclever, Arthur Brussee, Francesco Nori, Tuomas Haarnoja, Ben Moran, Steven Bohez, Fereshteh Sadeghi, Bojan Vujatovic, and Nicolas Heess. NeRF2Real: Sim2real Transfer of Vision-guided Bipedal Motion Skills using Neural Radiance Fields, October 2022.

79. Aldrich A. Cabrera-Ponce, Manuel Martin-Ortiz, and José Martínez-Carranza. Continual learning for multi-camera relocalisation. In *Mexican International Conference on Artificial Intelligence*, pages 289–302. Springer, 2021.

80. Aldrich A. Cabrera-Ponce and Jose Martinez-Carranza. Aerial geo-localisation for mavs using posenet. In *2019 Workshop on Research, Education and Development of Unmanned Aerial Systems (RED UAS)*, pages 192–198. IEEE, 2019.

81. Aldrich A. Cabrera-Ponce, Jose Martinez-Carranza, and Caleb Rascon. Detection of nearby uavs using a multi-microphone array on board a uav. *International Journal of Micro Air Vehicles*, 12:1756829320925748, 2020.

82. Aldrich A. Cabrera-Ponce and Jose Martinez-Carranza. Convolutional neural networks for geo-localisation with a single aerial image. *Journal of Real-Time Image Processing*, 19(3):565–575, 2022.

83. Aldrich A. Cabrera-Ponce and José Martínez-Carranza. Onboard cnn-based processing for target detection and autonomous landing for mavs. In *Mexican Conference on Pattern Recognition*, pages 195–208. Springer, 2020.

84. Aldrich A. Cabrera-Ponce, Leticia Oyuki Rojas-Perez, Jesus Ariel Carrasco-Ochoa, Jose Francisco Martinez-Trinidad, and Jose Martinez-Carranza. Gate detection for micro aerial vehicles using a single shot detector. *IEEE Latin America Transactions*, 17(12):2045–2052, 2019.

85. Aldrich Alfredo Cabrera-Ponce, Manuel Isidro Martin-Ortiz, and Jose Martinez-Carranza. Multi-model continual learning for camera localisation from aerial images. In G. de Croon and C. De Wagter, editors, *13th International Micro Air Vehicle Conference*, pages 103–109, Delft, the Netherlands, Sep 2022. Paper no. IMAV2022-12.

86. Ming Cai, Chunhua Shen, and Ian D. Reid. A hybrid probabilistic model for camera relocalization. In *BMVC*, volume 1, page 8, 2018.

87. S. A. Campbell. The science of engineering of microelectronic fabrication. In *2nd ed. Oxford University in electrical and computer engineering*, pages 326–350. Oxford University Press, 2001.

88. Carlos Campos, Richard Elvira, Juan J. Gómez Rodríguez, José M. M. Montiel, and Juan D. Tardós. Orb-slam3: An accurate open-source library for visual, visual–inertial, and multimap slam. *IEEE Transactions on Robotics*, 37(6):1874–1890, 2021.

89. Leobardo Campos-Macías, Rodrigo Aldana-López, Rafael de la Guardia, José I. Parra-Vilchis, and David Gómez-Gutiérrez. Autonomous navigation of mavs in unknown cluttered environments. *Journal of Field Robotics*, 38(2):307–326, 2021.

90. Ang Cao and Justin Johnson. HexPlane: A Fast Representation for Dynamic Scenes, January 2023. arXiv:2301.09632 [cs].

91. Anh-Quan Cao and Raoul de Charette. SceneRF: Self-Supervised Monocular 3D Scene Reconstruction with Radiance Fields, January 2023. arXiv:2212.02501 [cs].

92. Junli Cao, Huan Wang, Pavlo Chemerys, Vladislav Shakhrai, Ju Hu, Yun Fu, Denys Makoviichuk, Sergey Tulyakov, and Jian Ren. Real-Time Neural Light Field on Mobile Devices, December 2022. arXiv:2212.08057 [cs].

93. Alexandra Carlson, Manikandasriram Srinivasan Ramanagopal, Nathan Tseng, Matthew Johnson-Roberson, Ram Vasudevan, and Katherine A. Skinner. CLONeR: Camera-Lidar Fusion for Occupancy Grid-aided Neural Representations, September 2022. arXiv:2209.01194 [cs].

94. Rita Carter. *The Human Brain Book: An Illustrated Guide to its Structure, Function, and Disorders*. Penguin, January 2019. Google-Books-ID: s8bhDwAAQBAJ.

95. Guilherme Carvalho, Ihannah Guedes, Milena Pinto, Alessandro Zachi, Luciana Almeida, Fabio Andrade, and Aurelio G. Melo. Hybrid PID-Fuzzy controller for autonomous UAV stabilization. In *2021 14th IEEE International Conference on Industry Applications (INDUSCON)*, pages 1296–1302, São Paulo, Brazil, August 2021. IEEE.

96. Pedro Castillo, Rogelio Lozano, and Alejandro Dzul. Stabilization of a mini rotorcraft with four rotors. *IEEE control systems magazine*, 25(6):45–55, 2005.

97. Pedro Castillo, Rogelio Lozano, and Alejandro E. Dzul. *Modelling and Control of Mini-Flying Machines*. Advances in Industrial Control. Springer-Verlag, London, 1 edition, 2005. https://doi.org/10.1007/1-84628-179-2.

98. Dylan Cawthorne and Peter Møller Juhl. Designing for calmness: Early investigations into drone noise pollution management. In *2022 International Conference on Unmanned Aircraft Systems (ICUAS)*, pages 839–848. IEEE, 2022.

99. Cómputo Comunicaciones y Contacto Ciudadano de la Ciudad de México Centro de Comando, Control and C5 CDMX. Gobierno de la Ciudad de México, 2023.

100. Fabio Cermelli, Massimiliano Mancini, Samuel Rota Bulo, Elisa Ricci, and Barbara Caputo. Modeling the background for incremental learning in semantic segmentation. In *Proceedings of the IEEE/CVF Conference on Computer Vision and Pattern Recognition*, pages 9233–9242, 2020.

101. Ender Çetin, Alicia Cano, Robin Deransy, Sergi Tres, and Cristina Barrado. Implementing mitigations for improving societal acceptance of urban air mobility. *Drones*, 6(2):28, 2022.

102. T. Chai and R. R. Draxler. Root mean square error (RMSE) or mean absolute error (cao)? â arguments against avoiding RMSE in the literature. *Geoscientific Model Development*, 7(3):1247–1250, 2014.

103. U. Chatterjee, R. S. Chakraborty, J. Mathew, and Dhiraj K. P. Memristor based arbiter PUF: cryptanalysis threat and its mitigation. In *IEEE 29th Int. Conf. on VLSI Design and IEEE 15th Int. Conference on Embedded Systems (VLSID)*, pages 535–540, 2016.

104. Aggelina Chatziagapi, ShahRukh Athar, Abhinav Jain, Rohith MV, Vimal Bhat, and Dimitris Samaras. LipNeRF: What is the right feature space to lip-sync a NeRF? In *2023 IEEE 17th International Conference on Automatic Face and Gesture Recognition (FG)*, pages 1–8, January 2023.

105. Nived Chebrolu, Philipp Lottes, Thomas Läbe, and Cyrill Stachniss. Robot localization based on aerial images for precision agriculture tasks in crop fields. In *2019 International Conference on Robotics and Automation (ICRA)*, pages 1787–1793. IEEE, 2019.

106. M. L. Chen, Doddi A. Royer J., Freschi L., Schito M., Ezewudo M., Kohane I. S., Beam A., and Farhat M. Beyond multidrug resistance: Leveraging rare variants with machine and statistical learning models in mycobacterium tuberculosis resistance prediction. *EBioMedicine*, 43:356–369, 2019.

107. Shilang Chen, Junjun Wu, Qinghua Lu, Yanran Wang, and Zeqin Lin. Cross-scene loop-closure detection with continual learning for visual simultaneous localization and mapping. *International Journal of Advanced Robotic Systems*, 18(5): 17298814211050560, 2021. SAGE Publications Sage UK: London, England.

108. X. Chen, O. Schulz-Trieglaff, R. Shaw, B. Barnes, F. Schlesinger, M. Källberg, A. Cox, S. Kruglyak, and S. Saunders. Manta: Rapid detection of structural variants and indels for germline and cancer sequencing applications. *Bioinformatics*, 32:1220–1222, 2016.

109. Zetao Chen, Adam Jacobson, Niko Sünderhauf, Ben Upcroft, Lingqiao Liu, Chunhua Shen, Ian Reid, and Michael Milford. Deep learning features at scale for visual

place recognition. In *2017 IEEE International Conference on Robotics and Automation (ICRA)*, pages 3223–3230. IEEE, 2017.

110. Luca Chiaraviglio, Marco Fiore, Edouard Rossi, Marco Ajmone Marsan, Nicola Blefari Melazzi, and Stefano Buzzi. 5G technology: Which risks from the health perspective. *The 5G Italy Book*, 2019.

111. Elisabetta Chicca, Fabio Stefanini, Chiara Bartolozzi, and Giacomo Indiveri. Neuromorphic electronic circuits for building autonomous cognitive systems. *Proceedings of the IEEE*, 102(9):1367–1388, 2014.

112. Ioannis Chiotellis, Franziska Zimmermann, Daniel Cremers, and Rudolph Triebel. Incremental semi-supervised learning from streams for object classification. In *2018 IEEE/RSJ International Conference on Intelligent Robots and Systems (IROS)*, pages 5743–5749. IEEE, 2018.

113. Hyeon Cho, Dongyi Kim, Junho Park, Kyungshik Roh, and Wonjun Hwang. 2d barcode detection using images for drone-assisted inventory management. In *2018 15th International Conference on Ubiquitous Robots (UR)*, pages 461–465. IEEE, 2018.

114. Francois Chollet et al. *Deep learning with Python*, volume 361. Manning New York, 2018.

115. Yong Shean Chong and Yong Haur Tay. Abnormal event detection in videos using spatiotemporal autoencoder. In *Advances in Neural Networks-ISNN 2017: 14th International Symposium, ISNN 2017, Sapporo, Hakodate, and Muroran, Hokkaido, Japan, June 21–26, 2017, Proceedings, Part II 14*, pages 189–196. Springer, 2017.

116. N. Chowdhury, M. Kabir, M. Rahman, and S. Islam. Machine learning for detecting covid-19 from cough sounds: An ensemble-based mcdm method. *Comput Biol Med*, 145:105405, 2022.

117. L. Chua. Memristor-the missing circuit element. *IEEE Transactions on circuit theory*, 18(5):507–519, 1971.

118. Jaeyoung Chung, Kanggeon Lee, Sungyong Baik, and Kyoung Mu Lee. MEIL-NeRF: Memory-Efficient Incremental Learning of Neural Radiance Fields, December 2022. arXiv:2212.08328 [cs].

119. Claudio Cimarelli, Dario Cazzato, Miguel A. Olivares-Mendez, and Holger Voos. Faster visual-based localization with mobile-posenet. In *International Conference on Computer Analysis of Images and Patterns*, pages 219–230. Springer, 2019.

120. Mircea Cimpoi, Subhransu Maji, Iasonas Kokkinos, Sammy Mohamed, and Andrea Vedaldi. Describing textures in the wild. In *Proceedings of the IEEE Conference on Computer Vision and Pattern Recognition*, pages 3606–3613, 2014.

121. Petr Cisar, Dinara Bekkozhayeva, Oleksandr Movchan, Mohammadmehdi Saberioon, and Rudolf Schraml. Computer vision based individual fish identification using skin dot pattern. *Scientific Reports*, 11, 12 2021.

122. CISCO. Cisco annual internet report (2018–2023) white paper. Technical Report C11-741490-01, CISCO Public, Mar 2020.

123. Charles Clancy, Joe Hecker, Erich Stuntebeck, and Tim O'Shea. Applications of machine learning to cognitive radio networks. *IEEE Wireless Communications*, 14(4):47–52, 2007.

124. J. Arturo Cocoma-Ortega and Jose Martinez-Carranza. A compact cnn approach for drone localisation in autonomous drone racing. *Journal of Real-Time Image Processing*, pages 1–14, 2021.

125. J. Arturo Cocoma-Ortega, L. Oyuki Rojas-Perez, Aldrich A. Cabrera-Ponce, and Jose Martinez-Carranza. Overcoming the blind spot in cnn-based gate detection for

autonomous drone racing. In *2019 Workshop on Research, Education and Development of Unmanned Aerial Systems (RED UAS)*, pages 253–259. IEEE, 2019.

126. Jos Arturo Cocoma-Ortega and Jose Martinez-Carranza. A cnn based drone localisation approach for autonomous drone racing. In *11th International Micro Air Vehicle Competition and Conference*, 2019.

127. José Arturo Cocoma-Ortega and José Martínez-Carranza. Towards high-speed localisation for autonomous drone racing. In *Mexican International Conference on Artificial Intelligence*, pages 740–751. Springer, 2019.

128. The CRyPTIC Consortium. A data compendium associating the genomes of 12,289 mycobacterium tuberculosis isolates with quantitative resistance phenotypes to 13 antibiotics. *PLOS Biology*, 20:e3001721, 2022.

129. The CRyPTIC Consortium. Genome-wide association studies of global mycobacterium tuberculosis resistance to 13 antimicrobials in 10,228 genomes identify new resistance mechanisms. *PLOS Biology*, 20:e3001755, 2022.

130. Eduardo Enrique Contreras-Luján, Enrique Efrén García-Guerrero, Oscar Roberto López-Bonilla, Esteban Tlelo-Cuautle, Didier López-Mancilla, and Everardo Inzunza-González. Evaluation of machine learning algorithms for early diagnosis of deep venous thrombosis. *Mathematical and Computational Applications*, 27(2):24, 2022.

131. F. Cornejo-Granados, G. López-Leal, D. Mata-Espinosa, J. Barrios-Payán, B. Marquina-Castillo, E. Equihua-Medina, Z. Zatarain-Barrón, C. Molina-Romero, R. Hernández-Pando, and A. Ochoa-Leyva. Targeted rna-seq reveals the m. tuberculosis transcriptome from an in vivo infection model. *Biology*, 10:1–18, 2021.

132. Abril Corona-Figueroa, Jonathan Frawley, Sam Bond-Taylor, Sarath Bethapudi, Hubert P. H. Shum, and Chris G. Willcocks. MedNeRF: Medical Neural Radiance Fields for Reconstructing 3D-aware CT-Projections from a Single X-ray, April 2022. arXiv:2202.01020 [cs, eess].

133. M. Coyle, D. Keenan, L. Henderson, M. Watkins, B. Haumann, D. Mayleben, and M. Wilson. Evaluation of an ambulatory system for the quantification of cough frequency in patients with chronic obstructive pulmonary disease. *Cough*, 1:3, 2005.

134. Davide Cristiani, Filippo Bottonelli, Angelo Trotta, and Marco Di Felice. Inventory management through mini-drones: Architecture and proof-of-concept implementation. In *2020 IEEE 21st International Symposium on "A World of Wireless, Mobile and Multimedia Networks"(WoWMoM)*, pages 317–322. IEEE, 2020.

135. Ernesto Cruz-Esquivel and Zobeida J Guzman-Zavaleta. An examination on autoencoder designs for anomaly detection in video surveillance. *IEEE Access*, 10:6208–6217, 2022.

136. J. R. Cárdenas-Valdez, J. A. Galaviz-Aguilar, C. Vargas-Rosales, E. Inzunza-González, and L. Flores-Hernández. A crest factor reduction technique for lte signals with target relaxation in power amplifier linearization. *Sensors*, 22(3), 2022.

137. Qiyu Dai, Yan Zhu, Yiran Geng, Ciyu Ruan, Jiazhao Zhang, and He Wang. GraspNeRF: Multiview-based 6-DoF Grasp Detection for Transparent and Specular Objects Using Generalizable NeRF. October 2022.

138. P. Danecek, J. Bonfield, J. Liddle, J. Marshall, V. Ohan, M. Pollard, A. Whitwham, T. Keane, S. McCarthy, and R. Davies. Twelve years of samtools and bcftools. *GigaScience*, 10, 2021.

139. Shaveta Dargan, Munish Kumar, Maruthi Rohit Ayyagari, and Gulshan Kumar. A survey of deep learning and its applications: a new paradigm to machine learning. *Archives of Computational Methods in Engineering*, 27:1071–1092, 2020.

140. Andrew J Davison, Ian D Reid, Nicholas D Molton, and Olivier Stasse. Monoslam: Real-time single camera slam. *IEEE transactions on pattern analysis and machine intelligence*, 29(6):1052–1067, 2007.

141. Luca Davoli, Emanuele Pagliari, and Gianluigi Ferrari. Hybrid LoRa-IEEE 802.11s Opportunistic Mesh Networking for Flexible UAV Swarming. *Drones*, 5(2):26, June 2021. Number: 2 Publisher: Multidisciplinary Digital Publishing Institute.

142. Antonio De Falco, Fabio Narducci, and Alfredo Petrosino. An uav autonomous warehouse inventorying by deep learning. In *International Conference on Image Analysis and Processing*, pages 443–453. Springer, 2019.

143. Luca De Luigi, Damiano Bolognini, Federico Domeniconi, Daniele De Gregorio, Matteo Poggi, and Luigi Di Stefano. ScanNeRF: a Scalable Benchmark for Neural Radiance Fields. In *2023 IEEE/CVF Winter Conference on Applications of Computer Vision (WACV)*, pages 816–825, January 2023. ISSN: 2642-9381.

144. A. De Ruijter and F. Guldenmund. The bowtie method: A review. *Safety Science*, 88:211–218, 2015.

145. U. De Silva, T. Koike-Akino, R. Ma, A. Yamashita, and H. Nakamizo. A Modular 1D-CNN Architecture for Real-time Digital Pre-distortion. In *2022 IEEE Topical Conference on RF/Microwave Power Amplifiers for Radio and Wireless Applications (PAWR)*, pages 79–81. IEEE, 2022.

146. W. Deelder, S. Christakoudi, J. Phelan, E. Diez Benavente, S. Campino, R. McNerney, L. Palla, and T.G. Clark. Machine learning predicts accurately mycobacterium tuberculosis drug resistance from whole genome sequencing data. *Frontiers in Genetics*, 10:1–9, 2019.

147. M. V. Deepak Nair, R. Giofre, P. Colantonio, and F. Giannini. NARMA based novel closed loop digital predistortion using Moore-Penrose inverse technique. *2016 11th European Microwave Integrated Circuits Conference (EuMIC)*, pages 405–408, 2016.

148. M. Delavar, S. Mirzakuchaki, and J. Mohajeri. A ring oscillator-based PUF with enhanced challenge-response pairs. *Canadian Journal of Electrical and Computer Engineering*, 39(2):174–180, 2016.

149. Jeffrey Delmerico and Davide Scaramuzza. A benchmark comparison of monocular visual-inertial odometry algorithms for flying robots. In *2018 IEEE International Conference on Robotics and Automation (ICRA)*, pages 2502–2509, 2018.

150. Mehmet Ozgun Demir, Ali Emre Pusane, Guido Dartmann, Gerd Ascheid, and Gunes Karabulut Kurt. A garden of cyber physical systems: Requirements, challenges, and implementation aspects. *IEEE Internet of Things Magazine*, 3(3):84–89, 2020.

151. Congyue Deng, Chiyu "Max" Jiang, Charles R. Qi, Xinchen Yan, Yin Zhou, Leonidas Guibas, and Dragomir Anguelov. NeRDi: Single-View NeRF Synthesis with Language-Guided Diffusion as General Image Priors, December 2022. arXiv:2212.03267 [cs].

152. Dawa Derksen and Dario Izzo. Shadow Neural Radiance Fields for Multi-view Satellite Photogrammetry. In *2021 IEEE/CVF Conference on Computer Vision and Pattern Recognition Workshops (CVPRW)*, pages 1152–1161, June 2021. ISSN: 2160-7516.

153. S. Dikmese, L. Anttila, P. P. Campo, M. Valkama, and M. Renfors. Behavioral Modeling of Power Amplifiers With Modern Machine Learning Techniques. *IEEE MTT-S International Microwave Conference on Hardware and Systems for 5G and Beyond (IMC-5G)*, pages 1–3, 2019.

154. Sinahi Dionisio-Ortega, L. Oyuki Rojas-Perez, Jose Martinez-Carranza, and Israel Cruz-Vega. A deep learning approach towards autonomous flight in forest environments. In *2018 International Conference on Electronics, Communications and Computers (CONIELECOMP)*, pages 139–144. IEEE, 2018.

155. Na Dong, Yongqiang Zhang, Mingli Ding, and Gim Hee Lee. Open world detr: Transformer based open world object detection. *arXiv preprint arXiv:2212.02969*, 2022.

156. Songlin Dong, Haoyu Luo, Yuhang He, and Xing Wei Yihong Gong. Knowledge restore and transfer for multi-label class-incremental learning. *arXiv preprint arXiv:2302.13334*, 2023.

157. Con Doolan, Yendrew Yauwenas, and Danielle Moreau. Drone propeller noise under static and steady inflow conditions. In *Flinovia—Flow Induced Noise and Vibration Issues and Aspects-III*, pages 45–60. Springer, 2021.

158. Alexey Dosovitskiy, Philipp Fischer, Eddy Ilg, Philip Hausser, Caner Hazirbas, Vladimir Golkov, Patrick Van Der Smagt, Daniel Cremers, and Thomas Brox. Flownet: Learning optical flow with convolutional networks. In *Proceedings of the IEEE international conference on computer vision*, pages 2758–2766, 2015.

159. DroneScan. Physical inventory: Stock take, physical inventory, cycle counting. `https://www.dronescan.co/info`. Accessed on June 1, 2023.

160. Hualong Du, Pengfei Liu, Qiuyu Cui, Xin Ma, and He Wang. PID Controller Parameter Optimized by Reformative Artificial Bee Colony Algorithm. *Journal of Mathematics*, 2022:e3826702, February 2022. Publisher: Hindawi.

161. Yilun Du, Yinan Zhang, Hong-Xing Yu, Joshua B. Tenenbaum, and Jiajun Wu. Neural Radiance Flow for 4D View Synthesis and Video Processing, September 2021. arXiv:2012.09790 [cs].

162. D. Dua and C. Graff. Uci machine learning repository, 2017. `https://archive.ics.uci.edu/ml/datasets/`.

163. Hugh Durrant-Whyte and Tim Bailey. Simultaneous localization and mapping: part i. *IEEE robotics & automation magazine*, 13(2):99–110, 2006.

164. Malay Kishore Dutta, Namita Sengar, Narayan Kamble, Kaushik Banerjee, Navroj Minhas, and Biplab Sarkar. Image processing based technique for classification of fish quality after cypermethrine exposure. *LWT*, 68:408–417, 5 2016.

165. Hinnerk Eißfeldt and Marcus Biella. The public acceptance of drones–challenges for advanced aerial mobility (aam). *Transportation Research Procedia*, 66:80–88, 2022.

166. Oussama EL JIATI. The use of drones in e-commerce logistics and supply chain managament. 2021.

167. O. F. Erturul. A novel randomized machine learning approach: Reservoir computing extreme learning machine. *Applied Soft Computing*, 94:106433, 2020.

168. Z. Xu et al. A tangent approximation coefficient extraction method for instantaneous sample indexed magnitude-selective affine behavioral model. *IEEE Microwave and Wireless Components Letters*, 32(11):1375–1378, 2022.

169. Yaxiang Fan, Gongjian Wen, Deren Li, Shaohua Qiu, Martin D. Levine, and Fei Xiao. Video anomaly detection and localization via gaussian mixture fully convolutional variational autoencoder. *Computer Vision and Image Understanding*, 195:102920, 2020.

170. Shuo Feng, Peyman Setoodeh, and Simon Haykin. Smart home: Cognitive interactive people-centric internet of things. *IEEE Communications Magazine*, 55(2):34–39, 2017.

171. Jose R. Fermin and Jairo Beltrán. Modelos de propagacion electromagnetica para telefonia movil. 2012.

172. Arthur F. A. Fernandes, Eduardo M. Turra, Érika R. de Alvarenga, Tiago L. Passafaro, Fernando B. Lopes, Gabriel F. O. Alves, Vikas Singh, and Guilherme J. M. Rosa. Deep learning image segmentation for extraction of fish body measurements and prediction of body weight and carcass traits in nile tilapia. *Computers and Electronics in Agriculture*, 170:105274, 2020.

173. Tiago M Fernández-Caramés, Oscar Blanco-Novoa, Iván Froiz-Míguez, and Paula Fraga-Lamas. Towards an autonomous industry 4.0 warehouse: A uav and blockchain-based system for inventory and traceability applications in big data-driven supply chain management. *Sensors*, 19(10):2394, 2019.

174. Tiago M. Fernández-Caramés, Oscar Blanco-Novoa, Manuel Suárez-Albela, and Paula Fraga-Lamas. A uav and blockchain-based system for industry 4.0 inventory and traceability applications. In *Multidisciplinary Digital Publishing Institute Proceedings*, volume 4, page 26, 2018.

175. Greg Foderaro, Pingping Zhu, Hongchuan Wei, Thomas A. Wettergren, and Silvia Ferrari. Distributed optimal control of sensor networks for dynamic target tracking. *IEEE Transactions on Control of Network Systems*, 5(1):142–153, 2018.

176. Philipp Foehn, Dario Brescianini, Elia Kaufmann, Titus Cieslewski, Mathias Gehrig, Manasi Muglikar, and Davide Scaramuzza. Alphapilot: Autonomous drone racing. *arXiv preprint arXiv:2005.12813*, 2020.

177. Philipp Foehn, Angel Romero, and Davide Scaramuzza. Time-optimal planning for quadrotor waypoint flight. *Science Robotics*, 6(56):eabh1221, 2021.

178. Cesar Augusto Fontanillo Lopez, Guangye Li, and Dingguo Zhang. Beyond Technologies of Electroencephalography-Based Brain-Computer Interfaces: A Systematic Review From Commercial and Ethical Aspects. *Frontiers in Neuroscience*, 14, 2020.

179. Azade Fotouhi, Haoran Qiang, Ming Ding, Mahbub Hassan, Lorenzo Galati Giordano, Adrian Garcia-Rodriguez, and Jinhong Yuan. Survey on UAV cellular communications: Practical aspects, standardization advancements, regulation, and security challenges. *IEEE Communications Surveys & Tutorials*, 21(4):3417–3442, 2019.

180. Jonas Frey, Hermann Blum, Francesco Milano, Roland Siegwart, and Cesar Cadena. Continual adaptation of semantic segmentation using complementary 2d-3d data representations. *IEEE Robotics and Automation Letters*, 7(4):11665–11672, 2022.

181. Guihong Fu and Yun Yuna. Phenotyping and phenomics in aquaculture breeding, 3 2022.

182. Shin Fujieda, Kohei Takayama, and Toshiya Hachisuka. Wavelet convolutional neural networks. *arXiv preprint arXiv:1805.08620*, 2018.

183. Taku Fujitomi, Ken Sakurada, Ryuhei Hamaguchi, Hidehiko Shishido, Masaki Onishi, and Yoshinari Kameda. LB-NERF: Light Bending Neural Radiance Fields for Transparent Medium. In *2022 IEEE International Conference on Image Processing (ICIP)*, pages 2142–2146, October 2022. ISSN: 2381-8549.

184. Fadri Furrer, Michael Burri, Markus Achtelik, and Roland Siegwart. RotorS—A Modular Gazebo MAV Simulator Framework. In *Robot operating system (ROS)*, pages 595–625. Springer, 2016.

185. J. M. Fuster. *Cortex and Mind: Unifying Cognition*. Oxford University Press, 2003.

186. Joaquin M. Fuster. The prefrontal cortex makes the brain a preadaptive system. *Proceedings of the IEEE*, 102(4):417–426, 2014.

187. Guy Gafni, Justus Thies, Michael Zollhöfer, and Matthias Nießner. Dynamic Neural Radiance Fields for Monocular 4D Facial Avatar Reconstruction, December 2020. arXiv:2012.03065 [cs].

188. J. Galagan, K. Minch, M. Peterson, A. Lyubetskaya, E. Azizi, L. Sweet, A. Gomes, T. Rustad, G. Dolganov, I. Glotova, T. Abeel, C. Mahwinney, A. Kennedy, R. Allard, W. Brabant, A. Krueger, S. Jaini, B. Honda, W. Yu, M. Hickey, J. Zucker, C. Garay, B. Weiner, P. Sisk, C. Stolte, J. Winkler, Y. Van De Peer, P. Iazzetti, D. Camacho, J. Dreyfuss, Y. Liu, A. Dorhoi, H. Mollenkopf, P. Drogaris, J. Lamontagne, Y. Zhou, J. Piquenot, S. Park, S. Raman, S. Kaufmann, R. Mohney, D. Chelsky, D. Moody,

D. Sherman, and G. Schoolnik. The mycobacterium tuberculosis regulatory network and hypoxia. *Nature*, 499:178–183, 2013.

189. J. A. Galaviz-Aguilar, C. Vargas-Rosales, J. R. Cárdenas-Valdez, D. S. Aguila-Torres, and L. Flores-Hernández. A Comparison of Surrogate Behavioral Models for Power Amplifier Linearization under High Sparse Data. *Sensors*, 22(19), 2022.

190. Chen Gao, Ayush Saraf, Johannes Kopf, and Jia-Bin Huang. Dynamic View Synthesis from Dynamic Monocular Video. In *2021 IEEE/CVF International Conference on Computer Vision (ICCV)*, pages 5692–5701, October 2021. ISSN: 2380-7504.

191. Dasong Gao, Chen Wang, and Sebastian Scherer. Airloop: Lifelong loop closure detection. In *2022 International Conference on Robotics and Automation (ICRA)*, pages 10664–10671. IEEE, 2022.

192. Sicun Gao. Nonlinearity, automation, and reliable cyberphysical systems. *Computer*, 54(7):94–96, 2021.

193. Y. Gao, C. Jin, J. Kim, H. Nili, X. Xu, W. Burleson, O. Kavehi, Ma. van Dijk, D. Chinthana Ranasinghe, and U. Rührmair. Efficient erasable PUFs from programmable logic and memristors. *IACR Cryptology ePrint Archive*, 2018:358, 2018.

194. Y. Gao, D. C. Ranasinghe, S. F. Al-Sarawi, O. Kavehi, and D. Abbott. Emerging physical unclonable functions with nanotechnology. *IEEE Access*, 4:61–80, 2016.

195. Y. Gao, D. C. Ranasinghe, S.F. Al-Sarawi, O. Kavehi, and D. Abbott. Mrpuf: A novel memristive device based physical unclonable function. In *Int. Conf. on Applied Cryptography and Network Security*, pages 595–615. Springer, 2015.

196. Yunyuan Gao, Bo Gao, Qiang Chen, Jia Liu, and Yingchun Zhang. Deep convolutional neural network-based epileptic electroencephalogram (eeg) signal classification. *Frontiers in Neurology*, 11, 2020.

197. Stephan J. Garbin, Marek Kowalski, Matthew Johnson, Jamie Shotton, and Julien Valentin. FastNeRF: High-Fidelity Neural Rendering at 200FPS, April 2021. arXiv:2103.10380 [cs].

198. M. Garcia-Bosque, G. Díez-Señorans, C. Sánchez-Azqueta, and S. Celma. Proposal and analysis of a novel class of PUFs based on Galois ring oscillators. *IEEE Access*, 8:157830–157839, 2020.

199. K. Garikapati, S. Turnbull, R. Bennett, T. Campbell, J. Kanawati, M. Wong, S. Thomas, C. Chow, and S. Kumar. The role of contemporary wearable and handheld devices in the diagnosis and management of cardiac arrhythmias. *Heart Lung Circ*, 31:1432–1449, 2022.

200. B. Gassend, D. Clarke, M. Van Dijk, and S. Devadas. Silicon physical random functions. In *Proceedings of the 9th ACM conf. on Computer and communications security*, pages 148–160. ACM, 2002.

201. Hasan Genc, Yazhou Zu, Ting-Wu Chin, Matthew Halpern, and Vijay Janapa Reddi. Flying iot: Toward low-power vision in the sky. *IEEE Micro*, 37(6):40–51, 2017.

202. J. Giudicessi, M. Schram, J. Bos, C. Galloway, J. Shreibati, P. Johnson, R. Carter, L. Disrud, R. Kleiman, Z. Attia, P. Noseworthy, P. Friedman, D. Albert, and M. Ackerman. Artificial intelligence-enabled assessment of the heart rate corrected qt interval using a mobile electrocardiogram device. *Circulation*, 143:1274–1286, 2021.

203. Lorenzo Giusti, Josue Garcia, Steven Cozine, Darrick Suen, Christina Nguyen, and Ryan Alimo. MaRF: Representing Mars as Neural Radiance Fields, December 2022. arXiv:2212.01672 [cs].

204. Payam Gohari, Hossein Mohammadi, and Sajjad Taghvaei. Using chaotic maps for 3D boundary surveillance by quadrotor robot. *Applied Soft Computing*, 76, December 2018.

205. Ary L. Goldberger, Luis A. N. Amaral, Leon Glass, Jeffrey M. Hausdorff, Plamen Ch. Ivanov, Roger G. Mark, Joseph E. Mietus, George B. Moody, Chung-Kang Peng, and H. Eugene Stanley. Physiobank, physiotoolkit, and physionet. *Circulation*, 101(23):e215–e220, 2000.

206. I. Goldenberg, A. Moss, and W. Zareba. Qt interval: How to measure it and what is "normal". *Journal of Cardiovascular Electrophysiology*, 17:333–336, 2006.

207. Bingchen Gong, Yuehao Wang, Xiaoguang Han, and Qi Dou. RecolorNeRF: Layer Decomposed Radiance Fields for Efficient Color Editing of 3D Scenes, February 2023. arXiv:2301.07958 [cs].

208. Cheng Gong, Jianwei Gong, Chao Lu, Zhe Liu, and Zirui Li. Life-long multi-task learning of adaptive path tracking policy for autonomous vehicle. *arXiv preprint arXiv:2109.07210*, 2021.

209. Fabio Alfonso Gonzalez, M. E. Afanador Cristancho, and E. F. Niño López. Modelamiento y simulación de un quadrotor mediante la integración de simulink y solidworks. *Maskay*, 9(1):15–24, 2019.

210. Ramon Gonzalez, Francisco Rodriguez, Jose Luis Guzman, Cedric Pradalier, and Roland Siegwart. Combined visual odometry and visual compass for off-road mobile robots localization. *Robotica*, 30(6):865–878, 2012.

211. Ian Goodfellow, Yoshua Bengio, and Aaron Courville. *Deep Learning*. MIT Press, 2016. http://www.deeplearningbook.org.

212. Ian Goodfellow, Jean Pouget-Abadie, Mehdi Mirza, Bing Xu, David Warde-Farley, Sherjil Ozair, Aaron Courville, and Yoshua Bengio. Generative adversarial networks. *Communications of the ACM*, 63(11):139–144, 2020.

213. Ramesh A Gopinath, Haitao Guo, C Sidney Burrus, and L Sidney Burrus. Introduction to wavelets and wavelet transforms: a primer, 1997.

214. K. S. Goud and S. R. Gidituri. Security challenges and related solutions in software defined networks: A survey. *International Journal of Computer Networks and Applications*, 9(1):22–37, 2022.

215. Maximilian Grobbelaar, Souvik Phadikar, Ebrahim Ghaderpour, Aaron F. Struck, Nidul Sinha, Rajdeep Ghosh, and Md. Zaved Iqubal Ahmed. A survey on denoising techniques of electroencephalogram signals using wavelet transform. *Signals*, 3(3):577–586, 2022.

216. Tianhao Gu, Zhe Wang, Ziqiu Chi, Yiwen Zhu, and Wenli Du. Unsupervised cycle optimization learning for single-view depth and camera pose with kalman filter. *Engineering Applications of Artificial Intelligence*, 106:104488, 2021.

217. François Guérin, Frédéric Guinand, Jean-François Brethé, Hervé Pelvillain, et al. Towards an autonomous warehouse inventory scheme. In *2016 IEEE Symposium Series on Computational Intelligence (SSCI)*, pages 1–8. IEEE, 2016.

218. M. Guerrero-Chevannier. Modelo basado en deep learning para el diagnostico de tuberculosis pulmonar utilizando radiografias de torax y perfiles clinicos. Master's thesis, Universidad Autonoma de Baja California, 2022.

219. Ente Guo, Zhifeng Chen, Yanlin Zhou, and Dapeng Oliver Wu. Unsupervised learning of depth and camera pose with feature map warping. *Sensors*, 21(3):923, 2021.

220. Yuan-Chen Guo, Di Kang, Linchao Bao, Yu He, and Song-Hai Zhang. NeRFReN: Neural Radiance Fields with Reflections, April 2022. arXiv:2111.15234 [cs].

221. Yudong Guo, Keyu Chen, Sen Liang, Yong-Jin Liu, Hujun Bao, and Juyong Zhang. AD-NeRF: Audio Driven Neural Radiance Fields for Talking Head Synthesis, August 2021. arXiv:2103.11078 [cs].

222. Ohad Gur and Aviv Rosen. Design of quiet propeller for an electric mini unmanned air vehicle. *Journal of Propulsion and Power*, 25(3):717–728, 2009.

223. V. Guryanova. Transfer learning for tuberculosis screening by single-channel ecg. *Data Science*, pages 149–154, 2020.

224. Marco Antonio Gómez-Guzmán, Laura Jiménez-Beristaín, Enrique Efren García-Guerrero, Oscar Roberto López-Bonilla, Ulises Jesús Tamayo-Perez, José Jaime Esqueda-Elizondo, Kenia Palomino-Vizcaino, and Everardo Inzunza-González. Classifying brain tumors on magnetic resonance imaging by using convolutional neural networks. *Electronics*, 12:955, 2, 2023.

225. Teuku Mohd Ichwanul Hakim and Ony Arifianto. Implementation of dryden continuous turbulence model into simulink for LSA-02 flight test simulation. In *Journal of Physics: Conference Series*, volume 1005, page 012017. IOP Publishing, 2018.

226. Bohyung Han, Jack Sim, and Hartwig Adam. Branchout: Regularization for online ensemble tracking with convolutional neural networks. In *Proceedings of the IEEE conference on computer vision and pattern recognition*, pages 3356–3365, 2017.

227. E. Harausz, H. Cox, M. Rich, C. Mitnick, P. Zimetbaum, and J. Furin. Qtc prolongation and treatment of multidrug-resistant tuberculosis. *The International Journal of Tuberculosis and Lung Disease*, 19:385–391, 2015.

228. Mahmudul Hasan, Jonghyun Choi, Jan Neumann, Amit K. Roy-Chowdhury, and Larry S. Davis. Learning temporal regularity in video sequences. In *Proceedings of the IEEE conference on computer vision and pattern recognition*, pages 733–742, 2016.

229. Yasunari Hashimoto and Junichi Ushiba. EEG-based classification of imaginary left and right foot movements using beta rebound. *Clinical Neurophysiology*, 124(11):2153–2160, November 2013.

230. S. Haykin, A.O. Hero, and E. Moulines. Modeling, identification, and control of large-scale dynamical systems. In *Proceedings. (ICASSP '05). IEEE International Conference on Acoustics, Speech, and Signal Processing, 2005.*, volume 5, pages v/945–v/948 Vol. 5, 2005.

231. Simon Haykin. *Cognitive Dynamic Systems: Perception-action Cycle, Radar and Radio*. Cambridge University Press, 2012.

232. Simon Haykin, Mehdi Fatemi, Peyman Setoodeh, and Yanbo Xue. Cognitive control. *Proceedings of the IEEE*, 100(12):3156–3169, 2012.

233. Simon Haykin, Joaquin M. Fuster, David Findlay, and Shuo Feng. Cognitive risk control for physical systems. *IEEE Access*, 5:14664–14679, 2017.

234. Simon Haykin and Eric Moulines. Special issue on large-scale dynamic systems. *Proceedings of the IEEE*, 95(5):849–852, 2007.

235. Simon Haykin, Yanbo Xue, and Peyman Setoodeh. Cognitive radar: Step toward bridging the gap between neuroscience and engineering. *Proceedings of the IEEE*, 100(11):3102–3130, 2012.

236. Jiangpeng He, Runyu Mao, Zeman Shao, and Fengqing Zhu. Incremental learning in online scenario. In *Proceedings of the IEEE/CVF conference on computer vision and pattern recognition*, pages 13926–13935, 2020.

237. Kaiming He, Xiangyu Zhang, Shaoqing Ren, and Jian Sun. Deep residual learning for image recognition, 2015.

238. Peter Hedman, Pratul P. Srinivasan, Ben Mildenhall, Jonathan T. Barron, and Paul Debevec. Baking Neural Radiance Fields for Real-Time View Synthesis, March 2021. arXiv:2103.14645 [cs].

239. Coanda Henri. Device for deflecting a stream of elastic fluid projected into an elastic fluid, September 1 1936. US Patent 2,052,869.

240. C. Herder, M. Yu, F. Koushanfar, and S. Devadas. Physical unclonable functions and applications: A tutorial. *Proceedings of the IEEE*, 102(8):1126–1141, 2014.

241. J. M. Hernández-Ontiveros, E. Inzunza-González, E. E. García-Guerrero, O. R. López-Bonilla, S. O. Infante-Prieto, J. R. Cárdenas-Valdez, and E. Tlelo-Cuautle. Development and implementation of a fish counter by using an embedded system. *Computers and Electronics in Agriculture*, 145:53–62, 2 2018.

242. Gert Herold. In-flight directivity and sound power measurement of small-scale unmanned aerial systems. *Acta Acustica*, 6:58, 2022.

243. Mario Patricio Herrera Holguín and Juan Carlos Inclán Luna. Estudio y metodología de diseño de antenas utilizando geometría fractal (antenas fractales). B.S. thesis, Quito: Escuela Politécnica Nacional, 2004.

244. T. Hill and R. Unckless. A deep learning approach for detecting copy number variation in next-generation sequencing data. *G3 (Bethesda, Md.)*, 9:3575–3582, 2019.

245. Eißfeldt Hinnerk and Verena Vogelpohl. Drone acceptance and noise concerns-some findings. In *53rd International Symposium on Aviation Psychology. Wright State University, Dayton, OH, USA*, pages 199–204, 2019.

246. K. Honda and H. Tamukoh. A hardware-oriented echo state network and its FPGA implementation. *Journal of Robotics, Networking and Artificial Life*, 71:58–62, 2020.

247. Kunlong Hong, Hongguang Wang, and Bingbing Yuan. Inspection-Nerf: Rendering Multi-Type Local Images for Dam Surface Inspection Task Using Climbing Robot and Neural Radiance Field. *Buildings*, 13(1):213, January 2023. Number: 1 Publisher: Multidisciplinary Digital Publishing Institute.

248. Yang Hong, Bo Peng, Haiyao Xiao, Ligang Liu, and Juyong Zhang. HeadNeRF: A Real-time NeRF-based Parametric Head Model, April 2022. arXiv:2112.05637 [cs].

249. Csaba Horváth, Bence Fenyvesi, and Bálint Kocsis. Drone noise reduction via radiation efficiency considerations. 2018.

250. Mohammad-Parsa Hosseini, Amin Hosseini, and Kiarash Ahi. A review on machine learning for eeg signal processing in bioengineering. *IEEE Reviews in Biomedical Engineering*, 14:204–218, 2021.

251. Xiangwang Hou, Jingjing Wang, Zhengru Fang, Yong Ren, Kwang-Cheng Chen, and Lajos Hanzo. Edge intelligence for mission-critical 6g services in space-air-ground integrated networks. *IEEE Network*, 36(2):181–189, 2022.

252. Kerstin Howe, Matthew D. Clark, Carlos F. Torroja, James Torrance, et al. The zebrafish reference genome sequence and its relationship to the human genome. *Nature*, 496:498–503, 4 2013.

253. Alexander E. Hramov, Vladimir A. Maksimenko, and Alexander N. Pisarchik. Physical principles of brain–computer interfaces and their applications for rehabilitation, robotics and control of human brain states. *Physics Reports*, 918:1–133, 2021. Physical principles of brain–computer interfaces and their applications for rehabilitation, robotics and control of human brain states.

254. Benran Hu, Junkai Huang, Yichen Liu, Yu-Wing Tai, and Chi-Keung Tang. Nerf-rpn: A general framework for object detection in nerfs. In *Proceedings of the IEEE/CVF Conference on Computer Vision and Pattern Recognition*, pages 23528–23538, 2023.

255. Jing Hu, Daoliang Li, Qingling Duan, Yueqi Han, Guifen Chen, and Xiuli Si. Fish species classification by color, texture and multi-class support vector machine using computer vision. *Computers and Electronics in Agriculture*, 88:133–140, 10 2012.

256. Junjie Hu, Chenyou Fan, Liguang Zhou, Qing Gao, Honghai Liu, and Tin Lun Lam. Lifelong-monodepth: Lifelong learning for multi-domain monocular metric depth estimation. *arXiv preprint arXiv:2303.05050*, 2023.

257. G.-B. Huang, Q-Y. Zhu, and C-K. Siew. Extreme learning machine: Theory and applications. *Neurocomputing*, 70:489–501, 2006.

258. Yi-Hua Huang, Yue He, Yu-Jie Yuan, Yu-Kun Lai, and Lin Gao. StylizedNeRF: Consistent 3D Scene Stylization as Stylized NeRF via 2D-3D Mutual Learning. In *2022 IEEE/CVF Conference on Computer Vision and Pattern Recognition (CVPR)*, pages 18321–18331, June 2022. ISSN: 2575-7075.

259. Zilong Huang, Wentian Hao, Xinggang Wang, Mingyuan Tao, Jianqiang Huang, Wenyu Liu, and Xian-Sheng Hua. Half-real half-fake distillation for class-incremental semantic segmentation. *arXiv preprint arXiv:2104.00875*, 2021.

260. José Manuel Huidobro. Antenas de telecomunicaciones. *Revista digital de acta*, 4, 2013.

261. Jeffrey Ichnowski, Yahav Avigal, Justin Kerr, and Ken Goldberg. Dex-NeRF: Using a Neural Radiance Field to Grasp Transparent Objects. October 2021.

262. A. Ijaz, M. Nabeel, U. Masood, T. Mahmood, M. Hashmi, I. Posokhova, A. Rizwan, and A. Imran. Towards using cough for respiratory disease diagnosis by leveraging artificial intelligence: A survey. *Informatics in Medicine Unlocked*, 29:100832, 2022.

263. Eddy Ilg, Nikolaus Mayer, Tonmoy Saikia, Margret Keuper, Alexey Dosovitskiy, and Thomas Brox. Flownet 2.0: Evolution of optical flow estimation with deep networks. In *Proceedings of the IEEE conference on computer vision and pattern recognition*, pages 2462–2470, 2017.

264. Kutalmis Gokalp Ince, Aybora Koksal, Arda Fazla, and A Aydin Alatan. Semi-automatic annotation for visual object tracking. In *Proceedings of the IEEE/CVF International Conference on Computer Vision*, pages 1233–1239, 2021.

265. Norimasa Iwanami. Zebrafish as a model for understanding the evolution of the vertebrate immune system and human primary immunodeficiency. *Experimental Hematology*, 42(8):697–706, 2014. Genomics and Model Organisms: New Horizons for Experimental Hematology.

266. S. Jaeger, A. Karargyris, S. Candemir, J. Siegelman, L. Folio, S. Antani, and G. Thoma. Automatic screening for tuberculosis in chest radiographs: a survey. *Quantitative Imaging in Medicine and Surgery*, 3:89, 2013.

267. Ajay Jain, Matthew Tancik, and Pieter Abbeel. Putting NeRF on a Diet: Semantically Consistent Few-Shot View Synthesis, April 2021. arXiv:2104.00677 [cs].

268. A. K. Jain, J. Mao, and K. M. Mohiuddin. Artificial neural networks: A tutorial. *IEEE Computer Magazine*, 29(3):31–44, 1996.

269. S. Jamal, M. Khubaib, R. Gangwar, S. Grover, A. Grover, and S. E. Hasnain. Artificial intelligence and machine learning based prediction of resistant and susceptible mutations in mycobacterium tuberculosis. *Scientific Reports*, 10:1–16, 2020.

270. Guolai Jiang, Lei Yin, Shaokun Jin, Chaoran Tian, Xinbo Ma, and Yongsheng Ou. A simultaneous localization and mapping (slam) framework for 2.5 d map building based on low-cost lidar and vision fusion. *Applied Sciences*, 9(10):2105, 2019.

271. Lei Jiang, Gerald Schaefer, and Qinggang Meng. An Improved Novel View Synthesis Approach Based on Feature Fusion and Channel Attention. In *2022 IEEE International Conference on Systems, Man, and Cybernetics (SMC)*, pages 2459–2464, October 2022. ISSN: 2577-1655.

272. Yizhang Jiang, Dongrui Wu, Zhaohong Deng, Pengjiang Qian, Jun Wang, Guanjin Wang, Fu-Lai Chung, Kup-Sze Choi, and Shitong Wang. Seizure classification from eeg signals using transfer learning, semi-supervised learning and tsk fuzzy system. *IEEE Transactions on Neural Systems and Rehabilitation Engineering*, 25(12):2270–2284, 2017.

273. Z. Jiang, Y. Lu, Z. Liu, W. Wu, X. Xu, A. Dinnyes, Z. Yu, L. Chen, and Q. Sun. Drug resistance prediction and resistance genes identification in mycobacterium tuberculosis based on a hierarchical attentive neural network utilizing genome-wide variants. *Briefings in Bioinformatics*, 23:bbac041, 2022.

274. A. Jimenez-Ruano, C. Madrazo-Moya, I. Cancino-Muñoz, P. Mejia-Ponce, C. Licona-Cassani, I. Comas, R. Muñiz Salazar, and R. Zenteno-Cuevas. Whole genomic sequencing based genotyping reveals a specific x3 sublineage restricted to mexico and related with multidrug resistance. *Scientific Reports*, 11:1870, 2021.

275. Glenn Jocher, Ayush Chaurasia, Alex Stoken, Jirka Borovec, NanoCode012, Yonghye Kwon, Kalen Michael, TaoXie, Jiacong Fang, imyhxy, Lorna, Zeng Yifu, Colin Wong, Abhiram V, Diego Montes, Zhiqiang Wang, Cristi Fati, Jebastin Nadar, Laughing, UnglvKitDe, Victor Sonck, tkianai, yxNONG, Piotr Skalski, Adam Hogan, Dhruv Nair, Max Strobel, and Mrinal Jain. ultralytics/yolov5: v7.0 – YOLOv5 SOTA Realtime Instance Segmentation, November 2022.

276. Y. N. Joglekar and Stephen J. Wolf. The elusive memristor: properties of basic electrical circuits. *European Journal of physics*, 30(4):661, 2009.

277. Sunggoo Jung, Sungwook Cho, Dasol Lee, Hanseob Lee, David Hyunchul Shim, et al. A direct visual servoing-based framework for the 2016 iros autonomous drone racing challenge. *Journal of Field Robotics*, 35(1):146–166, 2018.

278. Sunggoo Jung, Sunyou Hwang, Heemin Shin, David Hyunchul Shim, et al. Perception, guidance, and navigation for indoor autonomous drone racing using deep learning. *IEEE Robotics and Automation Letters*, 3(3):2539–2544, 2018.

279. Sunggoo Jung, Hanseob Lee, Sunyou Hwang, and David Hyunchul Shim. Real time embedded system framework for autonomous drone racing using deep learning techniques. In *2018 AIAA Information Systems-AIAA Infotech@ Aerospace*, page 2138. 2018.

280. Sohag Kabir. Internet of things and safety assurance of cooperative cyber-physical systems: Opportunities and challenges. *IEEE Internet of Things Magazine*, 4(2):74–78, 2021.

281. Ivan Kalinov, Alexander Petrovsky, Valeriy Ilin, Egor Pristanskiy, Mikhail Kurenkov, Vladimir Ramzhaev, Ildar Idrisov, and Dzmitry Tsetserukou. Warevision: Cnn barcode detection-based uav trajectory optimization for autonomous warehouse stocktaking. *IEEE Robotics and Automation Letters*, 5(4):6647–6653, 2020.

282. Ivan Kalinov, Daria Trinitatova, and Dzmitry Tsetserukou. Warevr: Virtual reality interface for supervision of autonomous robotic system aimed at warehouse stocktaking. *arXiv preprint arXiv:2110.11052*, 2021.

283. Bouzgou Kamel, Bestaoui Yasmina, Benchikh Laredj, Ibari Benaoumeur, and Ahmed-Foitih Zoubir. Dynamic modeling, simulation and PID controller of unmanned aerial vehicle UAV. In *2017 Seventh International Conference on Innovative Computing Technology (INTECH)*, pages 64–69, Luton, UK, August 2017. IEEE.

284. Rohan Kapoor, Nicola Kloet, Alessandro Gardi, Abdulghani Mohamed, and Roberto Sabatini. Sound propagation modelling for manned and unmanned aircraft noise assessment and mitigation: A review. *Atmosphere*, 12(11):1424, 2021.

285. Elia Kaufmann, Mathias Gehrig, Philipp Foehn, René Ranftl, Alexey Dosovitskiy, Vladlen Koltun, and Davide Scaramuzza. Beauty and the beast: Optimal methods meet learning for drone racing. In *2019 International Conference on Robotics and Automation (ICRA)*, pages 690–696. IEEE, 2019.

286. Elia Kaufmann, Antonio Loquercio, Rene Ranftl, Alexey Dosovitskiy, Vladlen Koltun, Davide Scaramuzza, et al. Deep drone racing: Learning agile flight in dynamic environments. In *Conference on Robot Learning*, pages 133–145. PMLR, 2018.

287. E.S. Kavvas, E. Catoiu, N. Mih, J. T. Yurkovich, Y. Seif, N. Dillon, D. Heckmann, A. Anand, L. Yang, V. Nizet, J. M. Monk, and B. O. Palsson. Machine learning and structural analysis of mycobacterium tuberculosis pan-genome identifies genetic signatures of antibiotic resistance. *Nature Communications*, 9, 2018.

288. Berk Kaya, Suryansh Kumar, Francesco Sarno, Vittorio Ferrari, and Luc Van Gool. Neural Radiance Fields Approach to Deep Multi-View Photometric Stereo, October 2021. arXiv:2110.05594 [cs].

289. Bob Kemp and Jesus Olivan. European data format "plus" (edf+), an edf alike standard format for the exchange of physiological data. *Clinical Neurophysiology*, 114(9):1755–1761, 2003.

290. Alex Kendall and Roberto Cipolla. Modelling uncertainty in deep learning for camera relocalization. In *2016 IEEE international conference on Robotics and Automation (ICRA)*, pages 4762–4769. IEEE, 2016.

291. Alex Kendall and Roberto Cipolla. Geometric loss functions for camera pose regression with deep learning. In *Proceedings of the IEEE conference on computer vision and pattern recognition*, pages 5974–5983, 2017.

292. Alex Kendall, Matthew Grimes, and Roberto Cipolla. Posenet: A convolutional network for real-time 6-dof camera relocalization. In *Proceedings of the IEEE international conference on computer vision*, pages 2938–2946, 2015.

293. John Kennedy, Simone Garruccio, and Kai Cussen. Modelling and mitigation of drone noise. *Vibroengineering Procedia*, 37:60–65, 2021.

294. Justin Kerr, Letian Fu, Huang Huang, Yahav Avigal, Matthew Tancik, Jeffrey Ichnowski, Angjoo Kanazawa, and Ken Goldberg. Evo-NeRF: Evolving NeRF for Sequential Robot Grasping of Transparent Objects. November 2022.

295. René Keßler, Christian Melching, Ralph Goehrs, and Jorge Marx Gómez. Using camera-drones and artificial intelligence to automate warehouse inventory. In *AAAI Spring Symposium: Combining Machine Learning with Knowledge Engineering*, 2021.

296. Muhammad Umar Karim Khan. Towards continual, online, self-supervised depth. *arXiv preprint arXiv:2103.00369*, 2021.

297. Wahab Khawaja, Ismail Guvenc, David W. Matolak, Uwe-Carsten Fiebig, and Nicolas Schneckenburger. A survey of air-to-ground propagation channel modeling for unmanned aerial vehicles. *IEEE Communications Surveys & Tutorials*, 21(3):2361–2391, 2019.

298. M. S. Khusro, A. Q. Hashmi, A. Q. Ansari, and M. Auyenur. A new and Reliable Decision Tree Based Small-Signal Behavioral Modeling of GaN HEMT. *IEEE 62nd International Midwest Symposium on Circuits and Systems (MWSCAS)*, pages 303–306, 2019.

299. Kristine Kiernan, Robert Joslin, and John Robbins. Standardization roadmap for unmanned aircraft systems, version 2.0. 2020.

300. Dae Han Kim, Chun Hyuk Park, and Young J Moon. Aerodynamic analyses on the steady and unsteady loading-noise sources of drone propellers. *International Journal of Aeronautical and Space Sciences*, 20:611–619, 2019.

301. Daewoon Kim and Kwanghee Ko. Camera localization with siamese neural networks using iterative relative pose estimation. *Journal of Computational Design and Engineering*, 9(4):1482–1497, 2022.

302. Jangho Kim, Jeesoo Kim, and Nojun Kwak. Stacknet: Stacking feature maps for continual learning. In *Proceedings of the IEEE/CVF Conference on Computer Vision and Pattern Recognition Workshops*, pages 242–243, 2020.

303. James Kirkpatrick, Razvan Pascanu, Neil Rabinowitz, Joel Veness, Guillaume Desjardins, Andrei A Rusu, Kieran Milan, John Quan, Tiago Ramalho, Agnieszka Grabska-Barwinska, et al. Overcoming catastrophic forgetting in neural networks. *Proceedings of the national academy of sciences*, 114(13):3521–3526, 2017.

304. Simon Klenk, Lukas Koestler, Davide Scaramuzza, and Daniel Cremers. E-NeRF: Neural Radiance Fields from a Moving Event Camera. August 2022.

305. Yejun Ko, Sunghoon Kim, Kwanghyun Shin, Youngmin Park, Sundo Kim, and Dongsuk Jeon. A 65nm 12.92nj/inference mixed-signal neuromorphic processor for image classification. *IEEE Transactions on Circuits and Systems II: Express Briefs*, pages 1–1, 2023.

306. D. Koboldt, Q. Zhang, D. Larson, D. Shen, M. McLellan, L. Lin, C. Miller, E. Mardis, L. Ding, and R. Wilson. Varscan 2: Somatic mutation and copy number alteration discovery in cancer by exome sequencing. *Genome Research*, 22:568–576, 2012.

307. D. C. Koboldt. Best practices for variant calling in clinical sequencing. *Genome Medicine*, 12:1–13, 2020.

308. T. Kohl, C. Utpatel, V. Schleusener, M. De Filippo, P. Beckert, D. Cirillo, and S. Niemann. Mtbseq: A comprehensive pipeline for whole genome sequence analysis of mycobacterium tuberculosis complex isolates. *PeerJ*, 2018, 2018.

309. O. Kömmerling and M. Kuhn. Design principles for tamper-resistant smartcard processors. *Smartcard*, 99:9–20, 1999.

310. Naruya Kondo, Yuya Ikeda, Andrea Tagliasacchi, Yutaka Matsuo, Yoichi Ochiai, and Shixiang Shane Gu. VaxNeRF: Revisiting the Classic for Voxel-Accelerated Neural Radiance Field, November 2021. arXiv:2111.13112 [cs].

311. I. Kononenko. Machine learning for medical diagnosis: History, state of the art and perspective. *Artificial Intelligence in Medicine*, 23:89–109, 2001.

312. Nicholas Blaise Konzel and Eric Greenwood. Ground-based acoustic measurements of small multirotor aircraft. In *Vertical Flight Society 78th Annual Forum & Technology Display, Fort Worth, TX*, pages 10–12, 2022.

313. J. Korpas, J. Sadlonova, and M. Vrabec. Analysis of the cough sound: an overview. *Pulm Pharmacol*, 9:261–268, 1996.

314. Adam R. Kosiorek, Heiko Strathmann, Daniel Zoran, Pol Moreno, Rosalia Schneider, Soňa Mokrá, and Danilo J. Rezende. NeRF-VAE: A Geometry Aware 3D Scene Generative Model, April 2021. arXiv:2104.00587 [cs, stat].

315. S. Kouchaki, Y. Yang, T. M. Walker, A. S. Walker, D. J. Wilson, T. E. A. Peto, D. W. Crook, and D. A. CRyPTIC Consortium; Clifton. Application of machine learning techniques to tuberculosis drug resistance analysis. *Bioinformatics*, 35, 2019.

316. Olga Krestinskaya, Aidana Irmanova, and Alex Pappachen James. Memristive nonidealities: Is there any practical implications for designing neural network chips?. In *IEEE Int. Symp. on Circuits and Systems (ISCAS)*, pages 1–5, 2019.

317. Alex Krizhevsky, Ilya Sutskever, and Geoffrey E Hinton. Imagenet classification with deep convolutional neural networks. *Advances in neural information processing systems*, 25:1097–1105, 2012.

318. Alex Krizhevsky, Ilya Sutskever, and Geoffrey E Hinton. Imagenet classification with deep convolutional neural networks. *Communications of the ACM*, 60(6):84–90, 2017.

319. F. Krueger. Trim galore: A wrapper tool around cutadapt and fastqc to consistently apply quality and adapter trimming to fastq files. *Babraham Institute*, pages undefined–undefined, 2015.

320. Zhengfei Kuang, Fujun Luan, Sai Bi, Zhixin Shu, Gordon Wetzstein, and Kalyan Sunkavalli. PaletteNeRF: Palette-based Appearance Editing of Neural Radiance Fields, January 2023. arXiv:2212.10699 [cs].

321. S Kulkarni and S Jha. Artificial intelligence, radiology, and tuberculosis: A review. *Academic radiology*, 27:71–75, 2020.

322. N. Kumar, S. Bhargava, C. Agrawal, K. George, P. Karki, and D. Baral. Chest radiographs and their reliability in the diagnosis of tuberculosis. *JNMA; journal of the Nepal Medical Association*, 44:138–142, 2005.

323. Minseop Kwak, Jiuhn Song, and Seungryong Kim. GeCoNeRF: Few-shot Neural Radiance Fields via Geometric Consistency, January 2023. arXiv:2301.10941 [cs].

324. Woong Kwon, Jun Ho Park, Minsu Lee, Jongbeom Her, Sang-Hyeon Kim, and Ja-Won Seo. Robust autonomous navigation of unmanned aerial vehicles (uavs) for warehouses? inventory application. *IEEE Robotics and Automation Letters*, 5(1):243–249, 2019.

325. Moussa Labbadi and Mohamed Cherkaoui. Robust adaptive nonsingular fast terminal sliding-mode tracking control for an uncertain quadrotor UAV subjected to disturbances. *ISA Transactions*, 99:290–304, April 2020.

326. Moussa Labbadi, Kamal Elyaalaoui, Mohamed Amine Dabachi, Soufian Lakrit, Mohamed Djemai, and Mohamed Cherkaoui. Robust flight control for a quadrotor under external disturbances based on predefined-time terminal sliding mode manifold. *Journal of Vibration and Control*, 0(0):1–13, April 2022. Publisher: SAGE Publications Ltd STM.

327. J. Laguarta, F. Hueto, and B. Subirana. Covid-19 artificial intelligence diagnosis using only cough recordings. *IEEE Open J Eng Med Biol*, 1:275–281, 2020.

328. N. F. Lahens, Ricciotti E., O. Smirnova, E. Toorens, E. J. Kim, G. Baruzzo, K. E. Hayer, T. Ganguly, J. Schug, and G. R. Grant. A comparison of illumina and ion torrent sequencing platforms in the context of differential gene expression. *BMC genomics*, 18:602, 2017.

329. Qiuxia Lai, Salman Khan, Yongwei Nie, Hanqiu Sun, Jianbing Shen, and Ling Shao. Understanding more about human and machine attention in deep neural networks. *IEEE Transactions on Multimedia*, 23:2086–2099, 2021.

330. G. Lanza, A. De Vita, S. Ravenna, A. D'Aiello, M. Covino, F. Franceschi, and F. Crea. Electrocardiographic findings at presentation and clinical outcome in patients with sars-cov-2 infection. *EP Europace*, 23:123–129, 2020.

331. Christian Lawrence. The husbandry of zebrafish (danio rerio): A review. *Aquaculture*, 269(1):1–20, 2007.

332. R. Layer, C. Chiang, A. Quinlan, and I. Hall. Lumpy: A probabilistic framework for structural variant discovery. *Genome Biology*, 15, 2014.

333. Verica Lazova, Vladimir Guzov, Kyle Olszewski, Sergey Tulyakov, and Gerard Pons-Moll. Control-NeRF: Editable Feature Volumes for Scene Rendering and

Manipulation. In *2023 IEEE/CVF Winter Conference on Applications of Computer Vision (WACV)*, pages 4329–4339, January 2023. ISSN: 2642-9381.

334. Tuan Le, Jon Glenn Omholt Gjevestad, and Pål Johan, From Online 3d mapping and localization system for agricultural robots. *IFAC-PapersOnLine*, 52(30):167–172, 2019.

335. Bruce W. Lee and Jason Hyung-Jong Lee. Traditional Readability Formulas Compared for English, January 2023. arXiv:2301.02975 [cs].

336. G. Lee, K. Wohlfahrt R. Gommers, F. Waselewski, and A. O'Leary. Pywavelets: A python package for wavelet analysis. In *Journal of Open Source Software*, volume 4, page 1237, 2019.

337. Kibok Lee, Kimin Lee, Jinwoo Shin, and Honglak Lee. Overcoming catastrophic forgetting with unlabeled data in the wild. In *Proceedings of the IEEE/CVF International Conference on Computer Vision*, pages 312–321, 2019.

338. Kyuhwa Lee, Dong Liu, Laetitia Perroud, Ricardo Chavarriaga, and José del R. Millán. A brain-controlled exoskeleton with cascaded event-related desynchronization classifiers. *Special Issue on New Research Frontiers for Intelligent Autonomous Systems*, 90(Supplement C):15–23, April 2017.

339. Soomin Lee, Le Chen, Jiahao Wang, Alexander Liniger, Suryansh Kumar, and Fisher Yu. Uncertainty Guided Policy for Active Robotic 3D Reconstruction Using Neural Radiance Fields. *IEEE Robotics and Automation Letters*, 7(4):12070–12077, October 2022. Conference Name: IEEE Robotics and Automation Letters.

340. Gaston Lenczner, Adrien Chan-Hon-Tong, Nicola Luminari, and Bertrand Le Saux. Weakly-supervised continual learning for class-incremental segmentation. In *IGARSS 2022-2022 IEEE International Geoscience and Remote Sensing Symposium*, pages 4843–4846. IEEE, 2022.

341. R. Lent, F. A. C. Azevedo, C. H. Andrade-Moraes, and A. V. Pinto. How many neurons do you have? Some dogmas of quantitative neuroscience under revision. *Eur J Neurosci*, 35(1):1–9, 2012.

342. Stanley Lewis, Jana Pavlasek, and Odest Chadwicke Jenkins. NARF22: Neural Articulated Radiance Fields for Configuration-Aware Rendering, October 2022.

343. Beibei Li, Jun Yue, Shixiang Jia, Qing Wang, Zhenbo Li, and Zhenzhong Li. Recognition of abnormal body surface characteristics of oplegnathus punctatus. *Information Processing in Agriculture*, 9:575–585, 12 2022.

344. H. Li and R. Durbin. Fast and accurate short read alignment with burrows-wheeler transform. *Bioinformatics*, 25:1754–1760, 2009.

345. Haifeng Li, Hao Jiang, Xin Gu, Jian Peng, Wenbo Li, Liang Hong, and Chao Tao. Clrs: Continual learning benchmark for remote sensing image scene classification. *Sensors*, 20(4):1226, 2020.

346. Honggen Li, Hongbo Chen, Wenke Jing, Yuwei Li, and Rui Zheng. 3D Ultrasound Spine Imaging with Application of Neural Radiance Field Method. In *2021 IEEE International Ultrasonics Symposium (IUS)*, pages 1–4, September 2021. ISSN: 1948-5727.

347. Jiadong Li, Zirui Lian, Zhelin Wu, Lihua Zeng, Liangliang Mu, Ye Yuan, Hao Bai, Zheng Guo, Kangsen Mai, Xiao Tu, and Jianmin Ye. Artificial intelligence?based method for the rapid detection of fish parasites (ichthyophthirius multifiliis, gyrodactylus kobayashii, and argulus japonicus). *Aquaculture*, 563, 1 2023.

348. Jianxun Li, Hao Liu, Kin Keung Lai, and Bhagwat Ram. Vehicle and UAV Collaborative Delivery Path Optimization Model. *Mathematics*, 10(20):3744, January 2022. Number: 20 Publisher: Multidisciplinary Digital Publishing Institute.

349. Jing Shuang Li, Anish A. Sarma, Terrence J. Sejnowski, and John C. Doyle. Internal feedback in the cortical perception-action loop enables fast and accurate behavior, 2022.

350. Ruihao Li, Sen Wang, Zhiqiang Long, and Dongbing Gu. Undeepvo: Monocular visual odometry through unsupervised deep learning. In *2018 IEEE International Conference on Robotics and Automation (ICRA)*, pages 7286–7291. IEEE, 2018.

351. S. Li, C. De Wagter, C. C. de Visser, Q. P. Chu, G. C. H. E. de Croon, et al. In-flight model parameter and state estimation using gradient descent for high-speed flight. *International Journal of Micro Air Vehicles*, 11:1756829319833685, 2019.

352. Shuo Li, Michaël MOI Ozo, Christophe De Wagter, Guido CHE de Croon, et al. Autonomous drone race: A computationally efficient vision-based navigation and control strategy. *Robotics and Autonomous Systems*, 133:103621, 2020.

353. Shuo Li, Erik van der Horst, Philipp Duernay, Christophe De Wagter, and Guido C. H. E. de Croon. Visual model-predictive localization for computationally efficient autonomous racing of a 72-g drone. *Journal of Field Robotics*, 37(4):667–692, 2020.

354. Shuo Li, Erik van der Horst, Philipp Duernay, Christophe De Wagter, Guido CHE de Croon, et al. Visual model-predictive localization for computationally efficient autonomous racing of a 72-g drone. *Journal of Field Robotics*, 37(4):667–692, 2020.

355. Shuo Li, Erik van der Horst, Philipp Duernay, Christophe De Wagter, and Guido C. H. E. de Croon. Visual model-predictive localization for computationally efficient autonomous racing of a 72-gram drone. *arXiv: Robotics*, 2019.

356. Sicheng Li, Hao Li, Yue Wang, Yiyi Liao, and Lu Yu. SteerNeRF: Accelerating NeRF Rendering via Smooth Viewpoint Trajectory, December 2022. arXiv:2212.08476 [cs].

357. Weixin Li, Vijay Mahadevan, and Nuno Vasconcelos. Anomaly detection and localization in crowded scenes. *IEEE transactions on pattern analysis and machine intelligence*, 36(1):18–32, 2013.

358. Wentong Li, Yijie Chen, Kaixuan Hu, and Jianke Zhu. Oriented reppoints for aerial object detection. In *Proceedings of the IEEE/CVF Conference on Computer Vision and Pattern Recognition*, pages 1829–1838, 2022.

359. Yuan Li, Zhi-Hao Lin, David Forsyth, Jia-Bin Huang, and Shenlong Wang. ClimateNeRF: Physically-based Neural Rendering for Extreme Climate Synthesis, November 2022. arXiv:2211.13226 [cs].

360. Zhi Li, Jinghao Xin, and Ning Li. Autonomous exploration and mapping for mobile robots via cumulative curriculum reinforcement learning. *arXiv preprint arXiv:2302.13025*, 2023.

361. Julien Li-Chee-Ming and Costas Armenakis. Uav navigation system using line-based sensor pose estimation. *Geo-spatial information science*, 21(1):2–11, 2018.

362. Susan Liang, Chao Huang, Yapeng Tian, Anurag Kumar, and Chenliang Xu. AV-NeRF: Learning Neural Fields for Real-World Audio-Visual Scene Synthesis, February 2023. arXiv:2302.02088 [cs, eess].

363. D. Lim, J. W. Lee, B. Gassend, G. E. Suh, M. Van Dijk, and S. Devadas. Extracting secret keys from integrated circuits. *IEEE Transactions on Very Large Scale Integration (VLSI) Systems*, 13(10):1200–1205, 2005.

364. Chen-Hsuan Lin, Wei-Chiu Ma, Antonio Torralba, and Simon Lucey. BARF: Bundle-Adjusting Neural Radiance Fields. In *2021 IEEE/CVF International Conference on Computer Vision (ICCV)*, pages 5721–5731, October 2021. ISSN: 2380-7504.

365. Ping-Ju Lin, Tianyu Jia, Chong Li, Tianyi Li, Chao Qian, Zhibin Li, Yu Pan, and Linhong Ji. Cnn-based prognosis of bci rehabilitation using eeg from first session

bci training. *IEEE Transactions on Neural Systems and Rehabilitation Engineering*, 29:1936–1943, 2021.

366. Yimin Lin, Zhaoxiang Liu, Jianfeng Huang, Chaopeng Wang, Guoguang Du, Jinqiang Bai, and Shiguo Lian. Deep global-relative networks for end-to-end 6-dof visual localization and odometry. In *Pacific Rim International Conference on Artificial Intelligence*, pages 454–467. Springer, 2019.

367. Yunzhi Lin, Thomas Müller, Jonathan Tremblay, Bowen Wen, Stephen Tyree, Alex Evans, Patricio A. Vela, and Stan Birchfield. Parallel Inversion of Neural Radiance Fields for Robust Pose Estimation, October 2022.

368. Jingwang Ling, Zhibo Wang, and Feng Xu. ShadowNeuS: Neural SDF Reconstruction by Shadow Ray Supervision, November 2022. arXiv:2211.14086 [cs].

369. Jing Shuang Lisa Li. Internal feedback in biological control: Locality and system level synthesis. In *2022 American Control Conference (ACC)*, pages 474–479, 2022.

370. Daniil Lisus and Connor Holmes. Towards Open World NeRF-Based SLAM, January 2023. arXiv:2301.03102 [cs].

371. Ta Kang Liu, Hwung Hweng Hwung, Jin Li Yu, and Ruey Chy Kao. Managing deep ocean water development in taiwan: Experiences and future challenges. *Ocean and Coastal Management*, 51:126–140, 2008.

372. Wei Liu, Dragomir Anguelov, Dumitru Erhan, Christian Szegedy, Scott Reed, Cheng-Yang Fu, and Alexander C Berg. Ssd: Single shot multibox detector. In *Computer Vision–ECCV 2016: 14th European Conference, Amsterdam, The Netherlands, October 11–14, 2016, Proceedings, Part I 14*, pages 21–37. Springer, 2016.

373. Y. Liu, D. Shukla, H. Newman, and Y. Zhu. Soft wearable sensors for monitoring symptoms of covid-19 and other respiratory diseases: a review. *Progress in Biomedical Engineering*, 4:12001, 2022.

374. Z. Liu, X. Hu, L. Xu, W. Weidong, and F. M. Ghannouchi. Low computational complexity digital predistortion based on convolutional neural network for wideband power amplifiers. *IEEE Transactions on Circuits and Systems II: Express Briefs*, 69(3):1702–1706, 2022.

375. K. Lohmann and C. Klein. Next generation sequencing and the future of genetic diagnosis. *Neurotherapeutics*, 11:699, 2014.

376. Vincenzo Lomonaco and Davide Maltoni. Core50: a new dataset and benchmark for continuous object recognition. *arXiv preprint arXiv:1705.03550*, 2017.

377. Vincenzo Lomonaco, Davide Maltoni, and Lorenzo Pellegrini. Fine-grained continual learning. *arXiv preprint arXiv:1907.03799*, 1, 2019.

378. Vincenzo Lomonaco, Davide Maltoni, and Lorenzo Pellegrini. Rehearsal-free continual learning over small non-iid batches. In *CVPR Workshops*, volume 1, page 3, 2020.

379. B. Long, W. Brady, R. Bridwell, M. Ramzy, T. Montrief, M. Singh, and M. Gottlieb. Electrocardiographic manifestations of covid-19. *Am J Emerg Med*, 41:96–103, 2021.

380. Shing Yan Loo, Moein Shakeri, Sai Hong Tang, Syamsiah Mashohor, and Hong Zhang. Online mutual adaptation of deep depth prediction and visual slam. *arXiv preprint arXiv:2111.04096*, 2021.

381. J. T. H. Loong, N. A. N. Hashim, M. S. Hamid, and F. A. Hamid. Performance analysis of CMOS-memristor hybrid ring oscillator physically unclonable function (RO-PUF). In *International Conference on Semiconductor Electronics (ICSE)*, pages 304–307. IEEE, 2016.

382. Manuel Lopez and Jose Martinez-Carranza. A cnn-based approach for cable-suspended load lifting with an autonomous mav. *Journal of Intelligent & Robotic Systems*, 105(2):32, 2022.

383. Cewu Lu, Jianping Shi, and Jiaya Jia. Abnormal event detection at 150 fps in matlab. In *Proceedings of the IEEE international conference on computer vision*, pages 2720–2727, 2013.

384. G. Lu, S. Huang, S. Ying, Y. Zeng, Y. Yao, F. Ning, and Feng. Z. Low-complexity power amplifier model driven wireless emitter identification using random forest. *IEEE Globecom Workshops (GC Wkshps)*, pages 1–12, 2019.

385. H. Lu, F. Giordano, and Z. Ning. Oxford nanopore minion sequencing and genome assembly. *Genomics, Proteomics and Bioinformatics*, 14:265–279, 2016.

386. Hanchen Lu, Hongming Shen, Bailing Tian, Xuewei Zhang, Zhenzhou Yang, and Qun Zong. Flight in gps-denied environment: Autonomous navigation system for micro-aerial vehicle. *Aerospace Science and Technology*, 124:107521, 2022.

387. Miguel ángel Luján, María Verónica Jimeno, Jorge Mateo Sotos, Jorge Javier Ricarte, and Alejandro L. Borja. A survey on eeg signal processing techniques and machine learning: Applications to the neurofeedback of autobiographical memory deficits in schizophrenia. *Electronics*, 10(23), 2021.

388. Weixin Luo, Wen Liu, and Shenghua Gao. A revisit of sparse coding based anomaly detection in stacked rnn framework. In *Proceedings of the IEEE international conference on computer vision*, pages 341–349, 2017.

389. Tianxiang Ma, Bingchuan Li, Qian He, Jing Dong, and Tieniu Tan. Semantic 3D-aware Portrait Synthesis and Manipulation Based on Compositional Neural Radiance Field, February 2023. arXiv:2302.01579 [cs].

390. J. Maarsingh, S. Yang, J. Park, and S. Haydel. Comparative transcriptomics reveals prrab-mediated control of metabolic, respiration, energy-generating, and dormancy pathways in mycobacterium smegmatis. *BMC Genomics*, 20:942, 2019.

391. Ratnesh Madaan, Nicholas Gyde, Sai Vemprala, Matthew Brown, Keiko Nagami, Tim Taubner, Eric Cristofalo, Davide Scaramuzza, Mac Schwager, Ashish Kapoor, et al. Airsim drone racing lab. pages 177–191, 2020.

392. C. Madrazo-Moya, I. Cancino-Muñoz, B. Cuevas-Cordoba, V. Gonzalez-Covarrubias, M. Barbosa-Amezcua, X. Soberón, R. Muñiz-Salazar, A. Martínez-Guarneros, C. Bäcker, J. Zarrabal-Meza, C. Sampieri-Ramirez, A. Enciso-Moreno, M. Lauzardo, I. Comas, and R. Zenteno-Cuevas. Whole genomic sequencing as a tool for diagnosis of drug and multidrug-resistance tuberculosis in an endemic region in mexico. *PLoS ONE*, 14:e0213046, 2019.

393. Dominic Maggio, Marcus Abate, Jingnan Shi, Courtney Mario, and Luca Carlone. Loc-NeRF: Monte Carlo Localization using Neural Radiance Fields. September 2022.

394. Trupti Mahendrakar, Basilio Caruso, Van Minh Nguyen, Ryan T. White, and Todd Steffen. 3D Reconstruction of Non-cooperative Resident Space Objects using Instant NGP-accelerated NeRF and D-NeRF, February 2023. arXiv:2301.09060 [cs].

395. A. Maiti, J. Casarona, L. McHale, and P. Schaumont. A large scale characterization of RO-PUF. In *International Symposium on Hardware-Oriented Security and Trust (HOST)*, pages 94–99. IEEE, 2010.

396. A. Maiti, V. Gunreddy, and P. Schaumont. A systematic method to evaluate and compare the performance of physical unclonable functions. In *Embedded systems design with FPGAs*, pages 245–267. Springer, 2013.

397. Stephane Mallat et al. A wavelet tour of signal processing: the sparce way. *AP Professional, Third Edition, London*, 2009.

398. Davide Maltoni and Vincenzo Lomonaco. Continuous learning in single-incremental-task scenarios. *Neural Networks*, 116:56–73, 2019.

399. Dipan Mandal, Abhilash Jain, and Sreenivas Subramoney. Unsupervised learning of depth, camera pose and optical flow from monocular video. *arXiv preprint arXiv:2205.09821*, 2022.

400. Dilip Mandloi and Rajeev Arya. Seamless connectivity with 5G enabled unmanned aerial vehicles base stations using machine programming approach. *Expert Systems*, 39(5):e12828, 2022.

401. Abhishek Manjrekar, Dr Jha, Pratiksha Jagtap, Vinay Yadav, et al. Warehouse inventory management with cycle counting using drones. *Siddhi and Jagtap, Pratiksha and Yadav, Vinay, Warehouse Inventory Management with Cycle Counting Using Drones (May 7, 2021)*, 2021.

402. A. Manson, K. Cohen, T. Abeel, C. Desjardins, D. Armstrong, C. Barry, J. Brand, S. Chapman, S. Cho, A. Gabrielian, J. Gomez, A. Jodals, M. Joloba, P. Jureen, J. Lee, L. Malinga, M. Maiga, D. Nordenberg, E. Noroc, E. Romancenco, A. Salazar, W. Ssengooba, A. Velayati, K. Winglee, A. Zalutskaya, L. Via, G. Cassell, S. Dorman, J. Ellner, P. Farnia, J. Galagan, A. Rosenthal, V. Crudu, D. Homorodean, P. Hsueh, S. Narayanan, A. Pym, A. Skrahina, S. Swaminathan, M. Van Der Walt, D. Alland, W. Bishai, T. Cohen, S. Hoffner, B. Birren, and A. Earl. Genomic analysis of globally diverse mycobacterium tuberculosis strains provides insights into the emergence and spread of multidrug resistance. *Nature Genetics*, 49:395–402, 2017.

403. José Martínez-Carranza, Richard Bostock, Simon Willcox, Ian Cowling, and Walterio Mayol-Cuevas. Indoor mav auto-retrieval using fast 6d relocalisation. *Advanced Robotics*, 30(2):119–130, 2016.

404. Jose Martinez-Carranza and Caleb Rascon. A review on auditory perception for unmanned aerial vehicles. *Sensors*, 20(24):7276, 2020.

405. Jose Martinez-Carranza and Leticia Oyuki Rojas-Perez. Warehouse inspection with an autonomous micro air vehicle. *Unmanned Systems*, pages 1–14, 2022.

406. Roger Marí, Gabriele Facciolo, and Thibaud Ehret. Sat-NeRF: Learning multi-view satellite photogrammetry with transient objects and shadow modeling using RPC cameras. In *2022 IEEE/CVF Conference on Computer Vision and Pattern Recognition Workshops (CVPRW)*, pages 1310–1320, June 2022. ISSN: 2160-7516.

407. Ruben Mascaro, Lucas Teixeira, Timo Hinzmann, Roland Siegwart, and Margarita Chli. Gomsf: Graph-optimization based multi-sensor fusion for robust uav pose estimation. In *2018 IEEE International Conference on Robotics and Automation (ICRA)*, pages 1421–1428. IEEE, 2018.

408. S. Matos, S. Birring, I. Pavord, and D. Evans. An automated system for 24-h monitoring of cough frequency: the leicester cough monitor. *IEEE Trans Biomed Eng*, 54:1472–1479, 2007.

409. A. McKenna, M. Hanna, E. Banks, A. Sivachenko, K. Cibulskis, A. Kernytsky, K. Garimella, D. Altshuler, S. Gabriel, M. Daly, and M.A. DePristo. The genome analysis toolkit: A mapreduce framework for analyzing next-generation dna sequencing data. *Genome Research*, 20:1297–1303, 2010.

410. E. Mehraeen, S. Seyed Alinaghi, A. Nowroozi, O. Dadras, S. Alilou, P. Shobeiri, F. Behnezhad, and A. Karimi. A systematic review of ecg findings in patients with covid-19. *Indian Heart J*, 72:500–507, 2020.

411. Ramin Mehran, Alexis Oyama, and Mubarak Shah. Abnormal crowd behavior detection using social force model. In *2009 IEEE conference on computer vision and pattern recognition*, pages 935–942. IEEE, 2009.

412. M. Melek. Diagnosis of covid-19 and non-covid-19 patients by classifying only a single cough sound. *Neural Computing and Applications*, 33:17621–17632, 2021.

413. Daniel Mellinger and Vijay Kumar. Minimum snap trajectory generation and control for quadrotors. In *2011 IEEE international conference on robotics and automation*, pages 2520–2525. IEEE, 2011.

414. Lili Meng, Jianhui Chen, Frederick Tung, James J. Little, and Clarence W. de Silva. Exploiting random rgb and sparse features for camera pose estimation. In *BMVC*, 2016.

415. Geison Pires Mesquita, Margarita Mulero-Pázmány, Serge A. Wich, and José Domingo Rodríguez-Teijeiro. Terrestrial megafauna response to drone noise levels in ex situ areas. *Drones*, 6(11):333, 2022.

416. Umberto Michieli and Pietro Zanuttigh. Incremental learning techniques for semantic segmentation. In *Proceedings of the IEEE/CVF International Conference on Computer Vision Workshops*, pages 0–0, 2019.

417. Ben Mildenhall, Pratul P. Srinivasan, Matthew Tancik, Jonathan T. Barron, Ravi Ramamoorthi, and Ren Ng. NeRF: Representing Scenes as Neural Radiance Fields for View Synthesis, August 2020. arXiv:2003.08934 [cs].

418. Michael J. Milford and Gordon F. Wyeth. Seqslam: Visual route-based navigation for sunny summer days and stormy winter nights. In *2012 IEEE International Conference on Robotics and Automation*, pages 1643–1649. IEEE, 2012.

419. Yuhang Ming, Weicai Ye, and Andrew Calway. idf-slam: End-to-end rgb-d slam with neural implicit mapping and deep feature tracking. *arXiv preprint arXiv:2209.07919*, 2022.

420. P. Miotto, F. Forti, A. Ambrosi, D. Pellin, D. Veiga, G. Balazsi, M. Gennaro, C. Di Serio, D. Ghisotti, and D. Cirillo. Genome-wide discovery of small rnas in mycobacterium tuberculosis. *PLoS ONE*, 7:e51950, 2012.

421. Roger Miranda-Colorado, Luis T. Aguilar, and José E. Herrero-Brito. Reduction of power consumption on quadrotor vehicles via trajectory design and a controller-gains tuning stage. *Aerospace Science and Technology*, 78:280–296, 2018.

422. F. Mkadem, A. Islam, and S. Boumaiza. Multi-band complexity-reduced generalized-memory-polynomial power-amplifier digital predistortion. *IEEE Transactions on Microwave Theory and Techniques*, 64(6):1763–1774, 2016.

423. Omid Mofid, Saleh Mobayen, Chunwei Zhang, and Balasubramanian Esakki. Desired tracking of delayed quadrotor UAV under model uncertainty and wind disturbance using adaptive super-twisting terminal sliding mode control. *ISA Transactions*, 123(2022):455–471, April 2022.

424. Abdirahman Mohamud and Ashwin Ashok. Drone noise reduction through audio waveguiding. In *Proceedings of the 4th ACM Workshop on Micro Aerial Vehicle Networks, Systems, and Applications*, pages 92–94, 2018.

425. Vikram Mohanty, Shubh Agrawal, Shaswat Datta, Arna Ghosh, Vishnu Dutt Sharma, and Debashish Chakravarty. Deepvo: A deep learning approach for monocular visual odometry. *arXiv preprint arXiv:1611.06069*, 2016.

426. M. Molnar and L. Ilie. Correcting illumina data. *Briefings in Bioinformatics*, 16:588–599, 2014.

427. Hyungpil Moon, Jose Martinez-Carranza, Titus Cieslewski, Matthias Faessler, Davide Falanga, Alessandro Simovic, Davide Scaramuzza, Shuo Li, Michael Ozo, Christophe

De Wagter, et al. Challenges and implemented technologies used in autonomous drone racing. *Intelligent Service Robotics*, 12(2):137–148, 2019.

428. Hyungpil Moon, Yu Sun, Jacky Baltes, Si Jung Kim, et al. The iros 2016 competitions [competitions]. *IEEE Robotics and Automation Magazine*, 24(1):20–29, 2017.

429. Koichiro Morihiro, Teijiro Isokawa, Nobuyuki Matsui, and Haruhiko Nishimura. Effects of chaotic exploration on reinforcement learning in target capturing task. *International Journal of Knowledge-based and Intelligent Engineering Systems*, 12(5-6):369–377, January 2008. Publisher: IOS Press.

430. Bashir I Morshed and Abdulhalim Khan. A brief review of brain signal monitoring technologies for bci applications: challenges and prospects. *Journal of Bioengineering & Biomedical Sciences*, 4(1):1, 2014.

431. Luca Mottola and Gian Pietro Picco. Programming wireless sensor networks: Fundamental concepts and state of the art. *ACM Comput. Surv.*, 43(3), apr 2011.

432. Sayed Hamid Mousavian and Hamid Reza Koofigar. Identification-based robust motion control of an AUV: Optimized by particle swarm optimization algorithm. *Journal of Intelligent & Robotic Systems*, 85(2):331–352, February 2017.

433. Muhammad F Mridha, Sujoy Chandra Das, Muhammad Mohsin Kabir, Aklima Akter Lima, Md Rashedul Islam, and Yutaka Watanobe. Brain-computer interface: Advancement and challenges. *Sensors*, 21(17):5746, 2021.

434. Shiv Kumar Mudgal, Suresh K Sharma, Jitender Chaturvedi, and Anil Sharma. Brain computer interface advancement in neurosciences: Applications and issues. *Interdisciplinary Neurosurgery*, 20:100694, 2020.

435. Mark W Mueller, Markus Hehn, and Raffaello D'Andrea. A computationally efficient motion primitive for quadrocopter trajectory generation. *IEEE Transactions on Robotics*, 31(6):1294–1310, 2015.

436. M. S. Müller, S. Urban, and B. Jutzi. Squeezeposenet: Image based pose regression with small convolutional neural networks for real time uas navigation. *ISPRS Annals of the Photogrammetry, Remote Sensing and Spatial Information Sciences*, 4:49, 2017.

437. Raul Mur-Artal, Jose Maria Martinez Montiel, and Juan D. Tardos. Orb-slam: a versatile and accurate monocular slam system. *IEEE Transactions on Robotics*, 31(5):1147–1163, 2015.

438. Raul Mur-Artal and Juan D. Tardós. Orb-slam2: An open-source slam system for monocular, stereo, and rgb-d cameras. *IEEE Transactions on Robotics*, 33(5):1255–1262, 2017.

439. Raul Mur-Artal and Juan D. Tardós. Orb-slam2: an open-source slam system for monocular, stereo and rgb-d cameras. *IEEE Transactions on Robotics*, 2017.

440. M. Murugesan and R. Sukanesh. Automated detection of brain tumor in eeg signals using artificial neural networks. In *2009 International Conference on Advances in Computing, Control, and Telecommunication Technologies*, pages 284–288, 2009.

441. M. Mustapa and M. Niamat. A comparative study of ring oscillator PUFs implementation on different fpga families. In *Journal of Physics: Conference Series*, volume 1962, page 012054. IOP Publishing, 2021.

442. P. Muñoz-Benavent, G. Andreu-García, José M. Valiente-González, V. Atienza-Vanacloig, V. Puig-Pons, and V. Espinosa. Enhanced fish bending model for automatic tuna sizing using computer vision. *Computers and Electronics in Agriculture*, 150:52–61, 2018.

443. A. Nachiappan, K. Rahbar, X. Shi, E. Guy, E. Barbosa, G. Shroff, D. Ocazionez, A. Schlesinger, S. Katz, and M. Hammer. Pulmonary tuberculosis: Role of radiology in

diagnosis and management. *Radiographics : a review publication of the Radiological Society of North America, Inc.*, 37:52–72, 2017.

444. Michael Narine. Active noise cancellation of drone propeller noise through waveform approximation and pitch-shifting. 2020.

445. C. Neuper and G. Pfurtscheller. Evidence for distinct beta resonance frequencies in human EEG related to specific sensorimotor cortical areas. *Clinical Neurophysiology*, 112(11):2084–2097, November 2001.

446. Christa Neuper, Gernot R. Müller, Andrea Kübler, Niels Birbaumer, and Gert Pfurtscheller. Clinical application of an eeg-based brain–computer interface: a case study in a patient with severe motor impairment. *Clinical neurophysiology*, 114(3):399–409, 2003.

447. Ngoc Phi Nguyen, Nguyen Xuan Mung, Ha Le Nhu Ngoc Thanh, Tuan Tu Huynh, Ngoc Tam Lam, and Sung Kyung Hong. Adaptive Sliding Mode Control for Attitude and Altitude System of a Quadcopter UAV via Neural Network. *IEEE Access*, 9:40076–40085, 2021. Conference Name: IEEE Access.

448. Truong Nguyen. *Wavelets and filter banks*. Wellesley-Cambridge Press, 1996.

449. S. Niemann, C. Köser, S. Gagneux, C. Plinke, S. Homolka, H. Bignell, R. Carter, R. Cheetham, A. Cox, N. Gormley, P. Kokko-Gonzales, L. Murray, R. Rigatti, V. Smith, F. Arends, H. Cox, G. Smith, and J. Archer. Genomic diversity among drug sensitive and multidrug resistant isolates of mycobacterium tuberculosis with identical dna fingerprints. *PLoS ONE*, 4:e7407, 2009.

450. David Nilsson, Aleksis Pirinen, Erik Gärtner, and Cristian Sminchisescu. Embodied learning for lifelong visual perception. *arXiv preprint arXiv:2112.14084*, 2021.

451. Sivapong Nilwong, Delowar Hossain, Shin-ichiro Kaneko, and Genci Capi. Deep learning-based landmark detection for mobile robot outdoor localization. *Machines*, 7(2):25, 2019.

452. C. Nimmo, L. Shaw, R. Doyle, R. Williams, K. Brien, C. Burgess, J. Breuer, F. Balloux, and A. Pym. Whole genome sequencing mycobacterium tuberculosis directly from sputum identifies more genetic diversity than sequencing from culture. *BMC Genomics*, 20:389, 2019.

453. Ryusuke Noda, Teruaki Ikeda, Toshiyuki Nakata, and Hao Liu. Characterization of the low-noise drone propeller with serrated gurney flap. *Frontiers in Aerospace Engineering*, page 8, 2022.

454. Ryusuke Noda, Toshiyuki Nakata, Teruaki Ikeda, Di Chen, Yuma Yoshinaga, Kenta Ishibashi, Chen Rao, and Hao Liu. Development of bio-inspired low-noise propeller for a drone. *Journal of Robotics and Mechatronics*, 30(3):337–343, 2018.

455. Nvidia Corporation . https://www.nvidia.com/, Noviembre 2020.

456. Shu Lih Oh, Jahmunah Vicnesh, Edward J. Ciaccio, Rajamanickam Yuvaraj, and U. Rajendra Acharya. Deep convolutional neural network model for automated diagnosis of schizophrenia using eeg signals. *Applied Sciences*, 9(14), 2019.

457. World Healt Organizartion. Catalogue of mutations in mycobacterium tuberculosis complex and their association with drug resistance. Technical report, World Healt Organizartion, 2021.

458. World Health Organization. The use of next-generation sequencing technologies for the detection of mutations associated with drug resistance in mycobacterium tuberculosis complex: technical guide. Technical documents, World Health Organization, 2018.

459. World Health Organization et al. *Environmental noise guidelines for the European region*. World Health Organization. Regional Office for Europe, 2018.

460. Joseph Ortiz, Alexander Clegg, Jing Dong, Edgar Sucar, David Novotny, Michael Zollhoefer, and Mustafa Mukadam. isdf: Real-time neural signed distance fields for robot perception. *arXiv preprint arXiv:2204.02296*, 2022.

461. Joseph Ortiz, Talfan Evans, Edgar Sucar, and Andrew J. Davison. Incremental abstraction in distributed probabilistic slam graphs. In *2022 International Conference on Robotics and Automation (ICRA)*, pages 7566–7572. IEEE, 2022.

462. Felix Ott, Tobias Feigl, Christoffer Loffler, and Christopher Mutschler. Vipr: visual-odometry-aided pose regression for 6dof camera localization. In *Proceedings of the IEEE/CVF Conference on Computer Vision and Pattern Recognition Workshops*, pages 42–43, 2020.

463. Natasha Padfield, Jaime Zabalza, Huimin Zhao, Valentin Masero, and Jinchang Ren. Eeg-based brain-computer interfaces using motor-imagery: Techniques and challenges. *Sensors*, 19(6), 2019.

464. M. Pahar, M. Klopper, B. Reeve, R. Warren, G. Theron, A. Diacon, and T. Niesler. Automatic tuberculosis and covid-19 cough classification using deep learning. In *2022 International Conference on Electrical, Computer and Energy Technologies (ICE-CET)*, pages 1–9, 2022.

465. Jen-I. Pan, Jheng-Wei Su, Kai-Wen Hsiao, Ting-Yu Yen, and Hung-Kuo Chu. Sampling Neural Radiance Fields for Refractive Objects. In *SIGGRAPH Asia 2022 Technical Communications*, pages 1–4, December 2022. arXiv:2211.14799 [cs].

466. Bai. F Li. Z Li. H Mao. K. Pang. M, Zhu. Q and Tian. Y. Height-dependent LoS probability model for A2G mmWave communications under built-up scenarios. *arXiv preprint arXiv:2109.02263*, 2021.

467. Prabin Kumar Panigrahi and Sukant Kishoro Bisoy. Localization strategies for autonomous mobile robots: A review. *Journal of King Saud University-Computer and Information Sciences*, 34(8):6019–6039, 2022.

468. A. D. Pano-Azucena, E. Tlelo-Cuautle, B. Ovilla-Martinez, L. G. de la Fraga, and R. Li. Pipeline FPGA-based implementations of ANNs for the prediction of up to 600-steps-ahead of chaotic time series. *Journal of Circuits, Systems, and Computers*, 30(9):2150164, 2021.

469. R. Pappu, B. Recht, J. Taylor, and N. Gershenfeld. Physical one-way functions. *Science*, 297(5589):2026–2030, 2002.

470. Chun Hyuk Park, Dae Han Kim, and Young J. Moon. Computational study on the steady loading noise of drone propellers: Noise source modeling with the lattice boltzmann method. *International Journal of Aeronautical and Space Sciences*, 20(4):858–869, 2019.

471. Keunhong Park, Utkarsh Sinha, Jonathan T. Barron, Sofien Bouaziz, Dan B. Goldman, Steven M. Seitz, and Ricardo Martin-Brualla. Nerfies: Deformable Neural Radiance Fields, September 2021. arXiv:2011.12948 [cs].

472. Keunhong Park, Utkarsh Sinha, Peter Hedman, Jonathan T. Barron, Sofien Bouaziz, Dan B. Goldman, Ricardo Martin-Brualla, and Steven M. Seitz. HyperNeRF: A Higher-Dimensional Representation for Topologically Varying Neural Radiance Fields, September 2021. arXiv:2106.13228 [cs].

473. Seonghun Park, Jisoo Ha, Jimin Park, Kyeonggu Lee, and Chang-Hwan Im. Brain-controlled, ar-based home automation system using ssvep-based brain-computer interface and eog-based eye tracker: A feasibility study for the elderly end user. *IEEE Transactions on Neural Systems and Rehabilitation Engineering*, 31:544–553, 2023.

474. Navneet Paul. TransNeRF – Improving Neural Radiance fields using transfer learning for efficient scene reconstruction, 2021. Publisher: University of Twente.

475. Dario Pavllo, David Joseph Tan, Marie-Julie Rakotosaona, and Federico Tombari. Shape, Pose, and Appearance from a Single Image via Bootstrapped Radiance Field Inversion, November 2022. arXiv:2211.11674 [cs].

476. Naama Pearl, Tali Treibitz, and Simon Korman. NAN: Noise-Aware NeRFs for Burst-Denoising. In *2022 IEEE/CVF Conference on Computer Vision and Pattern Recognition (CVPR)*, pages 12662–12671, June 2022. ISSN: 2575-7075.

477. Shaotong Pei, Rui Yang, Yunpeng Liu, Wenxuan Xu, and Gonghao Zhang. Research on 3D reconstruction technology of large-scale substation equipment based on NeRF. *IET Science, Measurement & Technology*, n/a(n/a). _eprint: https://onlinelibrary. wiley.com/doi/pdf/10.1049/smt2.12131.

478. Lorenzo Pellegrini, Gabrile Graffieti, Vincenzo Lomonaco, and Davide Maltoni. Latent replay for real-time continual learning. *arXiv preprint arXiv:1912.01100*, 2019.

479. Lorenzo Pellegrini, Vincenzo Lomonaco, Gabriele Graffieti, and Davide Maltoni. Continual learning at the edge: Real-time training on smartphone devices. *arXiv preprint arXiv:2105.13127*, 2021.

480. Fabian Peller-Konrad, Rainer Kartmann, Christian R. G. Dreher, Andre Meixner, Fabian Reister, Markus Grotz, and Tamim Asfour. A memory system of a robot cognitive architecture and its implementation in armarx. *Robotics and Autonomous Systems*, page 104415, 2023.

481. Rafael Perez-Torres, Cesar Torres-Huitzil, and Hiram Galeana-Zapien. An on-device cognitive dynamic systems inspired sensing framework for the iot. *IEEE Communications Magazine*, 56(9):154–161, 2018.

482. Anders Persson. The coriolis effect. *History of Meteorology*, 2:1–24, 2005.

483. Christian Pfeiffer and Davide Scaramuzza. Expertise affects drone racing performance. *arXiv preprint arXiv:2109.07307*, 2021.

484. Christian Pfeiffer and Davide Scaramuzza. Human-piloted drone racing: Visual processing and control. *IEEE Robotics and Automation Letters*, 6(2):3467–3474, 2021.

485. Huy Xuan Pham, Halil Ibrahim Ugurlu, Jonas Le Fevre, Deniz Bardakci, and Erdal Kayacan. Deep learning for vision-based navigation in autonomous drone racing. In *Deep Learning for Robot Perception and Cognition*, pages 371–406. Elsevier, 2022.

486. Yalong Pi, Nipun D Nath, and Amir H Behzadan. Convolutional neural networks for object detection in aerial imagery for disaster response and recovery. *Advanced Engineering Informatics*, 43:101009, 2020.

487. O. R. Pinheiro, L. R. G. Alves, and J. R. D. Souza. Eeg signals classification: Motor imagery for driving an intelligent wheelchair. *IEEE Latin America Transactions*, 16(1):254–259, 2018.

488. Maciej Podsędkowski, Rafał Konopiński, and Michał Lipian. Sound noise properties of variable pitch propeller for small uav. In *2022 International Conference on Unmanned Aircraft Systems (ICUAS)*, pages 1025–1029. IEEE, 2022.

489. S. Preethi, A. Revathi, and M. Murugan. *Exploration of Cough Recognition Technologies Grounded on Sensors and Artificial Intelligence*, pages 193–214. Springer Singapore, 2020.

490. A. Proaño, M. Bravard, B. Tracey, J. López, G. Comina, M. Zimic, J. Coronel, G. Lee, L. Caviedes, J. L. Cabrera, A. Salas, E. Ticona, D. Kirwan, J. Friedland, C. Evans, D. Moore, and R. Gilman. Protocol for studying cough frequency in people with pulmonary tuberculosis. *BMJ Open*, 6:e010365, 2016.

491. Jesus Pestana Puerta, Michael Maurer, Friedrich Fraundorfer, and Horst Bischof. Towards an autonomous vision-based inventory drone. In *Logistikwerkstatt Graz 2019: Solution day* – https://linkprotect.cudasvc.com/url?a=http%3a%2f%2fprisma.dieti.unina.it%2findex.php%2fevents%2f552-robotics-for-logistics-in-warehouses-and-environments-shared-with-humans&c=E,1,M6aTT_OnvTlDb1vsmtag2–RFc43ktHjOKvp5ArBq3c3AwB-S6ZZ8v9nNWj8QfyTkIvgp9NudiPGEnNk5gS-PGXgqiNi_AkOaM5BSyHjcUc6MsG9hQk,&typo=1, 2019.

492. Albert Pumarola, Enric Corona, Gerard Pons-Moll, and Francesc Moreno-Noguer. D-NeRF: Neural Radiance Fields for Dynamic Scenes. In *2021 IEEE/CVF Conference on Computer Vision and Pattern Recognition (CVPR)*, pages 10313–10322, June 2021. ISSN: 2575-7075.

493. C. Qin, D. Yao, Y. Shi, and Z. Song. Computer-aided detection in chest radiography based on artificial intelligence: A survey. *BioMedical Engineering Online*, 17:1–23, 2018.

494. Hongwei Qin, Xiu Li, Jian Liang, Yigang Peng, and Changshui Zhang. Deepfish: Accurate underwater live fish recognition with a deep architecture. *Neurocomputing*, 187:49–58, 4 2016.

495. Z. Qin. Screening and triage for tb using computer-aided detection (cad) technology and ultra-portable x-ray systems: A practical guide. Technical report, Stop TB Partnership, 2022.

496. Z. Qin, S. Ahmed, M. Sarker, K. Paul, A. Adel, T. Naheyan, R. Barrett, S. Banu, and J. Creswell. Tuberculosis detection from chest x-rays for triaging in a high tuberculosis-burden setting: an evaluation of five artificial intelligence algorithms. *The Lancet. Digital health*, 3:e543–e554, 2021.

497. M. A. Quail, M. Smith, P. Coupland, T. D. Otto, R. S. Harris, T. R. Connor, A. Bertoni, H. P. Swerdlow, and Y. Gu. A tale of three next generation sequencing platforms: comparison of ion torrent, pacific biosciences and illumina miseq sequencers. *BMC Genomics*, 13:1–13, 2012.

498. Morgan Quigley, Ken Conley, Brian Gerkey, Josh Faust, Tully Foote, Jeremy Leibs, Rob Wheeler, Andrew Y Ng, et al. Ros: an open-source robot operating system. In *ICRA workshop on open source software*, volume 3, page 5. Kobe, Japan, 2009.

499. Sadegh Rabiee and Joydeep Biswas. Iv-slam: Introspective vision for simultaneous localization and mapping. In *Conference on Robot Learning*, pages 1100–1109. PMLR, 2021.

500. T. Rahman, A. Akinbi, M. Chowdhury, T. Rashid, A. Şengür, A. Khandakar, K. Islam, and A. Ismael. Cov-ecgnet: Covid-19 detection using ecg trace images with deep convolutional neural network. *Health Information Science and Systems*, 10:1, 2022.

501. Bharathkumar Ramachandra and Michael Jones. Street scene: A new dataset and evaluation protocol for video anomaly detection. In *Proceedings of the IEEE/CVF Winter Conference on Applications of Computer Vision*, pages 2569–2578, 2020.

502. Rabie A. Ramadan and Athanasios V. Vasilakos. Brain computer interface: control signals review. *Neurocomputing*, 223(Supplement C):26–44, February 2017.

503. Soundarya Ramesh, Thomas Pathier, and Jun Han. Sounduav: Towards delivery drone authentication via acoustic noise fingerprinting. In *Proceedings of the 5th Workshop on Micro Aerial Vehicle Networks, Systems, and Applications*, pages 27–32, 2019.

504. Carlos Ramos-Romero, Nathan Green, Seth Roberts, Charlotte Clark, and Antonio J. Torija. Requirements for drone operations to minimise community noise impact. *International Journal of Environmental Research and Public Health*, 19(15):9299, 2022.

505. Caleb Rascon, Oscar Ruiz-Espitia, and Jose Martinez-Carranza. On the use of the aira-uas corpus to evaluate audio processing algorithms in unmanned aerial systems. *Sensors*, 19(18):3902, 2019.

506. Mamunur Rashid, Norizam Sulaiman, Mahfuzah Mustafa, Sabira Khatun, and Bifta Sama Bari. The classification of eeg signal using different machine learning techniques for bci application. In *Robot Intelligence Technology and Applications: 6th International Conference, RiTA 2018, Kuala Lumpur, Malaysia, December 16–18, 2018, Revised Selected Papers 6*, pages 207–221. Springer, 2019.

507. Nasir Rashid, Javaid Iqbal, Amna Javed, Mohsin I Tiwana, and Umar Shahbaz Khan. Design of embedded system for multivariate classification of finger and thumb movements using eeg signals for control of upper limb prosthesis. *BioMed Research International*, 2018, 2018.

508. T. Rausch, T. Zichner, A. Schlattl, A. Stütz, V. Benes, and J. Korbel. Delly: Structural variant discovery by integrated paired-end and split-read analysis. *Bioinformatics*, 28, 2012.

509. Mahdyar Ravanbakhsh, Enver Sangineto, Moin Nabi, and Nicu Sebe. Training adversarial discriminators for cross-channel abnormal event detection in crowds. In *2019 IEEE Winter Conference on Applications of Computer Vision (WACV)*, pages 1896–1904. IEEE, 2019.

510. Joseph Redmon, Santosh Divvala, Ross Girshick, and Ali Farhadi. You only look once: Unified, real-time object detection. In *Proceedings of the IEEE conference on computer vision and pattern recognition*, pages 779–788, 2016.

511. Christian Reiser, Songyou Peng, Yiyi Liao, and Andreas Geiger. KiloNeRF: Speeding up Neural Radiance Fields with Thousands of Tiny MLPs, August 2021. arXiv:2103.13744 [cs].

512. B. Rekadwad and J. M. Gonzalez. *New generation DNA sequencing (NGS): Mining for genes and the potential of extremophiles*, pages 255–268. Springer International Publishing, 2017.

513. M. Rivera. Reeuso de redes preentrenadas (cimat). [Online]. Available: http://personal.cimat.mx:8181/~mrivera/cursos/aprendizaje_profundo/preentrenadas/ preentrenadas.html.

514. Damian Roca, Rodolfo Milito, Mario Nemirovsky, and Mateo Valero. Advances in the hierarchical emergent behaviors (heb) approach to autonomous vehicles. *IEEE Intelligent Transportation Systems Magazine*, 12(4):57–65, 2020.

515. Damian Roca, Daniel Nemirovsky, Mario Nemirovsky, Rodolfo Milito, and Mateo Valero. Emergent behaviors in the internet of things: The ultimate ultra-large-scale system. *IEEE Micro*, 36(6):36–44, 2016.

516. D. Rodman, S. Lowenstein, and T. Rodman. The electrocardiogram in chronic obstructive pulmonary disease. *J Emerg Med*, 8:607–615, 1990.

517. Jean-Paul Rodrigue. *The geography of transport systems*. Routledge, 2020.

518. L. Oyuki Rojas-Perez and Jose Martinez-Carranza. Leveraging a neural pilot via automatic gain tuning using gate detection for autonomous drone racing. In *13th International Micro Air Vehicle Conference at Delft, The Netherlands*, pages 110–118.

519. L Oyuki Rojas-Perez and Jose Martinez-Carranza. A temporal cnn-based approach for autonomous drone racing. In *2019 Workshop on Research, Education and Development of Unmanned Aerial Systems (RED UAS)*, pages 70–77. IEEE, 2019.

520. L Oyuki Rojas-Perez and Jose Martinez-Carranza. Metric monocular slam and colour segmentation for multiple obstacle avoidance in autonomous flight. In *2017 Workshop*

on Research, Education and Development of Unmanned Aerial Systems (RED-UAS), pages 234–239. IEEE, 2017.

521. L Oyuki Rojas-Perez and José Martínez-Carranza. On-board processing for autonomous drone racing: an overview. *Integration*, 80:46–59, 2021.

522. L Oyuki Rojas-Perez and Jose Martinez-Carranza. Deeppilot4pose: a fast pose localisation for mav indoor flight using the oak-d camera. *Journal of Real-Time Image Processing*, 20(1):8, 2023.

523. Leticia Oyuki Rojas-Perez and Jose Martinez-Carranza. Deeppilot: A cnn for autonomous drone racing. *Sensors*, 20(16):4524, 2020.

524. Leticia Oyuki Rojas-Perez and Jose Martinez-Carranza. Towards autonomous drone racing without gpu using an oak-d smart camera. *Sensors*, 21(22):7436, 2021.

525. Leticia Oyuki Rojas-Perez and Jose Martinez-Carranza. Where are the gates: Discovering effective waypoints for autonomous drone racing. In *Advances in Artificial Intelligence–IBERAMIA 2022: 17th Ibero-American Conference on AI, Cartagena de Indias, Colombia, November 23–25, 2022, Proceedings*, pages 353–365. Springer, 2023.

526. Angel Romero, Robert Penicka, and Davide Scaramuzza. Time-optimal online replanning for agile quadrotor flight. *IEEE Robotics and Automation Letters*, 7(3):7730–7737, 2022.

527. Martin Rosalie, Grégoire Danoy, Serge Chaumette, and Pascal Bouvry. Chaos-enhanced mobility models for multilevel swarms of UAVs. *Swarm and Evolutionary Computation*, 41:36–48, August 2018.

528. Martin Rosalie, Jan E. Dentler, Grégoire Danoy, Pascal Bouvry, Somasundar Kannan, Miguel A. Olivares-Mendez, and Holger Voos. Area exploration with a swarm of UAVs combining deterministic chaotic ant colony mobility with position MPC. In *2017 International Conference on Unmanned Aircraft Systems (ICUAS)*, pages 1392–1397, Miami, FL, June 2017.

529. Antoni Rosinol, John J. Leonard, and Luca Carlone. NeRF-SLAM: Real-Time Dense Monocular SLAM with Neural Radiance Fields, October 2022. arXiv:2210.13641 [cs].

530. Somak Roy, Christopher Coldren, Arivarasan Karunamurthy, Nefize S. Kip, Eric W. Klee, Stephen E. Lincoln, Annette Leon, Mrudula Pullambhatla, Robyn L. Temple-Smolkin, Karl V. Voelkerding, Chen Wang, and Alexis B. Carter. Standards and guidelines for validating next-generation sequencing bioinformatics pipelines: A joint recommendation of the association for molecular pathology and the college of american pathologists. *Journal of Molecular Diagnostics*, 20:4–27, 2018.

531. Markovskii. V. S., Fominykh. A. A., and Matveev. N. V. Simulation of air-to-ground based on cellular network. In *2021 Wave Electronics and its Application in Information and Telecommunication Systems (WECONF)*, pages 1–4. IEEE, 2021.

532. Sara Sabour, Suhani Vora, Daniel Duckworth, Ivan Krasin, David J. Fleet, and Andrea Tagliasacchi. RobustNeRF: Ignoring Distractors with Robust Losses, February 2023. arXiv:2302.00833 [cs].

533. Ali Safa, Tim Verbelen, Ilja Ocket, André Bourdoux, Hichem Sahli, Francky Catthoor, and Georges Gielen. Learning to slam on the fly in unknown environments: A continual learning approach for drones in visually ambiguous scenes. *arXiv preprint arXiv:2208.12997*, 2022.

534. A.H. Safari, N. Sedaghat, H. Zabeti, A. Forna, L. Chindelevitch, and M. Libbrecht. Predicting drug resistance in m. tuberculosis using a long-term recurrent convolutional network. *Proceedings of the 12th ACM Conference on Bioinformatics, Computational Biology, and Health Informatics, BCB 2021*, 1, 2021.

535. Chiranjib Saha, Mehrnaz Afshang, and Harpreet S. Dhillon. Integrated mmwave access and backhaul in 5G: Bandwidth partitioning and downlink analysis. In *2018 IEEE International Conference on Communications (ICC)*, pages 1–6, 2018.
536. Simanto Saha, Khondaker A. Mamun, Khawza Ahmed, Raqibul Mostafa, Ganesh R. Naik, Sam Darvishi, Ahsan H. Khandoker, and Mathias Baumert. Progress in brain computer interface: Challenges and opportunities. *Frontiers in Systems Neuroscience*, 15, 2021.
537. S. Saini and R. Gupta. Artificial intelligence methods for analysis of electrocardiogram signals for cardiac abnormalities: state-of-the-art and future challenges. *Artificial Intelligence Review*, 55:1519–1565, 2022.
538. Alzayat Saleh, Marcus Sheaves, and Mostafa Rahimi Azghadi. Computer vision and deep learning for fish classification in underwater habitats: A survey. *Fish and Fisheries*, 23:977–999, 7 2022.
539. M. Salim-Maza. *Generación y Distribución de Señal de Reloj para Sistemas en Chip utilizando Anillos Interconectados Acoplados*. PhD thesis, Ph.D. dissertation, Instituto Nacional de Astrofísica, Óptica, y Electrónica, 2005.
540. F. Sana, E. Isselbacher, J. Singh, E. Heist, B. Pathik, and A. Armoundas. Wearable devices for ambulatory cardiac monitoring: Jacc state-of-the-art review. *Journal of the American College of Cardiology*, 75:1582–1592, 2020.
541. S. E. Sandoval-Azuara, R. Muñiz Salazar, R. Perea-Jacobo, S. Robbe-Austerman, A. Perera-Ortiz, G. Lopez-Valencia, D.M. Bravo, A. Sanchez-Flores, D. Miranda-Guzman, C.A. Flores-Lopez, R. Zenteno-Cuevas, R. Laniado-Laborin, F.L. de la Cruz, and T.P. Stuber. Whole genome sequencing of mycobacterium bovis to obtain molecular fingerprints in human and cattle isolates from baja california, mexico. *International Journal of Infectious Diseases*, 63:48–56, 2017.
542. A. Santos-Pereira, C. Magalhães, P. Araujo, and N. Osorio. Evolutionary genetics of mycobacterium tuberculosis and hiv-1: the tortoise and the hare. *Microorganisms*, 9:1–15, 2021.
543. Paul-Edouard Sarlin, Ajaykumar Unagar, Mans Larsson, Hugo Germain, Carl Toft, Viktor Larsson, Marc Pollefeys, Vincent Lepetit, Lars Hammarstrand, Fredrik Kahl, et al. Back to the feature: Learning robust camera localization from pixels to pose. In *Proceedings of the IEEE/CVF conference on computer vision and pattern recognition*, pages 3247–3257, 2021.
544. Anish A. Sarma, Jing Shuang Lisa Li, Josefin Stenberg, Gwyneth Card, Elizabeth S. Heckscher, Narayanan Kasthuri, Terrence Sejnowski, and John C. Doyle. Internal feedback in biological control: Architectures and examples. In *2022 American Control Conference (ACC)*, pages 456–461, 2022.
545. A. Sarmiento-Reyes, J. U. Franco, Y. Rodríguez-Velásquez, and J. P. Loyo. Development of an operator-based fully analytical charge-controlled memristor model. In *IEEE 15th Int. Conf. on Electrical Engineering, Computing Science and Automatic Control (CCE)*, pages 1–4, 2018.
546. G. Satta, M. Lipman, G. Smith, C. Arnold, O. Kon, and T. McHugh. Mycobacterium tuberculosis and whole-genome sequencing: how close are we to unleashing its full potential? *Clinical Microbiology and Infection*, 24:604–609, 2018.
547. Torsten Sattler, Qunjie Zhou, Marc Pollefeys, and Laura Leal-Taixe. Understanding the limitations of cnn-based absolute camera pose regression. In *Proceedings of the IEEE Conference on Computer Vision and Pattern Recognition*, pages 3302–3312, 2019.

548. Marzieh Savadkoohi, Timothy Oladunni, and Lara Thompson. A machine learning approach to epileptic seizure prediction using electroencephalogram (eeg) signal. *Biocybernetics and Biomedical Engineering*, 40(3):1328–1341, 2020.

549. Davide Scaramuzza and Friedrich Fraundorfer. Visual odometry [tutorial]. *IEEE robotics & automation magazine*, 18(4):80–92, 2011.

550. Beat Schäffer, Reto Pieren, Kurt Heutschi, Jean Marc Wunderli, and Stefan Becker. Drone noise emission characteristics and noise effects on humans - a systematic review. *International journal of environmental research and public health*, 18(11):5940, 2021.

551. Gerwin Schalk, Dennis J McFarland, Thilo Hinterberger, Niels Birbaumer, and Jonathan R Wolpaw. Bci2000: a general-purpose brain-computer interface (bci) system. *IEEE Transactions on biomedical engineering*, 51(6):1034–1043, 2004.

552. Fabrizio Schiano, Dominik Natter, Davide Zambrano, and Dario Floreano. Autonomous detection and deterrence of pigeons on buildings by drones. *IEEE Access*, 10:1745–1755, 2021.

553. Daniel Schleich, Marius Beul, Jan Quenzel, and Sven Behnke. Autonomous flight in unknown gnss-denied environments for disaster examination. In *2021 International Conference on Unmanned Aircraft Systems (ICUAS)*, pages 950–957. IEEE, 2021.

554. C. Schötz. Comparison of methods for learning differential equations from data. In *EGU General Assembly 2023, Vienna, Austria, 24-28 Apr 2023, EGU23-7368*, 2023.

555. Soroush Seifi and Tinne Tuytelaars. How to improve cnn-based 6-dof camera pose estimation. In *Proceedings of the IEEE International Conference on Computer Vision Workshops*, pages 0–0, 2019.

556. Tabassum. H. Sekander. S and Hossai. E. Multi-tier drone architecture for 5G/B5G cellular networks: Challenges, trends, and prospects. *IEEE Communications Magazine*, 56(3):96–103, 2018.

557. Dimitrios Serpanos. The cyber-physical systems revolution. *Computer*, 51(3):70–73, 2018.

558. Advaith Venkatramanan Sethuraman, Manikandasriram Srinivasan Ramanagopal, and Katherine A. Skinner. WaterNeRF: Neural Radiance Fields for Underwater Scenes, September 2022. arXiv:2209.13091 [cs, eess].

559. De Rosal Igantius Moses Setiadi. PSNR vs SSIM: imperceptibility quality assessment for image steganography. *Multimed Tools Appl*, 80(6):8423–8444, March 2021.

560. Syed Zakir Hussain Shah, Hafiz Tayyab Rauf, Muhammad IkramUllah, Malik Shahzaib Khalid, Muhammad Farooq, Mahroze Fatima, and Syed Ahmad Chan Bukhari. Fish-pak: Fish species dataset from pakistan for visual features based classification. *Data in Brief*, 27:104565, 2019.

561. Naresh R. Shanbhag, Subhasish Mitra, Gustavode de Veciana, Michael Orshansky, Radu Marculescu, Jaijeet Roychowdhury, Douglas Jones, and Jan M. Rabaey. The search for alternative computational paradigms. *IEEE Design & Test of Computers*, 25(4):334–343, 2008.

562. Amit Sheth. Internet of things to smart iot through semantic, cognitive, and perceptual computing. *IEEE Intelligent Systems*, 31(2):108–112, 2016.

563. Akshay Shetty and Grace Xingxin Gao. Uav pose estimation using cross-view geolocalization with satellite imagery. In *2019 International Conference on Robotics and Automation (ICRA)*, pages 1827–1833. IEEE, 2019.

564. Y. Shifman, A. Miller, Y. Weizman, A. Fish, and J. Shor. An SRAM PUF with 2 independent bits/cell in 65nm. In *Int. Symp. on Circuits and Systems (ISCAS)*, pages 1–5. IEEE, 2019.

565. Dongseok Shim, Seungjae Lee, and H. Jin Kim. SNeRL: Semantic-aware Neural Radiance Fields for Reinforcement Learning, January 2023. arXiv:2301.11520 [cs].

566. Dongyoon Shin, Hyeji Kim, Jihyuk Gong, Uijeong Jeong, Yeeun Jo, and Eric Matson. Stealth uav through coandă effect. In *2020 Fourth IEEE International Conference on Robotic Computing (IRC)*, pages 202–209. IEEE, 2020.

567. S. Shokralla, J. Spall, J. Gibson, and M. Hajibabaei. Next-generation sequencing technologies for environmental dna research. *Molecular Ecology*, 21:1794–1805, 2012.

568. Mark R Shortis, Mehdi Ravanbakhsh, Faisal Shafait, and Ajmal Mian. Progress in the automated identification, measurement and counting of fish in underwater image sequences, 2016.

569. Bruno Siciliano, Lorenzo Sciavicco, Luigi Villani, and Giuseppe Oriolo. *Robotics: Modelling, Planning and Control.* Springer London, 2008. DOI: 10.1007/978-1-84628-642-1.

570. Harun Siljak, John Kennedy, Stephen Byrne, and Karina Einicke. Noise mitigation of uav operations through a complex networks approach. In *INTER-NOISE and NOISE-CON Congress and Conference Proceedings*, volume 265, pages 2208–2214. Institute of Noise Control Engineering, 2023.

571. Karen Simonyan and Andrew Zisserman. Two-stream convolutional networks for action recognition in videos. *Advances in neural information processing systems*, 27, 2014.

572. Karen Simonyan and Andrew Zisserman. Very deep convolutional networks for large-scale image recognition. *arXiv preprint arXiv:1409.1556*, 2014.

573. Uriel Singer, Shelly Sheynin, Adam Polyak, Oron Ashual, Iurii Makarov, Filippos Kokkinos, Naman Goyal, Andrea Vedaldi, Devi Parikh, Justin Johnson, and Yaniv Taigman. Text-To-4D Dynamic Scene Generation, January 2023. arXiv:2301.11280 [cs].

574. Gurkirt Singh, Stephen Akrigg, Manuele Di Maio, Valentina Fontana, Reza Javanmard Alitappeh, Salman Khan, Suman Saha, Kossar Jeddisaravi, Farzad Yousefi, Jacob Culley, et al. Road: The road event awareness dataset for autonomous driving. *IEEE transactions on pattern analysis and machine intelligence*, 45(1):1036–1054, 2022.

575. J. Singh and B. Raj. "Temperature dependent analytical modeling and simulations of nanoscale memristor". *Engineering science and technology, an Int. journal*, 21(5):862–868, 2018.

576. Jason Sleight, Preeti J. Pillai, and S. Mohan. Classification of Executed and Imagined Motor Movement EEG Signals, 2009.

577. J. Smith and A. Woodcock. New developments in the objective assessment of cough. *Lung*, 186:48–54, 2008.

578. James E. Smith. Research agenda: Spacetime computation and the neocortex. *IEEE Micro*, 37(1):8–14, 2017.

579. Alex J Smola and Bernhard Schölkopf. A tutorial on support vector regression. *Statistics and computing*, 14:199–222, 2004.

580. C. Spaccarotella, S. Migliarino, A. Mongiardo, J. Sabatino, G. Santarpia, S. De Rosa, A. Curcio, and C. Indolfi. Measurement of the qt interval using the apple watch. *Scientific Report*, 11:10817, 2021.

581. Sara Spedicato and Giuseppe Notarstefano. Minimum-time trajectory generation for quadrotors in constrained environments. *IEEE Transactions on Control Systems Technology*, 26(4):1335–1344, 2017.

582. Igor Stancin, Mario Cifrek, and Alan Jovic. A review of eeg signal features and their application in driver drowsiness detection systems. *Sensors*, 21(11), 2021.

583. Josefin Stenberg, Jing Shuang Lisa Li, Anish A. Sarma, and John C. Doyle. Internal feedback in biological control: Diversity, delays, and standard theory. In *2022 American Control Conference (ACC)*, pages 462–467, 2022.

584. Daniel H. Stolfi, Matthias R. Brust, Grégoire Danoy, and Pascal Bouvry. A Cooperative Coevolutionary Approach to Maximise Surveillance Coverage of UAV Swarms. In *2020 IEEE 17th Annual Consumer Communications Networking Conference (CCNC)*, pages 528–535, Las Vegas, NV, USA, January 2020. ISSN: 2331-9860.

585. R. Strahan, T. Uppal, and S.C. Verma. Next-generation sequencing in the understanding of kaposi's sarcoma-associated herpesvirus (kshv) biology. *Viruses*, 8:1–27, 2016.

586. Jonathan Stroud, David Ross, Chen Sun, Jia Deng, and Rahul Sukthankar. D3d: Distilled 3d networks for video action recognition. In *Proceedings of the IEEE/CVF Winter Conference on Applications of Computer Vision*, pages 625–634, 2020.

587. D. B. Strukov, G. S. Snider, D. R. Stewart, and R. S. Williams. The missing memristor found. *Nature*, 453(7191):80–83, 2008.

588. Wang Su, Abhijeet Ravankar, Ankit A Ravankar, Yukinori Kobayashi, and Takanori Emaru. Uav pose estimation using ir and rgb cameras. In *2017 IEEE/SICE International Symposium on System Integration (SII)*, pages 151–156. IEEE, 2017.

589. Edgar Sucar, Shikun Liu, Joseph Ortiz, and Andrew J Davison. imap: Implicit mapping and positioning in real-time. In *Proceedings of the IEEE/CVF International Conference on Computer Vision*, pages 6229–6238, 2021.

590. G. E. Suh and S. Devadas. Physical unclonable functions for device authentication and secret key generation. In *44th ACM/IEEE Design Automation Conference*, pages 9–14, 2007.

591. C. Sun, M. Song, S. Hong, and H. Li. A review of designs and applications of echo state networks. *arXiv:2012.02974v1*, 2020.

592. Cheng Sun, Min Sun, and Hwann-Tzong Chen. Direct Voxel Grid Optimization: Superfast Convergence for Radiance Fields Reconstruction, June 2022. arXiv:2111.11215 [cs].

593. Cheng Sun, Min Sun, and Hwann-Tzong Chen. Improved Direct Voxel Grid Optimization for Radiance Fields Reconstruction, July 2022. arXiv:2206.05085 [cs].

594. Chunyi Sun, Yanbin Liu, Junlin Han, and Stephen Gould. NeRFEditor: Differentiable Style Decomposition for Full 3D Scene Editing, December 2022. arXiv:2212.03848 [cs, eess].

595. Jiankai Sun, Yan Xu, Mingyu Ding, Hongwei Yi, Jingdong Wang, Liangjun Zhang, and Mac Schwager. NeRF-Loc: Transformer-Based Object Localization Within Neural Radiance Fields, September 2022.

596. Jianwei Sun, Koichi Yonezawa, Eiji Shima, and Hao Liu. Integrated evaluation of the aeroacoustics and psychoacoustics of a single propeller. *International Journal of Environmental Research and Public Health*, 20(3):1955, 2023.

597. Shaohui Sun, Ramesh Sarukkai, Jack Kwok, and Vinay Shet. Accurate deep direct geo-localization from ground imagery and phone-grade gps. In *Proceedings of the IEEE Conference on Computer Vision and Pattern Recognition Workshops*, pages 1016–1023, 2018.

598. Zhe Sun, Xuejian Zhao, Zhixin Sun, Feng Xiang, and Chunjing Mao. Optimal Sliding Mode Controller Design Based on Dynamic Differential Evolutionary Algorithm for Under-Actuated Crane Systems. *IEEE Access*, 6:67469–67476, 2018. Conference Name: IEEE Access.

599. Y. Suo, W. Qiao, C. Jiang, B. Zhang, and F. Liu. A Residual-Fitting Modeling Method for Digital Predistortion of Broadband Power Amplifiers. *IEEE Microwave and Wireless Components Letters*, 32(9):1115–1118, 2022.

600. Shun Taguchi and Noriaki Hirose. Unsupervised simultaneous learning for camera re-localization and depth estimation from video. In *2022 IEEE/RSJ International Conference on Intelligent Robots and Systems (IROS)*, pages 6840–6847. IEEE, 2022.

601. Morgan B Talbot, Rushikesh Zawar, Rohil Badkundri, Mengmi Zhang, and Gabriel Kreiman. Lifelong compositional feature replays beat image replays in stream learning. *arXiv preprint arXiv:2104.02206*, 2021.

602. G. Tanaka, T. Yamane, J.B. Héroux, R. Nakane, N. Kanazawa, S. Takeda, H. Numata, D. Nakano, and A. Hirose. Recent advances in physical reservoir computing: A review. *Neural Networks*, 115:100–123, 2019.

603. Matthew Tancik, Vincent Casser, Xinchen Yan, Sabeek Pradhan, Ben Mildenhall, Pratul P. Srinivasan, Jonathan T. Barron, and Henrik Kretzschmar. Block-NeRF: Scalable Large Scene Neural View Synthesis, February 2022. arXiv:2202.05263 [cs].

604. Matthew Tancik, Ethan Weber, Evonne Ng, Ruilong Li, Brent Yi, Justin Kerr, Terrance Wang, Alexander Kristoffersen, Jake Austin, Kamyar Salahi, Abhik Ahuja, David McAllister, and Angjoo Kanazawa. Nerfstudio: A Modular Framework for Neural Radiance Field Development, February 2023. arXiv:2302.04264 [cs].

605. Shitao Tang, Chengzhou Tang, Rui Huang, Siyu Zhu, and Ping Tan. Learning camera localization via dense scene matching. In *Proceedings of the IEEE/CVF Conference on Computer Vision and Pattern Recognition*, pages 1831–1841, 2021.

606. Zhenggang Tang, Balakumar Sundaralingam, Jonathan Tremblay, Bowen Wen, Ye Yuan, Stephen Tyree, Charles Loop, Alexander Schwing, and Stan Birchfield. RGB-Only Reconstruction of Tabletop Scenes for Collision-Free Manipulator Control, October 2022.

607. Chen. S. Tang. Q, Wei. Z and Cheng. Z. Modeling and simulation of A2G channel based on UAV array. In *2020 IEEE 6th International Conference on Computer and Communications (ICCC)*, pages 500–506. IEEE, 2020.

608. Ryutaro Tanno, Melanie F Pradier, Aditya Nori, and Yingzhen Li. Repairing neural networks by leaving the right past behind. *Advances in Neural Information Processing Systems*, 35:13132–13145, 2022.

609. Yi Tao, Weiwei Xu, Guangming Wang, Ziwen Yuan, Maode Wang, Michael Houston, Yingchun Zhang, Badong Chen, Xiangguo Yan, and Gang Wang. Decoding multi-class eeg signals of hand movement using multivariate empirical mode decomposition and convolutional neural network. *IEEE Transactions on Neural Systems and Rehabilitation Engineering*, 30:2754–2763, 2022.

610. Zachary Teed and Jia Deng. Droid-slam: Deep visual slam for monocular, stereo, and rgb-d cameras. *Advances in neural information processing systems*, 34:16558–16569, 2021.

611. Michal Teplan. Fundamental of EEG measurement. *Measurement Science Review*, 2, 01 2002.

612. M. Thilagaraj, S. Ramkumar, N. Arunkumar, A. Durgadevi, K. Karthikeyan, S. Hariharasitaraman, M. Pallikonda Rajasekaran, and Petchinathan Govindan. Classification of electroencephalogram signal for developing brain-computer interface using bioinspired machine learning approach. *Computational Intelligence and Neuroscience*, 2022, 2022.

613. Esteban Tlelo-Cuautle, Astrid Maritza González-Zapata, Jonathan Daniel Dáz-Muñoz, Luis Gerardo de la Fraga, and Israel Cruz-Vega. Optimization of fractional-order chaotic cellular neural networks by metaheuristics. *The European Physical Journal Special Topics*, 231(10):2037–2043, August 2022.

614. Shengbang Tong, Xili Dai, Ziyang Wu, Mingyang Li, Brent Yi, and Yi Ma. Incremental learning of structured memory via closed-loop transcription. *arXiv preprint arXiv:2202.05411*, 2022.

615. Antonio J Torija and Rory K Nicholls. Investigation of metrics for assessing human response to drone noise. *International Journal of Environmental Research and Public Health*, 19(6):3152, 2022.

616. A. J. Torija Martinez et al. Drone noise, a new public health challenge? In *Quiet Drones International e-Symposium on UAV/UAS Noise 2020*, pages 448–454. International Institute of Noise Control Engineering, 2020.

617. A. J. Torija Martinez, Zhengguang Li, et al. Metrics for assessing the perception of drone noise. In *Proceedings of Forum Acusticum 2020*, pages 3163–3168. European Acoustics Association (EAA)/HAL, 2020.

618. R. Torrance and D. James. The state-of-the-art in IC reverse engineering. In *Int. Workshop on Cryptographic Hardware and Embedded Systems*, pages 363–381. Springer, 2009.

619. Cesar Torres-Huitzil and Bernard Girau. Fault and error tolerance in neural networks: A review. *IEEE Access*, 5:17322–17341, 2017.

620. B. Tracey, G. Comina, S. Larson, M. Bravard, J. López, and R. Gilman. Cough detection algorithm for monitoring patient recovery from pulmonary tuberculosis. In *2011 Annual International Conference of the IEEE Engineering in Medicine and Biology Society*, pages 6017–6020, 2011.

621. Finnish Transport and Communications Agency. Definitions for communications services and networks used in traficom's statistics and requests for information.

622. Thanh-Dat Truong, Pierce Helton, Ahmed Moustafa, Jackson David Cothren, and Khoa Luu. Conda: Continual unsupervised domain adaptation learning in visual perception for self-driving cars. *arXiv preprint arXiv:2212.00621*, 2022.

623. Wei-Cheng Tseng, Hung-Ju Liao, Lin Yen-Chen, and Min Sun. CLA-NeRF: Category-Level Articulated Neural Radiance Field, March 2022. arXiv:2202.00181 [cs].

624. Haithem Turki, Deva Ramanan, and Mahadev Satyanarayanan. Mega-nerf: Scalable construction of large-scale nerfs for virtual fly-throughs. In *Proceedings of the IEEE/CVF Conference on Computer Vision and Pattern Recognition*, pages 12922–12931, 2022.

625. Aruna Tyagi, Sunil Semwal, and Gautam Shah. A review of eeg sensors used for data acquisition. *Journal of Computer Applications (IJCA)*, pages 13–17, 2012.

626. Monika Ullrich, Haider Ali, Maximilian Durner, Zoltán-Csaba Márton, and Rudolph Triebel. Selecting cnn features for online learning of 3d objects. In *2017 IEEE/RSJ International Conference on Intelligent Robots and Systems (IROS)*, pages 5086–5091. IEEE, 2017.

627. Balemir Uragun and Ibrahim N. Tansel. The noise reduction techniques for unmanned air vehicles. In *2014 International Conference on Unmanned Aircraft Systems (ICUAS)*, pages 800–807. IEEE, 2014.

628. Dragan B. Đurđević, Nikola Pavlov, and Mladen Božić. The role of warehouse in e-business. *E-Business Technologies*, page 7, 2021.

629. R. Urtasun, P. Lenz, and A. Geiger. Are we ready for autonomous driving? The KITTI vision benchmark suite. In *2012 IEEE Conference on Computer Vision and Pattern Recognition*, pages 3354–3361. IEEE Computer Society, 2012.

630. Abhinav Valada, Noha Radwan, and Wolfram Burgard. Deep auxiliary learning for visual localization and odometry. In *2018 IEEE international conference on robotics and automation (ICRA)*, pages 6939–6946. IEEE, 2018.

631. Leslie G. Valiant. A neuroidal architecture for cognitive computation. *Journal of ACM*, 47(5):854–882, sep 2000.

632. Andrea Vallone, Frederik Warburg, Hans Hansen, Søren Hauberg, and Javier Civera. Danish airs and grounds: A dataset for aerial-to-street-level place recognition and localization. *arXiv preprint arXiv:2202.01821*, 2022.

633. M. Van Cleeff, L. Kivihya-Ndugga, H. Meme, J. Odhiambo, and P. Klatser. The role and performance of chest x-ray for the diagnosis of tuberculosis: A cost-effective analysis in nairobi, kenya. *BMC Infectious Diseases*, 5:1–9, 2005.

634. G. Van der Auwera, M. Carneiro, C. Hartl, R. Poplin, G. Del Angel, A. Levy-Moonshine, T. Jordan, K. Shakir, D. Roazen, J. Thibault, E. Banks, K. Garimella, D. Altshuler, S. Gabriel, and M. DePristo. From fastq data to high-confidence variant calls: The genome analysis toolkit best practices pipeline. *Current Protocols in Bioinformatics*, 43:11.10.1–11.10.33, 2013.

635. Jan van Erp, Fabien Lotte, and Michael Tangermann. Brain-computer interfaces: Beyond medical applications. *Computer*, 45(4):26–34, 2012.

636. B. Vandenberk, E. Vandael, T. Robyns, J. Vandenberghe, C. Garweg, V. Foulon, J. Ector, and R. Willems. Which qt correction formulae to use for qt monitoring? *Journal of the American Heart Association*, 5:e003264, 2016.

637. Cecilia Vargas-Olmos. Procesamiento de imágenes con métodos de ondeleta. M.s. thesis, Instituto de Investigación en Comunicación Óptica, UASLP, San Luis Potosí, México, jun 2010.

638. Cecilia Vargas-Olmos. *Procesamiento de señales y solución de problemas con la transformada wavelet*. Ph.d. dissertation, Instituto de Investigación en Comunicación Óptica, UASLP, San Luis Potosí, México, 2017.

639. Natalia S Vassilieva. Content-based image retrieval methods. *Programming and Computer Software*, 35(3):158–180, 2009.

640. Francisco Velasco-Álvarez, Álvaro Fernández-Rodríguez, and Ricardo Ron-Angevin. Brain-computer interface (bci)-generated speech to control domotic devices. *Neurocomputing*, 509:121–136, 2022.

641. S. Vhaduri, T. Kessel, B. Ko, D. Wood, S. Wang, and T. Brunschwiler. Nocturnal cough and snore detection in noisy environments using smartphone-microphones. In *2019 IEEE International Conference on Healthcare Informatics (ICHI)*, pages 1–7, 2019.

642. R. Vijayanandh, M. Ramesh, G. Raj Kumar, U. K. Thianesh, K. Venkatesan, and M Senthil Kumar. Research of noise in the unmanned aerial vehicle's propeller using cfd. *International Journal of Engineering and Advanced Technology, ISSN*, pages 2249–8958, 2019.

643. Andrés Villa, Juan León Alcázar, Motasem Alfarra, Kumail Alhamoud, Julio Hurtado, Fabian Caba Heilbron, Alvaro Soto, and Bernard Ghanem. Pivot: Prompting for video continual learning. *arXiv preprint arXiv:2212.04842*, 2022.

644. Niclas Vödisch, Daniele Cattaneo, Wolfram Burgard, and Abhinav Valada. Continual slam: Beyond lifelong simultaneous localization and mapping through continual learning. In *Robotics Research*, pages 19–35. Springer, 2023.

645. Niclas Vödisch, Daniele Cattaneo, Wolfram Burgard, and Abhinav Valada. Covio: Online continual learning for visual-inertial odometry. *arXiv preprint arXiv:2303.10149*, 2023.

646. Niclas Vödisch, Kürsat Petek, Wolfram Burgard, and Abhinav Valada. Codeps: Online continual learning for depth estimation and panoptic segmentation. *arXiv preprint arXiv:2303.10147*, 2023.

647. Georgios Volanis, Angelos Antonopoulos, Alkis A. Hatzopoulos, and Yiorgos Makris. Toward silicon-based cognitive neuromorphic ics—a survey. *IEEE Design & Test*, 33(3):91–102, 2016.

648. Trieu-Duc Vu, Yuki Iwasaki, Kenshiro Oshima, Ming-Tzu Chiu, Masato Nikaido, and Norihiro Okada. Data of rna-seq transcriptomes in the brain associated with aggression in males of the fish betta splendens. *Data in Brief*, 38:107448, 2021.

649. C. A. Vázquez-Chacón, F. J. Rodríguez-Gaxiola, C. F. López-Carrera, M. Cruz-Rivera, A. Martínez-Guarneros, R. Parra-Unda, E. Arámbula-Meraz, S. Fonseca-Coronado, G. Vaughan, and P. A. López-Durán. Identification of drug resistance mutations among mycobacterium bovis lineages in the americas. *PLoS Neglected Tropical Diseases*, 15:1–12, 2021.

650. Kaido Värbu, Naveed Muhammad, and Yar Muhammad. Past, present, and future of eeg-based bci applications. *Sensors*, 22(9), 2022.

651. Florian Walch, Caner Hazirbas, Laura Leal-Taixe, Torsten Sattler, Sebastian Hilsenbeck, and Daniel Cremers. Image-based localization using lstms for structured feature correlation. In *Proceedings of the IEEE International Conference on Computer Vision*, pages 627–637, 2017.

652. James S Walker. *A primer on wavelets and their scientific applications*. Broken Sound Parkway NW: Chapman and hall/CRC, 2nd ed. edition, 2008.

653. T. Walker, C. Ip, R. Harrell, J. Evans, G. Kapatai, M. Dedicoat, D. Eyre, D. Wilson, P. Hawkey, D. Crook, J. Parkhill, D. Harris, A. Walker, R. Bowden, P. Monk, E. Smith, and T. Peto. Whole-genome sequencing to delineate mycobacterium tuberculosis outbreaks: A retrospective observational study. *The Lancet Infectious Diseases*, 13:137–146, 2013.

654. Matthew Wallingford, Aditya Kusupati, Alex Fang, Vivek Ramanujan, Aniruddha Kembhavi, Roozbeh Mottaghi, and Ali Farhadi. Neural radiance field codebooks. *arXiv preprint arXiv:2301.04101*, 2023.

655. Can Wang, Ruixiang Jiang, Menglei Chai, Mingming He, Dongdong Chen, and Jing Liao. NeRF-Art: Text-Driven Neural Radiance Fields Stylization, December 2022. arXiv:2212.08070 [cs].

656. J. Wang, S. Lu, S-H. Wang, and Y-D. Zhang. A review on extreme learning machine. *Multimedia Tools and Applications*, 2021.

657. J. Wang, I. Tsapakis, and C. Zhong. A space–time delay neural network model for travel time prediction. *Engineering Applications of Artificial Intelligence*, 52:145–160, 2016.

658. Jue Wang, Anoop Cherian, and Fatih Porikli. Ordered pooling of optical flow sequences for action recognition. In *2017 IEEE Winter Conference on Applications of Computer Vision (WACV)*, pages 168–176. IEEE, 2017.

659. Lina Wang, Weining Xue, Yang Li, Meilin Luo, Jie Huang, Weigang Cui, and Chao Huang. Automatic epileptic seizure detection in eeg signals using multi-domain feature extraction and nonlinear analysis. *Entropy*, 19(6):222, 2017.

660. Liyuan Wang, Xingxing Zhang, Hang Su, and Jun Zhu. A comprehensive survey of continual learning: Theory, method and application. *arXiv preprint arXiv:2302.00487*, 2023.

661. Sen Wang, Ronald Clark, Hongkai Wen, and Niki Trigoni. Deepvo: Towards end-to-end visual odometry with deep recurrent convolutional neural networks. In *2017 IEEE International Conference on Robotics and Automation (ICRA)*, pages 2043–2050. IEEE, 2017.

662. Shuzhe Wang, Zakaria Laskar, Iaroslav Melekhov, Xiaotian Li, and Juho Kannala. Continual learning for image-based camera localization. In *Proceedings of the IEEE/CVF International Conference on Computer Vision*, pages 3252–3262, 2021.

663. Yuehao Wang, Yonghao Long, Siu Hin Fan, and Qi Dou. Neural Rendering for Stereo 3D Reconstruction of Deformable Tissues in Robotic Surgery, June 2022.

664. Zhongshu Wang, Lingzhi Li, Zhen Shen, Li Shen, and Liefeng Bo. 4K-NeRF: High Fidelity Neural Radiance Fields at Ultra High Resolutions, December 2022. arXiv:2212.04701 [cs].

665. Zirui Wang, Shangzhe Wu, Weidi Xie, Min Chen, and Victor Adrian Prisacariu. NeRF–: Neural Radiance Fields Without Known Camera Parameters, April 2022. arXiv:2102.07064 [cs].

666. J. Webster. *Medical Instrumentation: Application and Design*. John Wiley & Sons, 2009.

667. Yuliang Wei, Feng Xu, Shiyuan Bian, and Deyi Kong. Noise reduction of uav using biomimetic propellers with varied morphologies leading-edge serration. *Journal of Bionic Engineering*, 17:767–779, 2020.

668. Chung-Yi Weng, Brian Curless, Pratul P. Srinivasan, Jonathan T. Barron, and Ira Kemelmacher-Shlizerman. HumanNeRF: Free-viewpoint Rendering of Moving People from Monocular Video, June 2022. arXiv:2201.04127 [cs].

669. Travis Williams, Robert Li, et al. An ensemble of convolutional neural networks using wavelets for image classification. *Journal of Software Engineering and Applications*, 11(02):69, 2018.

670. Travis Williams, Robert Li, et al. Wavelet pooling for convolutional neural networks. *Proc. Int. Conf. on Learning Representations*, 2018.

671. Klaudia Witte, Katharina Baumgärtner, Corinna Röhrig, and Sabine Nöbel. Test of the deception hypothesis in atlantic mollies poecilia mexicana–does the audience copy a pretended mate choice of others? *Biology*, 7(3), 2018.

672. David Wong, Daisuke Deguchi, Ichiro Ide, and Hiroshi Murase. Single camera vehicle localization using feature scale tracklets. *IEICE Transactions on Fundamentals of Electronics, Communications and Computer Sciences*, 100(2):702–713, 2017.

673. Sanghyun Woo, Kwanyong Park, Seoung Wug Oh, In So Kweon, and Joon-Young Lee. Bridging images and videos: A simple learning framework for large vocabulary video object detection. In *Computer Vision–ECCV 2022: 17th European Conference, Tel Aviv, Israel, October 23–27, 2022, Proceedings, Part XXV*, pages 238–258. Springer, 2022.

674. Zirui Wu, Yuantao Chen, Runyi Yang, Zhenxin Zhu, Chao Hou, Yongliang Shi, Hao Zhao, and Guyue Zhou. Asyncnerf: Learning large-scale radiance fields from asynchronous rgb-d sequences with time-pose function. *arXiv preprint arXiv:2211.07459*, 2022.

675. Jean Marc Wunderli, Jonas Meister, Oliver Boolakee, and Kurt Heutschi. A method to measure and model acoustic emissions of multicopters. *International Journal of Environmental Research and Public Health*, 20(1):96, 2023.

676. Magdalena Wysocki, Mohammad Farid Azampour, Christine Eilers, Benjamin Busam, Mehrdad Salehi, and Nassir Navab. Ultra-NeRF: Neural Radiance Fields for Ultrasound Imaging, January 2023. arXiv:2301.10520 [cs, eess].

677. Yuanbo Xiangli, Linning Xu, Xingang Pan, Nanxuan Zhao, Anyi Rao, Christian Theobalt, Bo Dai, and Dahua Lin. Bungeenerf: Progressive neural radiance field for extreme multi-scale scene rendering. In *Computer Vision–ECCV 2022: 17th European Conference, Tel Aviv, Israel, October 23–27, 2022, Proceedings, Part XXXII*, pages 106–122. Springer, 2022.

678. Christopher Xie, Keunhong Park, Ricardo Martin-Brualla, and Matthew Brown. FiGNeRF: Figure-Ground Neural Radiance Fields for 3D Object Category Modelling, April 2021. arXiv:2104.08418 [cs].

679. Yiheng Xie, Towaki Takikawa, Shunsuke Saito, Or Litany, Shiqin Yan, Numair Khan, Federico Tombari, James Tompkin, Vincent Sitzmann, and Srinath Sridhar. Neural fields in visual computing and beyond. In *Computer Graphics Forum*, volume 41, pages 641–676. Wiley Online Library, 2022.

680. Sheng. M Zhao. N. Xie. Z, Liu. J and Li. J. Exploiting aerial computing for air-to-ground coverage enhancement. *IEEE Wireless Communications*, 28(5):50–58, 2021.

681. H. Xin, L. Zhijun, Y. Xiaofei, Z. Yulong, C. Wenhua, H. Biao, D. Xuekun, L. Xiang, H. Mohamed, W. Weidong, and F. M. Ghannouchi. Convolutional neural network for behavioral modeling and predistortion of wideband power amplifiers. *IEEE Transactions on Neural Networks and Learning Systems*, 33(8):3923–3937, 2022.

682. Dan Xu, Elisa Ricci, Yan Yan, Jingkuan Song, and Nicu Sebe. Learning deep representations of appearance and motion for anomalous event detection. *arXiv preprint arXiv:1510.01553*, 2015.

683. Fangzhou Xu, Fenqi Rong, Yunjing Miao, Yanan Sun, Gege Dong, Han Li, Jincheng Li, Yuandong Wang, and Jiancai Leng. Representation learning for motor imagery recognition with deep neural network. *Electronics*, 10(2), 2021.

684. G. Xu, H. Yu, C. Hua, and T. Liu. Chebyshev Polynomial-LSTM Model for 5G Millimeter-Wave Power Amplifier Linearization. *IEEE Microwave and Wireless Components Letters*, 32(6):611–614, 2022.

685. J. Xu, W. Jiang, L. Ma, M. Li, Z. Yu, and Z. Geng. Augmented Time-Delay Twin Support Vector Regression-Based Behavioral Modeling for Digital Predistortion of RF Power Amplifier. *IEEE Access*, 7:59832–59843, 2019.

686. Renjie Xu, Haifeng Lin, Kangjie Lu, Lin Cao, and Yunfei Liu. A forest fire detection system based on ensemble learning. *Forests*, 12(2), 2021.

687. Xiaoling Xu, Wensheng Li, and Qingling Duan. Transfer learning and se-resnet152 networks-based for small-scale unbalanced fish species identification. *Computers and Electronics in Agriculture*, 180, 1 2021.

688. Yifeng Xu, Yang Zhang, Huigang Wang, and Xing Liu. Underwater image classification using deep convolutional neural networks and data augmentation. *2017 IEEE International Conference on Signal Processing, Communications and Computing, IC-SPCC 2017*, 2017-Janua:1–5, 2017.

689. LC Yan, B Yoshua, and H Geoffrey. Deep learning. *nature*, 521(7553):436–444, 2015.

690. Zike Yan, Yuxin Tian, Xuesong Shi, Ping Guo, Peng Wang, and Hongbin Zha. Continual neural mapping: Learning an implicit scene representation from sequential

observations. In *Proceedings of the IEEE/CVF International Conference on Computer Vision*, pages 15782–15792, 2021.

691. Guanci Yang, Zhanjie Chen, Yang Li, and Zhidong Su. Rapid relocation method for mobile robot based on improved orb-slam2 algorithm. *Remote Sensing*, 11(2):149, 2019.

692. Ling Yang, Huihui Yu, Yuelan Cheng, Siyuan Mei, Yanqing Duan, Daoliang Li, and Yingyi Chen. A dual attention network based on efficientnet-b2 for short-term fish school feeding behavior analysis in aquaculture. *Computers and Electronics in Agriculture*, 187:106316, 8 2021.

693. Xingrui Yang, Yuhang Ming, Zhaopeng Cui, and Andrew Calway. Fd-slam: 3-d reconstruction using features and dense matching. In *2022 International Conference on Robotics and Automation (ICRA)*, pages 8040–8046. IEEE, 2022.

694. Y. Yang, T. M. Walker, A. S. Walker, D. J. Wilson, T. E. A. Peto, D. W. Crook, F. Shamout, CRyPTIC Consortium, T. Zhu, and D. A. Clifton. Deepamr for predicting co-occurrent resistance of mycobacterium tuberculosis. *Bioinformatics*, 35:3240–3249, 2019.

695. Shuquan Ye, Dongdong Chen, Songfang Han, and Jing Liao. Robust point cloud segmentation with noisy annotations. *IEEE Transactions on Pattern Analysis and Machine Intelligence*, 2022.

696. Zhenhui Ye, Ziyue Jiang, Yi Ren, Jinglin Liu, JinZheng He, and Zhou Zhao. GeneFace: Generalized and High-Fidelity Audio-Driven 3D Talking Face Synthesis, January 2023. arXiv:2301.13430 [cs].

697. Chi Tsai Yeh, Tzuo Ming Chen, and Zhong Jie Liu. Flexible iot cloud application for ornamental fish recognition using yolov3 model. *Sensors and Materials*, 34:1229–1240, 2021.

698. Lin Yen-Chen, Pete Florence, Jonathan T. Barron, Tsung-Yi Lin, Alberto Rodriguez, and Phillip Isola. NeRF-Supervision: Learning Dense Object Descriptors from Neural Radiance Fields, March 2022.

699. Lin Yen-Chen, Pete Florence, Jonathan T. Barron, Alberto Rodriguez, Phillip Isola, and Tsung-Yi Lin. iNeRF: Inverting Neural Radiance Fields for Pose Estimation. In *2021 IEEE/RSJ International Conference on Intelligent Robots and Systems (IROS)*, pages 1323–1330, September 2021. ISSN: 2153-0866.

700. Lin Yen-Chen, Pete Florence, Andy Zeng, Jonathan T. Barron, Yilun Du, Wei-Chiu Ma, Anthony Simeonov, Alberto Rodriguez Garcia, and Phillip Isola. MIRA: Mental Imagery for Robotic Affordances, December 2022.

701. Peng Yin, Abulikemu Abuduweili, Shiqi Zhao, Changliu Liu, and Sebastian Scherer. Bioslam: A bio-inspired lifelong memory system for general place recognition. *arXiv preprint arXiv:2208.14543*, 2022.

702. Pengwei Yin, Jiawu Dai, Jingjing Wang, Di Xie, and Shiliang Pu. NeRF-Gaze: A Head-Eye Redirection Parametric Model for Gaze Estimation, December 2022. arXiv:2212.14710 [cs].

703. Bohan Yoon, Hyeonha Kim, Geonsik Youn, and Jongtae Rhee. 3d position estimation of drone and object based on qr code segmentation model for inventory management automation. In *2021 IEEE International Symposium on Safety, Security, and Rescue Robotics (SSRR)*, pages 223–229. IEEE, 2021.

704. Alex Yu, Vickie Ye, Matthew Tancik, and Angjoo Kanazawa. pixelNeRF: Neural Radiance Fields from One or Few Images. In *2021 IEEE/CVF Conference on Computer*

Vision and Pattern Recognition (CVPR), pages 4576–4585, June 2021. ISSN: 2575-7075.

705. Lu Yu, Xialei Liu, and Joost Van de Weijer. Self-training for class-incremental semantic segmentation. *IEEE Transactions on Neural Networks and Learning Systems*, 2022.

706. Wentao Yuan, Zhaoyang Lv, Tanner Schmidt, and Steven Lovegrove. STaR: Self-supervised Tracking and Reconstruction of Rigid Objects in Motion with Neural Rendering. In *2021 IEEE/CVF Conference on Computer Vision and Pattern Recognition (CVPR)*, pages 13139–13147, June 2021. ISSN: 2575-7075.

707. Yu-Jie Yuan, Yang-Tian Sun, Yu-Kun Lai, Yuewen Ma, Rongfei Jia, and Lin Gao. NeRF-Editing: Geometry Editing of Neural Radiance Fields, May 2022. arXiv:2205.04978 [cs].

708. Amad Zafar, Shaik Javeed Hussain, Muhammad Umair Ali, and Seung Won Lee. Meta-heuristic optimization-based feature selection for imagery and arithmetic tasks: An fnirs study. *Sensors*, 23(7), 2023.

709. C. Y. Zhang, Y. Y. Zhu, Q. F. Cheng, H. P. Fu, J. G. Ma, and Q. J. Zhang. Extreme learning machine for the behavioral modeling of RF power amplifiers. *IEEE MTT-S International Microwave Symposium (IMS), Honolulu, HI, USA*, pages 558–561, 2017.

710. Ce Zhang and Azim Eskandarian. A computationally efficient multiclass time-frequency common spatial pattern analysis on eeg motor imagery. In *2020 42nd Annual International Conference of the IEEE Engineering in Medicine & Biology Society (EMBC)*, pages 514–518, 2020.

711. Kai Zhang, Gernot Riegler, Noah Snavely, and Vladlen Koltun. NeRF++: Analyzing and Improving Neural Radiance Fields, October 2020. arXiv:2010.07492 [cs].

712. Ray Zhang, Zongtai Luo, Sahib Dhanjal, Christopher Schmotzer, and Snigdhaa Hasija. Posenet++: A cnn framework for online pose regression and robot re-localization.

713. Richard Zhang, Phillip Isola, Alexei A. Efros, Eli Shechtman, and Oliver Wang. The unreasonable effectiveness of deep features as a perceptual metric. In *2018 IEEE/CVF Conference on Computer Vision and Pattern Recognition*, pages 586–595, June 2018. ISSN: 2575-7075.

714. Xiaodong Zhang, Xiaoli Li, Kang Wang, Yanjun Lu, et al. A survey of modelling and identification of quadrotor robot. In *Abstract and Applied Analysis*, volume 2014. Hindawi, 2014.

715. Yin Zhang, Xiao Ma, Jing Zhang, M. Shamim Hossain, Ghulam Muhammad, and Syed Umar Amin. Edge intelligence in the cognitive internet of things: Improving sensitivity and interactivity. *IEEE Network*, 33(3):58–64, 2019.

716. Bowen Zhao, Xi Xiao, Guojun Gan, Bin Zhang, and Shu-Tao Xia. Maintaining discrimination and fairness in class incremental learning. In *Proceedings of the IEEE/CVF conference on computer vision and pattern recognition*, pages 13208–13217, 2020.

717. Chunhui Zhao, Bin Fan, Jinwen Hu, Limin Tian, Zhiyuan Zhang, Sijia Li, and Quan Pan. Pose estimation for multi-camera systems. In *2017 IEEE International Conference on Unmanned Systems (ICUS)*, pages 533–538. IEEE, 2017.

718. Shili Zhao, Song Zhang, Jincun Liu, He Wang, Jia Zhu, Daoliang Li, and Ran Zhao. Application of machine learning in intelligent fish aquaculture: A review, 7 2021.

719. Henghui Zhi, Chenyang Yin, Huibin Li, and Shanmin Pang. An unsupervised monocular visual odometry based on multi-scale modeling. *Sensors*, 22(14):5193, 2022.

720. Shaohong Zhong, Alessandro Albini, Oiwi Parker Jones, Perla Maiolino, and Ingmar Posner. Touching a NeRF: Leveraging Neural Radiance Fields for Tactile Sensory Data Generation. November 2022.

721. Xingguang Zhong, Yue Pan, Jens Behley, and Cyrill Stachniss. Shine-mapping: Large-scale 3d mapping using sparse hierarchical implicit neural representations. *arXiv preprint arXiv:2210.02299*, 2022.

722. Allan Zhou, Moo Jin Kim, Lirui Wang, Pete Florence, and Chelsea Finn. NeRF in the Palm of Your Hand: Corrective Augmentation for Robotics via Novel-View Synthesis, January 2023.

723. Da-Wei Zhou, Fu-Yun Wang, Han-Jia Ye, and De-Chuan Zhan. Pycil: A python toolbox for class-incremental learning. *arXiv preprint arXiv:2112.12533*, 2021.

724. Peng Zhou, Long Mai, Jianming Zhang, Ning Xu, Zuxuan Wu, and Larry S Davis. M 2 kd: Incremental learning via multi-model and multi-level knowledge distillation. *Memory*, 1000(3000):4000, 2000.

725. Yongchao Zhou, Ehsan Nezhadarya, and Jimmy Ba. Dataset distillation using neural feature regression. *arXiv preprint arXiv:2206.00719*, 2022.

726. Zewei Zhou and Martim Brandão. Noise and environmental justice in drone fleet delivery paths: A simulation-based audit and algorithm for fairer impact distribution. In *IEEE International Conference on Robotics and Automation*, 2023.

727. Haoyi Zhu, Hao-Shu Fang, and Cewu Lu. X-NeRF: Explicit Neural Radiance Field for Multi-Scene 360° $Insufficient RGB-D Views, October 2022. arXiv:2210.05135 [cs].

728. R. Zhu, S. Chang, H. Wang, Q. Huang, J. He, and F. Yi. A versatile and accurate compact model of memristor with equivalent resistor topology. *IEEE Electron Device Letters*, 38(10):1367–1370, 2017.

729. Zhenxin Zhu, Yuantao Chen, Zirui Wu, Chao Hou, Yongliang Shi, Chuxuan Li, Pengfei Li, Hao Zhao, and Guyue Zhou. Latitude: Robotic global localization with truncated dynamic low-pass filter in city-scale nerf. *arXiv preprint arXiv:2209.08498*, 2022.

730. Jialin Zhu, Zhiqing Wei, Huici Wu, Chen Qiu, and Zhiyong Feng. Capacity of uav-assisted air-to-ground communication with random perturbation of uav platform. In *2020 International Conference on Wireless Communications and Signal Processing (WCSP)*, pages 275–279, 2020.

731. Bingbing Zhuang and Manmohan Chandraker. Fusing the old with the new: Learning relative camera pose with geometry-guided uncertainty. In *Proceedings of the IEEE/CVF Conference on Computer Vision and Pattern Recognition*, pages 32–42, 2021.

732. Konstantinos C. Zikidis. Early warning against stealth aircraft, missiles and unmanned aerial vehicles. *Surveillance in Action: Technologies for Civilian, Military and Cyber Surveillance*, pages 195–216, 2018.

733. A. Zimmer, C. Ugarte-Gil, R. Pathri, P. Dewan, D. Jaganath, A. Cattamanchi, M. Pai, and S. Lapierre. Making cough count in tuberculosis care. *Communications Medicine*, 2:83, 2022.

734. M. Zimmermann, M. Kogadeeva, M. Gengenbacher, G. McEwen, H. Mollenkopf, N. Zamboni, S. Ernts, and U. Sauer. Integration of metabolomics and transcriptomics reveals a complex diet of mycobacterium tuberculosis during early macrophage infection. *mSystems*, 2:e00057–17, 2017.

735. R. Zimmermann and L. Thiede. Python library for reservoir computing using echo state networks, 2021. `https://github.com/kalekiu/easyesn`.

736. Daniel Zwillinger and Stephen Kokoska. *CRC standard probability and statistics tables and formulae*. Crc Press, 1999.

737. L. Álvarez López, J. A. Becerra, M. J. Madero-Ayora, and C. Crespo-Cadenas. Determining a Digital Predistorter Model Structure for Wideband Power Amplifiers through Random Forest. *IEEE Topical Conference on RF/Microwave Power Amplifiers for Radio and Wireless Applications (PAWR)*, pages 50–52, 2020.

Index

For Product Safety Concerns and Information please contact our EU
representative GPSR@taylorandfrancis.com
Taylor & Francis Verlag GmbH, Kaufingerstraße 24, 80331 München, Germany